LOGIC COLLOQUIUM '99

Lecture Notes in Logic

A Publication of

The Association for Symbolic Logic

LECTURE NOTES IN LOGIC 17

LOGIC COLLOQUIUM '99

Proceedings of the Annual European Summer Meeting of the Association for Symbolic Logic, held in Utrecht, Netherlands August 1–6, 1999

Edited by

Jan van Eijck
Centre for Mathematics and Computer Science (CWI)
Amsterdam

Vincent van Oostrom
Department of Philosophy
Utrecht University

Albert Visser
Department of Philosophy
Utrecht University

ASSOCIATION FOR SYMBOLIC LOGIC

A K Peters, Ltd. • Natick, Massachusetts

Addresses of the Editors of Lecture Notes in Logic and a Statement of Editorial Policy may be found at the back of this book.

Sales and Customer Service:
A K Peters, Ltd.
888 Worcester Street, Suite 230
Wellesley, Massachusetts 02482, USA

Association for Symbolic Logic:
C. Ward Henson, Publisher
Mathematics Department
University of Illinois
1409 West Green Street
Urbana, Illinois 61801, USA

Library of Congress Cataloging-in-Publication Data

Logic Colloquium (1999 : Utrecht, Netherlands)
   Logic Colloquium '99 : proceedings of the Annual European Summer Meeting of the Association for Symbolic Logic, held in Utrecht, Netherlands, August 1–6, 1999 / edited by Jan van Eijck, Vincent van Oostrom, Albert Visser.
      p. cm. – (Lecture notes in logic ; 17)
   Includes bibliographical references.
   ISBN 1-56881-199-3 (acid-free paper) – ISBN 1-56881-203-5 (pbk. : acid-free paper)
      1. Logic, Symbolic and mathematical–Congresses. I. Eijck, J. van (Jan) II. Oostrom, Vincent van, 1966- III. Visser, Albert. IV. Association for Symbolic Logic. V. Title. VI. Series.

QA9.A1L64 1999
511.3–dc22   2003061666

Publisher's note: This book was typeset in LATEX, by the ASL Typesetting Office, from electronic files produced by the authors, using the ASL documentclass `asl.cls`. The fonts are Monotype Times Roman. This book was printed by Friesens, of Manitoba, Canada, on acid-free paper. The cover design is by Richard Hannus, Hannus Design Associates, Boston, Massachusetts.

13 12 11 10 09 08 07 06 05 04   5 4 3 2 1

PREFACE

Logic Colloquium '99, the 1999 European Meeting of the Association for Symbolic Logic, took place in Utrecht, Netherlands, from Sunday August 1st (arrival day) until Friday August 6th, 1999. The colloquium sessions started on Monday August 2nd.

The main programme consisted of twelve invited lectures, four series of 3-lecture tutorials and three series of parallel sessions for contributed talks. There were also two evening lectures arranged by members of the Dutch logic community.

The Programme Committee consisted of Erik Barendsen, Jan van Eijck, Sergey Goncharov, Wilfrid Hodges (Chair), Dick de Jongh, Alexander Kechris, Peter Koepke, Manuel Lerman, David Marker, Jaap van Oosten, Andrew Pitts and Albert Visser. The Organising Committee consisted of Jan van Eijck (chair), Peter Blok, Vincent van Oostrom, Simone Panka, Frans Snijders.

The Logic Colloquium was generously supported by the following organizations: Ercim, CWI, the City of Utrecht, NWO (Exact Sciences), OzsL, VvL, Spinoza Logic in Action, the departments of mathematics and computer science of the Dutch universities, and ASL (with a contribution to the travel aid fund).

All papers in this volume are based on talks or tutorials given at the colloquium. The editors received reviewing help from Zoe Chatzidakis, Rosalie Iemhoff, Salma Kuhlmann, Quintijn Puite, Bas Terwijn, Simon Thomas, Anne Troelstra, and Yde Venema. Roman Kuznets helped us with the LaTeX formatting.

Just before this volume went to press we received the sad news that Dr. Andreja Prijatelj died on April 1, 2002, in Ljubljana, after a six-month illness. She was 48 years old. Andreja was a fine colleague and we will miss her enthusiasm and warmth.

The Editors
Jan van Eijck, *Amsterdam*
Vincent van Oostrom, *Utrecht*
Albert Visser, *Utrecht*

TABLE OF CONTENTS

SURVEY ARTICLES

RESEARCH ARTICLES

SURVEY ARTICLES

GROUP ACTIONS AND COUNTABLE MODELS

GREG HJORTH

CONTENTS

§1. **Introduction.** The field of Polish group actions has emerged in the last couple of years as a kind of sub-sub-discipline in its own right.

To some extent one can think of the study of countable models as being a special case of a Polish group action — since the infinite symmetric group, consisting of all permutations of the natural numbers, is a Polish group whose

Logic Colloquium '99
Edited by J. van Eijck, V. van Oostrom, and A. Visser
Lecture Notes in Logic, 17

action on countable models gives rise to the isomorphism relation as its orbit equivalence relation. In these talks I will try to emphasize the connections between recent work on Polish groups and some basic concepts one might see in a first course in logic or model theory.

I will begin the discussion at a very general level, which hopefully might make some sense to almost any kind of logician. Of course this means that the initial pace will seem tedious to an expert. On the other hand as the paper progresses I want to increasingly mention connections to other areas and recent results and include stick figure sketches of their proofs. Perhaps the later parts of the paper will only be completely clear for someone who already knows much of this material or is willing to spend hours in the library digging up references. Thus the criticism may be raised that the level of exposition is inconsistent; and by the end perhaps no one will be completely happy. But short of turning this paper into a book that problem strikes me as an inevitable consequence of its goals.

My own outlook on the history of this sub-area would really be to place *The descriptive set theory of Polish group actions*, Becker-Kechris [1996], as being central. This book represented the intersection of two seemingly unrelated schools of thought.

On the one hand was a sequence of papers treating topics in what one might call "invariant descriptive set theory." Perhaps the most well known papers here are those by Vaught, in which the development of the notion unified proofs for results by Ryll-Nardzewski (every orbit is Borel) and Lopez-Escobar (invariant Borel sets in the space of countable structures are described by $\mathcal{L}_{\omega_1,\omega}$ sentences). For this Vaught was awarded the first Karp prize. A decade later, in work that was completed in the 1980's but not published until 1994, Sami used Vaught's work to obtain results on changing topologies for the "logic action" of S_∞ and showed that the topological Vaught conjecture holds for abelian Polish groups. (More on this below.)

The second sequence of papers comes from researchers in operator algebras and the study of infinite dimensional group representations. Although the topic of Polish group actions for its own sake was somewhat incidental to the main direction of their research, the group responsible for conjugacy on infinite dimensional representations is the infinite dimensional unitary group and thus Polish but not locally compact. By the time of Effros [1982] one has a completely general result for Polish groups acting on Polish spaces, largely divorced from the original context of Effros [1965], Glimm [1961], and Mackey [1957].

Both these traditions were very much in the mind of Becker and Kechris by the time of *The descriptive set theory of Polish group actions*. I have drawn up a small flow chart to try to illustrate how these papers might be viewed as interrelating, but it goes without saying that this interpretation is personal

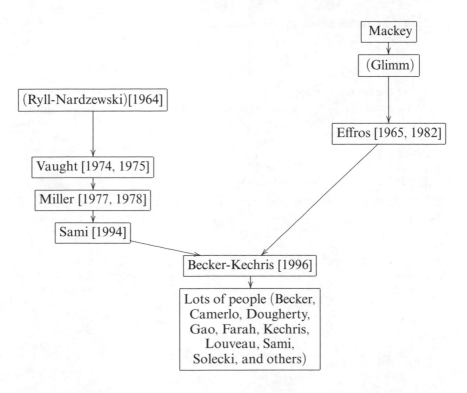

and subjective. Certainly there are papers off to one side of this chart (such as Friedman-Stanley [1989] and Burgess [1978]) which have proved extremely influential, but do not easily fall into the cartoon like summary of the history I have tried to give above.

§2. **Dynamic changes in topologies.** The main issue here is to discuss some of the tricks that enable us to change topologies and remain in the category of Polish spaces.

2.1. Polish spaces and Polish groups [definitions; the statement of the topological Vaught conjecture].

DEFINITION 2.1. A topological space is said to be *Polish* if it is separable and there is some complete metric which generates its topology.

EXAMPLES

(i) $\mathbb{R}$, the reals equipped with the topology generated by the openintervals.(The usual euclidean distance gives a complete metric.)

(ii) $C([0, 1])$, the continuous functions from the unit interval to $\mathbb{R}$ with the topology generated by the sup norm, $d(f_1, f_2) = \sup_{0 \leq x \leq 1} |f_1(x) - f_2(x)|$.

(iii) $\mathbb{N}$ in the discrete topology. As a metric,

$$d(x_1, x_2) = \begin{cases} 1 & \text{if } x_1 \neq x_2, \\ 0 & \text{if } x_1 = x_2. \end{cases}$$

(iv) $\mathbb{N}^{\mathbb{N}}$ in the product topology.

(v) More generally if $(X_i)_{i \in \mathbb{N}}$ is a sequence of Polish spaces, then their infinite product

$$\prod_{\mathbb{N}} X_i$$

is a Polish space in the product topology.

PROOF. After possibly replacing each metric d_i with

$$\hat{d}_i(x, y) = \frac{d_i(x, y)}{d_i(x, y) + 1}$$

we may assume that each d_i is bounded by 1.

Then we can obtain a compatible complete metric on the product given by

$$d(\vec{x}, \vec{y}) = \sum_{i \in \mathbb{N}} 2^{-i} d_i(x_i, y_i)$$

where

$$\vec{x} = (x_0, x_1, x_2, \ldots, x_i, \ldots),$$
$$\vec{y} = (y_0, y_1, y_2, \ldots, y_i, \ldots)$$

are both in $\prod_{\mathbb{N}} X_i$. ⊣

LEMMA 2.2. *Let X be a Polish space and $A \subset X$ a G_δ subset — that is to say, defined by a countable intersection of open sets. Then A in the subspace topology is again a Polish space.*

PROOF. First for open $\mathcal{O} \subset X$. Let d be complete metric for X, and find continuous

$$f : \mathcal{O} \to \mathbb{R}$$

with $f(x) \to \infty$ as $x \to X \setminus \mathcal{O}$. (For instance, $f(x) = \inf\{\frac{1}{d(x,c)} : c \notin \mathcal{O}\}$.) Then define $d_{\mathcal{O}}$ on $\mathcal{O}$ by

$$d_{\mathcal{O}}(x, y) = d(x, y) + |f(x) - f(y)|.$$

Suppose (x_i) is a sequence in $\mathcal{O}$ that is $d_{\mathcal{O}}$-Cauchy. From the definition of $d_{\mathcal{O}}$ it must be d-Cauchy, and hence by assumption that d witnesses X Polish, there is a limit x_∞. But we must also have that $(f(x_i))_{i \in \mathbb{N}}$ will be Cauchy in $\mathbb{R}$, and thus $x_\infty \in \mathcal{O}$.

For the general case

$$A = \bigcap_{i \in \mathbb{N}} \mathcal{O}_i$$

we can take

$$f_n \colon \mathcal{O}_n \to \mathbb{R}$$

as above and let

$$d_A(x, y) = d(x, y) + \sum_{n \in \mathbb{N}} 2^{-n} |f_n(x) - f_n(y)|.$$

⊣

DEFINITION 2.3. A topological group is said to be *Polish* if it is Polish as a space.

EXAMPLES 2.4.
  (i) $(\mathbb{R}, +)$, the reals equipped with its usual additive structure.
 (ii) Any discrete group. Countable products of discrete groups, and more generally the class of Polish groups is closed under countable products.
(iii) S_∞ the group of all permutations of $\mathbb{N}$, equipped with the topology of pointwise convergence. So as a basis we may take all sets of the form

$$\{\pi \in S_\infty \colon \pi(\ell_1) = k_1, \ldots, \pi(\ell_n) = k_n\}.$$

 (iv) U_∞ the unitary group of ℓ^2, infinite dimensional (separable) Hilbert space. Thus U_∞ is the group of all bijective linear operators

$$T \colon \ell^2 \to \ell^2$$

such that for all $\xi, \zeta \in \ell^2$

$$\langle \xi, \zeta \rangle = \langle T(\xi), T(\zeta) \rangle.$$

We give this the topology generated by the subbasis consisting of sets of the form

$$\{T \in U_\infty \colon |\langle T(\xi), \zeta \rangle - \langle T_0(\xi), \zeta \rangle| < \varepsilon\}.$$

DEFINITION 2.5. Let G be a Polish group. We say that a Polish space X is a *Polish G-space* if it is a Polish space which is equipped with a continuous action by G. We then let E_G denote the orbit equivalence relation: $x_1 E_G x_2$ if

$$\exists g \in G \ (g \cdot x_1 = x_2).$$

For $x \in X$ let $[x]_G$ indicate the orbit of x:

$$\{y \in X \colon \exists g \in G \ (g \cdot x = y)\}.$$

X/G indicates the set of orbits.

EXAMPLES 2.6.

(i) Let $\mathbb{Z}$ act on the complex unit circle $= \{e^{ix} : x \in \mathbb{R}\}$ in the following manner: For e^{ix} and $k \in \mathbb{Z}$ we let

$$k \cdot e^{ix} = e^{i(x+2\pi k \sqrt{2})}.$$

In other words the action is induced by a rotation of $\sqrt{2}$ radians. In this way, among many others, we could view the complex unit circle as a Polish $\mathbb{Z}$-space.

(ii) Given a countable set X we can consider the space $(U_\infty)^X$ of function from X to the unitary group of Hilbert space; this space is Polish since it is naturally isomorphic to a countable product of U_∞ with itself. We can let U_∞ act on this space by point-wise conjugation:

$$(T \cdot f)(x) = T \circ f(x) \circ T^{-1}$$

for any $x \in X$, $f \in (U_\infty)^X$.

U_∞ is a Polish group and under this action $(U_\infty)^X$ is a Polish U_∞-space.

(iii) Perhaps the most natural example for a logician of a Polish group acting on a Polish space is the action of S_∞ on the space of countable models. There are a few fiddley points here related to precisely which topology we would want to place on this space, so I will postpone the details to section 2.2.

CONJECTURE 2.7 (Topological Vaught Conjecture[1]). Whenever G is a Polish group and X a Polish G-space, then either

1. $X/G \leq \aleph_0$ (countably many orbits[2]), or
2. $X/G = 2^{\aleph_0}$ (continuum many orbits).

This conjecture should be compared with its counterparts for first order logic and countably infinitary logic.

CONJECTURE 2.8 (Original Vaught Conjecture). Let T be a first order theory. Then either T has countably many countable models up to isomorphism or it has continuum many countable models up to isomorphism.

DEFINITION 2.9. For $\mathcal{L}$ a language we let $\mathcal{L}_{\omega_1,\omega}$ be the collection of formulas obtained by closing under countable disjunctions and conjunctions, as well as the usual first operations of existential and universal quantification, substitutions, and negation.

CONJECTURE 2.10 (Vaught Conjecture for $\mathcal{L}_{\omega_1,\omega}$). Let $\mathcal{L}$ be a countable language and σ a sentence in $\mathcal{L}_{\omega_1,\omega}$. Then either σ has countably many countable models up to isomorphism or it has continuum many countable models up to isomorphism.

---

[1] This conjecture is very much still open.
[2] In general I use "countably many" to mean either finitely many or exactly $\aleph_0$ many.

As stated here, the topological Vaught conjecture is trivially true under CH (the continuum hypothesis). Frequently people use a reformulation of this conjecture that is equivalent under the negation of CH by Burgess' theorem 2.2 below. And in fact this equivalence is local, action by action, not just global: For each Polish group G and Polish G-space X, if CH fails then there are perfectly many orbits if and only if there are continuum many.

In fact I will find it convenient to refer to the following as *the* topological Vaught conjecture.

CONJECTURE 2.11 (Topological Vaught Conjecture, *modern form*). Whenever G is a Polish group and X a Polish G-space, then either

1. $X/G \leq \aleph_0$ (countably many orbits), or
2. there is a perfect set $P \subset X$ (that is to say, a set which is closed, non-empty, and has no isolated points) such that any two points in P are orbit inequivalent under the action of G.

Since any perfect set necessarily has size $2^{\aleph_0}$ it is easily seen that this alternative form of the topological Vaught conjecture implies the earlier. In fact it turns out that these two conjectures are equiprovable over ZFC — one can prove the first using only the axioms of ZFC if and only if one can prove the second using only the axioms of ZFC. This equiprovability follows by a short absoluteness argument. If the stronger form of the topological Vaught conjecture failed, then we can add lots of reals to blow up the size of the continuum; in the resulting model there must still be uncountably many orbits, by Shoenfield absoluteness; but equally there cannot be a perfect set of orbit inequivalent points, again by Shoenfield absoluteness; and thus we arrive at a contradiction with the local equivalence of the two forms of the topological Vaught conjecture under ¬CH. We sketch this argument due to Sami at 4.2 below. The proof of this equivalence for the model theoretic versions of the Vaught conjecture has also been written up precisely in Steel [1978].

One way to think of this alternate form of the topological Vaught conjecture is this: It is the conjecture that there cannot be "exactly $\aleph_1$ many orbits", where we understand "exactly" to mean that there are $\aleph_1$ many orbits but not perfectly many orbits.

It might be helpful before continuing with the discussion of the painfully open topological Vaught conjecture to look at some parallel problems. One can pose similar kinds of questions for other classes of equivalence relations, and in almost every case the answer is known.

I will need a couple more definitions from descriptive set theory. If you have not seen these before they may appear technical, but the motivation is a natural one: We wish to describe certain classes of equivalence relations which are not intractably complicated and compare them with the kind that arise for Polish group actions.

DEFINITION 2.12. A subset B of a Polish space X is said to be *Borel* if it appears in the smallest collection $\mathcal{B}$ of subsets of X such that

1. $\mathcal{B}$ includes the open sets in X;
2. $\mathcal{B}$ is an algebra — that is to say it is closed under complementation in X and finite unions and intersections;
3. moreover $\mathcal{B}$ is a σ-algebra — that is to say, an algebra which is also closed under countable intersections.

(Thus the collection of *the Borel sets* is the smallest σ-algebra containing the open sets.)

An equivalence relation $E \subset X \times X$ is *Borel* if it is Borel in the product topology.

A subset A of a Polish space X is said to be *analytic* or Σ_1^1 if there is a Polish space Y and a Borel set $B \subset X \times Y$ (that is to say, Borel in the product topology) with

$$A = \{x \in X \mid \exists y \in Y((x, y) \in B)\}.$$

$P \subset X$ is Π_1^1 or *co-analytic* if its complement is analytic.

Clearly every Borel set is analytic, and then since the Borel sets are closed under complementation we equally have that every Borel set is co-analytic. It is not hard to see that if a Polish group G acts continuously on a Polish space X then the set

$$\{(x_1, g, x_2) \mid g \cdot x_1 = x_2\}$$

is closed in $X \times G \times X$, and thus the resulting orbit equivalence relation E_G is analytic.

In the context of model theory Borel sets have an intuitive meaning. They correspond roughly to sets which can be defined by sentences in $\mathcal{L}_{\omega_1,\omega}$. (See corollary 2.18 below.) Analytic sets can be thought of as like a generalization of the set of models in a language which admit an expansion to a model of some given theory in a larger language.

THEOREM 2.1 (Silver [1980]). *Let E be a Borel (or even Π_1^1) equivalence relation on a Polish space X. Then either*

1. $X/E \leq \aleph_0$, *or*
2. $X/E = 2^{\aleph_0}$ *(continuum many equivalence classes), and in fact in this case there is a perfect set of E-inequivalent points in X.*

This would be very nice for those of us who slave away trying to prove the topological Vaught conjecture if it were the case that the equivalence relations arising from Polish group actions were always Borel. Unfortunately while they must be analytic they can in general be non-Borel.

THEOREM 2.2 (Burgess [1978]). *Let E be a Σ_1^1 equivalence relation on a Polish space X.*

1. $X/E \leq \aleph_1$, *or*
2. $X/E = 2^{\aleph_0}$, *and in fact in this case there is a perfect set of E-inequivalent points in X.*

Burgess' theorem is optimal. If one considers the space $2^{\mathbb{N}}$ and sets $x_1 E x_2$ if $\omega_1^{\mathrm{ck}(x_1)} = \omega_1^{\mathrm{ck}(x_2)}$ (the supremum of the lengths of the x_1-recursive well orders of $\mathbb{N}$ equals the supremum of the lengths of the x_2-recursive well orders of $\mathbb{N}$) then we obtain an equivalence relation which has exactly $\aleph_1$ many classes and is in the complexity class Σ_1^1. By Cohen we know that $2^{\aleph_0}$ may be bigger than $\aleph_1$, and thus we cannot hope to extend Burgess' result in ZFC.

Thus any proof of the topological Vaught conjecture must use more than the simple fact that orbit equivalence relations being considered are analytic. It is known from Ryll-Nardzewski [1964] that the orbit equivalence relations arising from continuous actions of Polish groups on Polish spaces are a special kind of Σ_1^1 equivalence relation in which every orbit is equivalence class is Borel, but unfortunately the counterexample above with $\aleph_1$ many equivalence classes is also one in which every class is Borel.

It does not seem as if any easily stated fact about the kinds of equivalence relations arising from Polish group actions would allow some general descriptive set theoretic proof that there cannot be exactly $\aleph_1$ many orbits. The proof, if the conjecture is indeed true, would need to use very specific facts about the context of a Polish group action. It is unlikely to be soft.

2.2. The space of countable models [isomorphism of countable models viewed as a kind of Polish group action; different Polish topologies we may place on spaces of countable models].

DEFINITION 2.13. Let $\mathcal{L}$ be a countable language, and for notational simplicity assume that $\mathcal{L}$ is generated by a single binary relation R. We then let $\mathrm{Mod}(\mathcal{L})$ be the space of all $\mathcal{L}$-structures whose underlying set is $\mathbb{N}$, equipped with the topology generated by sets of the form

$$\{\mathcal{M} \in \mathrm{Mod}(\mathcal{L}) \colon \mathcal{M} \models \varphi(k_1, \ldots, k_n)\}$$

where φ is quantifier free and $k_1, \ldots, k_n$ are in $\mathbb{N}$. There is then a natural homeomorphism

$$\mathrm{Mod}(\mathcal{L}) \to \{0, 1\}^{\mathbb{N} \times \mathbb{N}},$$
$$\mathcal{M} \mapsto \chi_{R^{\mathcal{M}}}$$

associating to each $\mathcal{M}$ the characteristic function of its interpretation of R.

Thus $\mathrm{Mod}(\mathcal{L})$ is a Polish space.

We let S_∞ act on $\mathrm{Mod}(\mathcal{L})$ by

$$\pi \cdot \mathcal{M} \models R(k_1, k_2)$$

if and only if

$$\mathcal{M} \models R(\pi^{-1}(k_1), \pi^{-1}(k_2)).$$

Then $\mathrm{Mod}(\mathcal{L})$ is a Polish S_∞ space and the orbit equivalence relation is just the usual isomorphism relation.

Our next goal is to see that the topological Vaught conjecture really does generalize the model theoretic versions of the Vaught conjecture. To do this we need to consider alternative Polish topologies we may place on the space of countable models.

DEFINITION 2.14.

(i) Let τ_{FO} be the topology on $\mathrm{Mod}(\mathcal{L})$ whose basic open sets have the form

$$\{\mathcal{M} \in \mathrm{Mod}(\mathcal{L}): \mathcal{M} \models \varphi(k_1, \dots, k_n)\}$$

where φ is first order and $k_1, \dots, k_n$ are in $\mathbb{N}$.

(ii) For $F \subset \mathcal{L}_{\omega_1, \omega}$ a countable fragment (closed under subformulas, instantiations, and first order operations), let τ_F be the topology on $\mathrm{Mod}(\mathcal{L})$ whose basic open sets have the form

$$\{\mathcal{M} \in \mathrm{Mod}(\mathcal{L}): \mathcal{M} \models \varphi(k_1, \dots, k_n)\}$$

where $\varphi \in F$.

All these provide Polish topologies on $\mathrm{Mod}(\mathcal{L})$. (Proof for (i) below). They are all finer that the topology of quantifier free formulas. They all give rise to the same Borel structure as the topology of quantifier free formulas, which I will continue to use as my default choice of topology on the space.

Thus for any theory T the set

$$\{\mathcal{M} \in \mathrm{Mod}(\mathcal{L}): \mathcal{M} \models T\}$$

is a closed S_∞-invariant subspace of $(\mathrm{Mod}(\mathcal{L}), \tau_{\mathrm{FO}})$, and hence a Polish S_∞ space in its own right. Sets of the form $\{\mathcal{M} \in \mathrm{Mod}(\mathcal{L}): \mathcal{M} \models T\}$ correspond exactly to the closed invariant subspaces; the subspace is minimal iff T complete.

A key step in the proof of Becker's theorem (which is coming up in section 3.3) is to pass from a not so good topology, like the quantifier free topology, to a better topology, like τ_{FO}.

LEMMA 2.15 (Gregorczyk, Mostowski, Ryll-Nardzewski [1960]).
$(\mathrm{Mod}(\mathcal{L}), \tau_{\mathrm{FO}})$ is a Polish space.

PROOF. Let B be the set of substitutions of the form $\varphi(k_1, \dots, k_n)$ where φ is a first order $\mathcal{L}$ formula. Let $X = \{0, 1\}^B$, the space of all functions from B to $\{0, 1\}$. We may naturally identify $(\mathrm{Mod}(\mathcal{L}), \tau_{\mathrm{FO}})$ with the functions in $f \in X$ which preserve the "inductive unraveling of truth" — for instance, $f(\varphi_1(k) \wedge \varphi_2(k)) = 1$ if and only if $f(\varphi_1(k)) = f(\varphi_2(k)) = 1$. Most of these requirements correspond to closed conditions in X, except for the rule governing existential quantifiers,

$$f(\exists x \varphi(k, x)) = 1 \Leftrightarrow \exists l \in \mathbb{N} \; f(\varphi(k, l)) = 1,$$

which is still no worse than G_δ.

Thus $(\text{Mod}(\mathcal{L}), \tau_{\text{FO}})$ is homeomorphic to a G_δ subspace of a Polish space and therefore Polish. $\dashv$

In many respects τ_{FO} would seem to be a nicer topology than the topology generated by quantifier free formulas. This becomes more apparent after we introduce the notion of *Vaught transform* below and note that τ_{FO} has a basis that is closed under the Vaught transforms.

The next subsection discusses an important theorem due to Becker and Kechris which states that among other things that for any Polish G-space there are "cofinally" many Polish topologies in which that action is still continuous and that are closed under the Vaught transforms. This was arguably the main result of Becker-Kechris [1996]. Sami [1994] had previously shown the same thing for a general class of continuous actions of S_∞, including its natural action on $(\text{Mod}(\mathcal{L}), \tau_{\text{FO}})$.

2.3. Becker-Kechris theorem [Vaught transforms; adding to the topology of a Polish G-space and keeping the action continuous].

DEFINITION 2.16. Let G be a Polish group and X a Polish G-space. For $V \subset G$ open and $B \subset X$ (preferably Borel), we let

$$B^{\Delta V}$$

be the set of $x \in X$ for which there is a non-meager[3] set of $g \in V$ for which $g \cdot x \in B$. We symmetrically define B^{*V} to be the set of $x \in X$ for which there is a relatively comeager set of $g \in V$ with $g \cdot x \in B$.

LEMMA 2.17 (Vaught [1973]). *For* $\varphi \in \mathcal{L}_{\omega_1, \omega}$, B *the set of* $\mathcal{M} \in \text{Mod}(\mathcal{L})$ *with* $\mathcal{M} \models \varphi(k_1, \ldots, k_n, a_1, \ldots, a_m)$, *and* $V = \{\pi \in S_\infty : \pi(a_1) = b_1, \ldots, \pi(a_m) = b_m\}$,

$$B^{\Delta V} = \{\mathcal{M} \in \text{Mod}(\mathcal{L}) : \mathcal{M} \models \exists x_1 \ldots \exists x_n \varphi(\vec{x}, \vec{b})\},$$

$$B^{*V} = \{\mathcal{M} \in \text{Mod}(\mathcal{L}) : \mathcal{M} \models \forall x_1 \ldots \forall x_n \varphi(\vec{x}, \vec{b})\}.$$

Using this one can show by an induction on the Borel complexity of B that:

COROLLARY 2.18 (Lopez-Escobar [1965]). *If* $B \subset \text{Mod}(\mathcal{L})$ *is Borel and invariant then there is a sentence* $\varphi \in \mathcal{L}_{\omega_1, \omega}$ *with*

$$B = \{\mathcal{M} \in \text{Mod}(\mathcal{L}) : \mathcal{M} \models \varphi\}.$$

---

[3]A set is meager if it is included in a countable union of closed sets each of which contain no non-empty open subset.

THEOREM 2.3 (Becker-Kechris [1996]). *Let G be a Polish group and X a Polish G-space. Let $B \subset X$ be Borel. Let $\mathcal{B}$ be a countable basis for G.*

Then there is a Polish topology τ on X such that:

1. *τ is richer than the original topology on X;*
2. *for any $V \in \mathcal{B}$*

$$B^{\Delta V} \in \tau;$$

3. *τ has a countable basis $\mathcal{C}$ such that for all $U_1, U_2 \in \mathcal{C}$ and $V_1, V_2 \in \mathcal{B}$,*

$$(U_1 \cap (X \setminus U_2)^{*V_1})^{\Delta V_2} \in \mathcal{C};$$

(This rather ominous looking condition plays a central role in the proof of Becker's theorem 3.6 below; it represents an approximation to the basis having the kind of closure conditions presented by the natural basis for τ_{FO} on $\mathrm{Mod}(\mathcal{L})$.)

4. *the action of G on X is still continuous with respect to τ (and thus (X, τ) is still a Polish G-space).*

In the case that B is invariant under the action of G we obtain that $B^{\Delta G} = B^{*G} = B$, and thus B is an invariant clopen subset of (X, τ). In particular, since for any sentence $\sigma \in \mathcal{L}_{\omega_1, \omega}$ we have that

$$\{\mathcal{M} : \mathcal{M} \models \sigma\}$$

is Borel and invariant, Vaught's conjecture for $\mathcal{L}_{\omega_1, \omega}$ is implied by the topological Vaught conjecture for S_∞.

It would take us too far afield to go into the details of their proof. Instead I will present a lemma, which by a transfinite induction can be used to prove the Becker Kechris theorem.

LEMMA 2.19. *X a Polish G-space, $V \subset G$ open and $C \subset X$ closed. Then there is a richer Polish topology on X for which the action is still continuous and $C^{\Delta V}$ is now open.*

PROOF. By a theorem of Birkhoff and Kakutani (see Becker-Kechris [1996], 1.1.1) we may find a compatible right invariant metric d on G that is bounded by 1. Let $\mathcal{L}(G)$ be the space of $f : G \to [0, 1]$ such that for all $g_1, g_2 \in G$

$$|f(g_1) - f(g_2)| \le d(g_1, g_2).$$

Then $\mathcal{L}(G)$ is a Polish G-space under the natural action $(g \cdot f)(h) = f(hg)$.

For $x \in X$ let $f_x \in \mathcal{L}(G)$ be given by

$$f_x(g) = \inf\{d(g, h) : h \cdot x \notin C\}.$$

$X_C = \{(x, f_x) : x \in X\}$ is an invariant subset of $X \times \mathcal{L}(G)$ and $\pi_C : X \to X_C$

$$x \mapsto (x, f_x)$$

is a bijection which respects the G-action.

One then verifies, as in Hjorth [1999], that X_C is a G_δ subset of $X \times \mathcal{L}(G)$ and thus a Polish G-space. The pullback of the topology on X_C to X is as required.   ⊣

§3. Smoothness and the Glimm-Effros dichotomy.

3.1. Smoothness [Effros lemma; G_δ orbits as the analogue of atomic models].

DEFINITION 3.1. An orbit equivalence relation E_G on Polish space X is said to be *smooth* if there is a Borel function[4] $f : X \to \mathbb{R}$ such that for all $x_1, x_2 \in X$

$$x_1 E_G x_2 \Leftrightarrow f(x_1) = f(x_2).$$

Burgess [1979] showed that for E_G given by the continuous action of a Polish group action we have E_G smooth if and only if there is a Borel set meeting every orbit in exactly one point.

EXAMPLE 3.2. For $\mathcal{M}, \mathcal{N} \in \mathrm{Mod}(\mathcal{L})$ set

$$\mathcal{M} \equiv \mathcal{N}$$

if they have the same (first order) theory.

Then $\equiv$ is smooth. To see this, let $(\varphi_n)_{n \in \mathbb{N}}$ enumerate the first order sentences of $\mathcal{L}$. For $\mathcal{M} \in \mathrm{Mod}(\mathcal{L})$ associate the real $x_\mathcal{M} \in (0, 1)$ in whose decimal expansion we have a 5 at the nth position if $\mathcal{M} \models \varphi_n$ and 6 if it does not. $\mathcal{M} \mapsto x_\mathcal{M}$ is a Borel function witnessing smoothness.

EXAMPLE 3.3. Let $\mathbb{Z}_2^{<\mathbb{N}}$ be the infinite binary sequences which are eventually constantly 0. This is a countable set, and we give it the discrete topological structure. We further introduce the group structure obtained by pointwise addition mod 2; thus for $\vec{g}, \vec{h} \in \mathbb{Z}_2^{<\mathbb{N}}$ and $n \in \mathbb{N}$ we have

$$(\vec{g} + \vec{h})(n) = \begin{cases} 0 & \text{if } \vec{h}(n) = \vec{g}(n), \\ 1 & \text{otherwise.} \end{cases}$$

We then let it act on $2^{\mathbb{N}}$ in the natural manner: For $\vec{x} \in 2^{\mathbb{N}} =_{df} \{0, 1\}^{\mathbb{N}}$, $\vec{g} \in \mathbb{Z}_2^{<\mathbb{N}}$, $n \in \mathbb{N}$, we again have $(\vec{g} \cdot \vec{x})(n) = 0$ if and only if $\vec{x}(n) = \vec{g}(n)$.

LEMMA 3.4. *The above orbit equivalence relation* $E_{\mathbb{Z}_2^{<\mathbb{N}}}$ *is non-smooth.*

PROOF. Suppose instead (by Burgess [1979]) B is a Borel selector. Let μ be the product ("coin flipping") measure on $2^{\mathbb{N}}$. Note that $(2^{\mathbb{N}}, \mu)$ is a probability space and $\mathbb{Z}_2^{<\mathbb{N}}$ acts by measure preserving transformations on $(2^{\mathbb{N}}, \mu)$.

Since countably many translates of B cover $2^{\mathbb{N}}$ we must have $\mu(B) > 0$. But then there are $\aleph_0$ many *disjoint* translates of B, each having the same measure, and hence a contradiction to $\mu(2^{\mathbb{N}}) = 1$.   ⊣

---

[4]That is to say, for any open $O \subset \mathbb{R}$ the set $f^{-1}(O)$ appears in the σ-algebra on X generated by the open sets — i.e., $f^{-1}[O]$ is Borel.

There are many theorems giving necessary conditions for non-smoothness in terms of being able to *embed* this orbit equivalence relation $E_{\mathbb{Z}_2^{<\mathbb{N}}}$. For a wide variety of groups (but not S_∞!) one can show that E_G is smooth if and only if we cannot so embed this canonical example of a non-smooth equivalence relation.

For instance:

THEOREM 3.1 (Becker-Kechris). *Let G be a Polish group and X a Polish G-space, $x \in X$. Suppose that no orbit $[y]_G$ with the same closure as $[x]_G$ is comeager in*

$$\overline{[x]_G},$$

the closure of the orbit.
    Then

1. *E_G is not smooth and*
2. *there is a continuous injection*

$$\rho \colon 2^{\mathbb{N}} \to X$$

such that for all $\vec{x}, \vec{y} \in 2^{\mathbb{N}}$

$$\vec{x} E_{\mathbb{Z}_2^{<\mathbb{N}}} \vec{y} \Leftrightarrow (\rho(\vec{x}) E_G \rho(\vec{y})).$$

In fact 2. implies 1. in the statement of the theorem.

After this Becker-Kechris theorem one might conclude that it would be nice to understand when an orbit is comeager in its own closure. Here a theorem was proved some 30 years earlier by Effros:

THEOREM 3.2 (Effros [1965]). *Let G be a Polish group and X a Polish G-space. Then for $x \in X$ the following are equivalent*:

1. *$[x]_G$ is comeager in $\overline{[x]_G}$;*
2. *$[x]_G$ is G_δ;*
3. *the map*

$$G \to [x]_G,$$
$$g \mapsto g \cdot x$$

    is open.

In the case of the logic action of S_∞ on $(\mathrm{Mod}(\mathcal{L}), \tau_{\mathrm{FO}})$ — the $\mathcal{L}$-structures on $\mathbb{N}$ with the topology generated by first order logic — one has that the G_δ orbits correspond to atomic models, see Miller [1978]. (The omitting types theorem shows that the realization of any non-principal type is meager.)

3.2. Examples [algebraic characterization of discrete groups with smooth dual (Thoma); Bernoulli shifts up to isomorphism are smooth (Ornstein); but not general mpt's (Feldman)].

EXAMPLE 3.5. As before let U_∞ be the infinite dimensional unitary group, consisting of all isomorphisms of Hilbert space. For H a countable (discrete) group we define $\mathrm{Rep}(H)$ to be the space of all homomorphism

$$\rho\colon H \to U_\infty.$$

This may naturally be identified with a closed subspace of the H-fold product of U_∞,

$$\prod_H U_\infty,$$

and thus is a Polish space. We let U_∞ act on $\mathrm{Rep}(H)$ by "pointwise conjugation",

$$(T \cdot \rho)(h) = T \circ \rho(h) \circ T^{-1}.$$

We then let $\mathrm{Irr}(H)$ be the *irreducible* representations in $\mathrm{Rep}(H)$ — that is to say those ρ for which there are no non-trivial closed subspaces of Hilbert space closed under the operators $\{\rho(h)\colon h \in H\}$. It turns out that $\mathrm{Irr}(H)$ is a G_δ subset of $\mathrm{Rep}(H)$, and thus a Polish U_∞-space in its own right.

THEOREM 3.3 (Thoma [1964]).   *The orbit equivalence E_{U_∞} on $\mathrm{Irr}(H)$ is smooth if and only if H is abelian-by-finite.*

One can extend the definition of $\mathrm{Rep}(H)$ and $\mathrm{Irr}(H)$ to a general locally compact group. Here one demands that the representations all be continuous, and we give these spaces the "compact-open" topology, generated by sets of the form

$$\{\rho\colon \rho[C] \subset O\},$$

where $C \subset H$ is compact and $O \subset U_\infty$ is open. Again we obtain that $\mathrm{Rep}(H)$ and $\mathrm{Irr}(H)$ are U_∞ Polish spaces.

THEOREM 3.4 (Harish-Chandra [1953], more or less[5]).   *If H is a semi-simple connected Lie group, then the orbit equivalence E_{U_∞} on $\mathrm{Irr}(H)$ is smooth.*

EXAMPLE 3.6. Let M_∞ be the group of measure preserving transformations of the unit interval, subject to identification of transformations agreeing almost everywhere. We give this the topology generated by sets of the form

$$\{\pi \in M_\infty\colon \mu(\pi(A)\Delta\pi_0(A)) < \varepsilon\},$$

for $\pi_0 \in M_\infty$, $A \subset [0, 1]$ measurable, $\varepsilon > 0$. In this topology and the operation of composition it becomes a Polish group.

It seems natural to think of measure preserving transformations as really being "isomorphic" if they are the same up to some isomorphism of the

---

[5]I fib terribly. In fact he proved that the representations of semi-simple connected Lie groups are all "type I". While at the time it was apparently felt that this should turn into a proof of the smoothness of $E_{U_\infty}|_{\mathrm{Irr}(H)}$, this does not seem to have been confirmed until the proof of the "Mackey conjecture" a few years later by Glimm.

underlying standard Borel probability space. Thus we are led to the orbit equivalence relation of M_∞ acting on itself by conjugation:

$$\pi \cdot \rho = \pi \circ \rho \circ \pi^{-1}.$$

THEOREM 3.5 (Feldman[6]). *The orbit equivalence relation E_{M_∞} obtained above is not smooth.*

For some special classes of measure preserving transformations one obtains smoothness of the restriction of E_{M_∞}. For instance Ornstein showed that the isomorphism relation on *Bernoulli shifts* is smooth and that *entropy* provides a complete invariant.

3.3. Becker's theorem [Glimm-Effros dichotomy, and hence Vaught's conjecture, for solvable groups].

DEFINITION 3.7 (Becker). A Polish group G satisfies the *Glimm-Effros dichotomy* if whenever G acts continuously on a Polish space X either:

1. E_G is smooth; or
2. there is a continuous injection

$$\rho: 2^{\mathbb{N}} \to X$$

such that for all $\vec{x}, \vec{y} \in 2^{\mathbb{N}}$

$$\vec{x} E_{\mathbb{Z}_2^{<\mathbb{N}}} \vec{y} \Leftrightarrow (\rho(\vec{x}) E_G \rho(\vec{y})).$$

It is not hard to see that the Glimm-Effros property for a group G implies that it does not provide a counterexample for the topological Vaught conjecture.

LEMMA 3.8. *Let G be a Polish group and X a Polish G-space. If G satisfies the Glimm-Effros dichotomy then X does not have exactly $\aleph_1$ many orbits.*

PROOF. Let X be a Polish G-space. We wish to show that either there are perfectly many orbits or (at most) countably many orbits.

Apply the definition of the Glimm-Effros dichotomy. Suppose first we are in case 2, and so we have a continuous embedding ρ of $E_{\mathbb{Z}_2^{<\mathbb{N}}}$ into E_G. We can consider the compact set

$$C = \{\vec{x} \mid \forall n \ (n \text{ not a prime power} \Rightarrow x(n) = 0)$$
$$\wedge \forall p \text{ prime } \forall k > 0 \ (x(p) = x(p^k))\}.$$

Any two elements of C are $E_{\mathbb{Z}_2^{<\mathbb{N}}}$-inequivalent. $\rho[C]$ is again compact and without isolated points. And thus $\rho[C]$ provides a perfect set of orbit inequivalent points.

---

[6]Actually something stronger and rather more specific is shown in Feldman [1974]. He obtains non-smoothness even for the property K transformations. See Foreman [to appear] for a thorough discussion of this and related results.

Alternatively suppose we are in case 1. Then there is Borel G-invariant

$$f : X \to \mathbb{R},$$

such that for all $x_1, x_2 \in X$

$$x_1 E_G x_2 \Leftrightarrow f(x_1) = f(x_2).$$

A routine calculation shows that E_G must consequently be Borel, and we can finish by applying Silver's theorem.   ⊣

Ever expanding classes of groups have been shown to satisfy the Glimm-Effros dichotomy:

1. Locally compact Polish groups (Effros, after Glimm).
2. Abelian Polish groups (Solecki).
3. Nilpotent (Hjorth).
4. Solvable (Becker).

Notably S_∞ fails this dichotomy.

THEOREM 3.6 (Becker). *Solvable Polish groups satisfy the Glimm-Effros dichotomy.*

PROOF. Let G be a solvable Polish group and X a Polish G-space. Assume that 2. fails, that is to say, we are unable to continuously inject $E_{\mathbb{Z}_2^{<\mathbb{N}}}$ into E_G. Since G is solvable it has a *complete* compatible right invariant metric (Gao [1998], Hjorth-Solecki [1999]), call it d_G.

Following theorem 2.3 we may assume that we have a basis $\mathcal{C}$ for the topology τ on X and $\mathcal{B}$ on G such that:

1. for all $U_1, U_2 \in \mathcal{C}$ and $V_1, V_2 \in \mathcal{B}$,

$$(U_1 \cap (X \setminus U_2)^{*V_1})^{\Delta V_2} \in \mathcal{C};$$

2. the action of G on X is still continuous with respect to τ (and thus (X, τ) is still a Polish G-space).

It is not hard to show that there is a Borel function

$$\theta : X \to \mathbb{R}$$

such that for all $x_1, x_2 \in X$

$$\theta(x_1) = \theta(x_2) \Leftrightarrow \overline{[x_1]_G} = \overline{[x_2]_G}.$$

(We may code the closure of an orbit by a real number, and the coding can be completed in a Borel manner.) Thus it suffices to show that $\overline{[x]_G}$ is a complete invariant of $[x]_G$; that is to say, $\overline{[x_1]_G} = \overline{[x_2]_G}$ if and only if $[x_1]_G = [x_2]_G$.

Fix some $x \in X$.

Now by theorem 3.1, the Becker-Kechris theorem on embedding $E_{\mathbb{Z}_2^{<\mathbb{N}}}$, we know that there must an $[y]_G$ with the same closure as $[x]_G$ which is comeager in $\overline{[x]_G} = \overline{[y]_G}$.

Using the Effros theorem, we can obtain for each open

$$V_n = \{g \in G : d_G(1_G, g) < 2^{-n}\}$$

some open $U_n \in \mathcal{C}$ containing y with diameter less than 2^{-n} such that

$$V_n \cdot y \supset U_n \cap [y]_G.$$

Then using the closure assumptions on the the basis $\mathcal{C}$ for the topology and a calculation which we do not present here, it can be argued that whenever $U \in \mathcal{C}$ and

$$U \cap U_n \cap [y]_G \neq \emptyset,$$
$$\hat{x} \in U_n \cap [x]_G$$

then there are g arbitrarily close to the identity with

$$g \cdot \hat{x} \in U^{\Delta V_n^2},$$

and thus in particular we may find $g, h \in V_n^2$ with $hg \cdot \hat{x} \in U$.

Repeating this observation infinitely often we may find a sequence $(h_n)_{n \in \mathbb{N}}$ in G and $(x_n)_{n \in \mathbb{N}}$ in $[x]_G$ with:

1. $x_n \in U_n$;
2. $h_n \in V_n^4$, and thus $d_G(h_n, 1_G) < 2^{-n+2}$;
3. $x_{n+1} = h_n \cdot x_n$.

We find h_{n+1} by applying the preceding observation with U_{n+1} taking the place of the set U.

Thus if we let $g_n = h_n h_{n-1} \ldots h_1 h_0$ then we have $x_n = g_n \cdot x_0 \in [x]_G$ and by 1. we have

$$x_n \rightarrow y.$$

On the other hand 2. and right invariance of the metric gives

$$d_G(g_n, g_{n-1}) < 2^{-n+2}$$

and thus (g_n) is Cauchy. But for g_∞ the limit we obtain from continuity of the action,

$$g_\infty \cdot x_0 = y.$$

Thus we have established that $[x]_G = [y]_G$ is comeager in $\overline{[x]_G}$.

Generalizing we therefore have that whenever $\overline{[x_1]_G} = \overline{[x_2]_G}$ we must have both $[x_1]_G$ and $[x_2]_G$ are comeager in this common closure. Hence the orbits overlap. Hence they are equal. ⊣

EXAMPLE 3.9. S_∞ does not satisfy the Glimm-Effros dichotomy.

Let $\mathcal{L}$ be generated by binary function $+$, unary function $-(\ )$, and constant symbol 0. Let AG be the subset of Mod($\mathcal{L}$) consisting of $\mathcal{G}$ in which

$(+^{\mathcal{G}}, -(\ )^{\mathcal{G}}, 0^{\mathcal{G}})$ defines an abelian group structure on $\mathbb{N}$. Certainly this corresponds to a first order definable property on $\mathcal{G}$, and so AG is a closed invariant subspace of $\mathrm{Mod}(\mathcal{L})$.

Then for p a prime let TAG_p be the set of all p-groups — that is to say, $\mathcal{G}$ such that for all $g \in \mathcal{G}$ there exists $n > 0$ with

$$\mathcal{G} \models p^n \cdot g = 0.$$

TAG_p is an invariant G_δ set, and hence a Polish S_∞-space in its own right.

The Ulm analysis of abelian p-groups shows that we may find a *reasonably nice* (e.g. absolutely Δ_2^1 in the codes, in the sense of Hjorth [2000]) function $U: \mathrm{TAG}_p \to 2^{<\omega_1}$ assigning bounded subsets of $\aleph_1$ as complete invariants.

CLAIM: $\mathrm{TAG}_p / S_\infty$ is not smooth.

Roughly this follows because in general there can be no *reasonably simply definable* (e.g. absolutely Δ_2^1!) ω_1 sequence of reals.

CLAIM: We cannot embed $E_{\mathbb{Z}_2^{<\mathbb{N}}}$ into $\mathrm{TAG}_p / S_\infty$.

Otherwise by postcomposing with U we would obtain an assignment $\hat{U}$ of bounded subsets of $\aleph_1$ as complete invariants for $2^{\mathbb{N}}/\mathbb{Z}_2^{<\mathbb{N}}$. Then we would obtain some fixed $\alpha < \omega_1$ and μ-measure[7] one set on which the reduction $\hat{U}$ takes values inside 2^α. (To see this, we can take $(M; \in)$ some countable transitive model containing the real parameter used in the embedding of $E_{\mathbb{Z}_2^{<\mathbb{N}}}$; then there will be a μ-measure 1 set of $x \in 2^{\mathbb{N}}$ that are "random" generic for M — and for any such x there will be some $\alpha < \omega_1^M = \omega_1^{M[x]} < \omega_1^V$ with $\hat{U}(x) \in 2^\alpha$).

But then since 2^α is naturally isomorphic to $2^{\mathbb{N}}$, and hence Borel isomorphic to $\mathbb{R}$, we obtain a contradiction to our proof before that $2^{\mathbb{N}}/\mathbb{Z}_2^{<\mathbb{N}}$ is non-smooth on any μ-measure one set.

(Here we have skipped over technicalities relating to the fact that $\hat{U}$ is not necessarily Borel. The issue here however is that it can be chosen to be universally measurable in the codes; hence on a measure one set it will equal a Borel function, or alternatively the proof of 3.4 can be seen to work in the more general context of μ-measurable functions.)

§4. Vaught's conjecture on analytic sets.

4.1. Variations on the conjecture.

NOTATION 4.1. For H a Polish group let $\mathrm{TVC}(H)$ be the statement that whenever X is a Polish H-space then either there are only countably many orbits or there is a perfect set $P \subset X$ of orbit inequivalent points:

$$\forall x_1, x_2 \in P \ (x_1 \neq x_2 \Rightarrow [x_1]_H \neq [x_2]_H).$$

---

[7] μ the "coin flipping" measure from before.

Let $\mathrm{WVC}(H)$ be the statement that when X is a Polish H-space then either there are only countably many orbits or there are $2^{\aleph_0}$ many orbits.

LEMMA 4.2 (Sami). *If H is a Polish group and X is a Polish H-space witnessing the failure of $\mathrm{TVC}(H)$, then X continues to witness the failure of $\mathrm{TVC}(H)$ through all generic extensions.*

CLAIM: In no generic extension can X contain a perfect set of orbits.

PROOF of Claim: The statement that there are perfectly many orbits may be expressed as:

$$\exists P \subset X \ (P \text{ perfect} \wedge \forall x_1, x_2 \in P \ \forall g \in G \ (x_1 \neq x_2 \Rightarrow g \cdot x_1 \neq x_2)).$$

This is Σ_2^1, and hence absolute by Shoenfield.[8]   ⊣

CLAIM: In every generic extension X/H is uncountable.

PROOF of Claim: For conceptual simplicity let us do the case that

$$X = \{\mathcal{M} \in \mathrm{Mod}(\mathcal{L}): \mathcal{M} \models T\}$$

and $H = S_\infty$. Then for any $\mathcal{M} \in \mathrm{Mod}(\mathcal{L})$ we have that $[\mathcal{M}]_{S_\infty}$ is uniformly Borel in (any code for) the Scott sentence of $\mathcal{M}$. Moreover, in some natural parameter space Y we have that

$$\mathcal{S} = \{(\mathcal{M}, y): y \text{ codes the Scott sentence of } \mathcal{M}\}$$

is Π_1^1. Then the statement that the models of T has uncountably many isomorphism types among its countable models becomes the statement that for all sequences $(\mathcal{M}_i), (y_i)$, either:

1. there exists i with $(\mathcal{M}_i, y_i) \notin \mathcal{S}$; or
2. there exists $\mathcal{N} \models T$ with $\mathcal{N}$ not satisfying any of the sentences coded by the $\{y_i: i \in \mathbb{N}\}$.

Thus it is Π_2^1 and subject to Shoenfield absoluteness.

The general case for arbitrary Polish group actions follows by working not with the Scott sentence, but the *stabilizer* of the points in question; again $[x]_H$ is uniformly Borel in (any code for) the stabilizer of x (see the proof of 1.2.4, 7.1.2 Becker-Kechris [1996]).   ⊣

COROLLARY 4.3. *It is provable in ZFC that every Polish group H satisfies $\mathrm{WVC}(H)$ (i.e. the original topological Vaught conjecture) if and only it is provable that every Polish group satisfies $\mathrm{TVC}(H)$.*

---

[8]This proof also works to show that if a Σ_1^1 A does not contain a perfect set of orbits then in no generic extension it will contain a perfect set of orbits. The point here is that if P is a perfect subset of A then by say an appeal to the Baire category theorem and the Jankov-von Neumann uniformization we may find a perfect $P_0 \subset P$ and a continuous function f on P_0 such that for all $x \in P_0$ we that $f(x)$ witnesses $x \in A$.

PROOF. ⇐ since any perfect set has size $2^{\aleph_0}$.

⇒ since if we have a group H and a Polish H-space X showing the failure of TVC(H) then we may go to a forcing extension in which $2^{\aleph_0} = \aleph_2$. In the forcing extension Sami's lemma and Burgess' theorem from 2.2 on analytic equivalence relations imply that we must have $|X/H| = \aleph_1$.   ⊣

NOTATION 4.4. For H a Polish group let TVC(H, Σ_1^1) be the statement that whenever X is a Polish H-space and $A \subset X$ is Σ_1^1 then either

$$A/H \leq \aleph_0$$

or there is perfect $P \subset A$ such that for all $x_1, x_2 \in P$

$$x_1 \neq x_2 \Rightarrow [x_1]_H \neq [x_2]_H.$$

Sami's argument also shows absoluteness of TVC(H, Σ_1^1).

LEMMA 4.5 (H. Friedman [1973]). *Up to isomorphism there are exactly $\aleph_1$ many possible order types for the ordinals in countable ω-models of Kripke-Platek set theory.*

COROLLARY 4.6. TVC(S_∞, Σ_1^1) *fails.*

4.2. Some sketches around the proof for characterizing Vaught's conjecture on analytic sets.

THEOREM 4.1. *A Polish group H satisfies* TVC(H, Σ_1^1) *if and only if S_∞ does not divide H.*

PROOF. ⇒: It was previously known by results of Mackey and Becker and Kechris (see 2.3.5 Becker-Kechris [1996]) that if K is a closed subgroup of a Polish group H and Y is a Polish K-space, then we can in some sense lift the action of K on Y up to an action of H on a Polish space X, and obtain many similarities between X/H and Y/K. In particular this "Mackey" lift suffices to show that ¬ TVC(K, Σ_1^1) implies ¬ TVC(H, Σ_1^1).

But now if we have that S_∞ is the continuous surjective image of K then certainly we can obtain that any orbit equivalence relation induced by a continuous action of S_∞ is induced by the continuous action of K.

⇐: We suppose that X is a Polish H-space which provides a counterexample to TVC(H, Σ_1^1). Let A be the Σ_1^1 set which has uncountably many but not perfectly many orbits. We need to find a closed subgroup of H which has S_∞ as its continuous homomorphic image.

Let us first consider the argument for the following special case. Let $\mathcal{L}_0$ be a countable language and $\mathcal{M}_0$ a $\mathcal{L}_0$-model with underlying set $\mathbb{N}$ and let us suppose that H is the automorphism group of $\mathcal{M}_0$. Then let us suppose for some countable $\mathcal{L}_1 \supset \mathcal{L}_0$ we have that X is space of all expansions of $\mathcal{M}_0$ to an $\mathcal{L}_1$-structure satisfying some given theory T.

(Actually this case is not quite as special as it seems. Any closed subgroup of S_∞ is the automorphism group of a countable structure, and Becker and

Kechris have shown that if $H = \text{Aut}(\mathcal{M}_0)$ then any Polish H-space may be naturally identified with a subspace of the expansions of $\mathcal{M}_0$ to some richer language $\mathcal{L}_1$.)

We need to show that the automorphism group of $\mathcal{M}_0$ is very complicated. In fact we will find some expansion $\mathcal{N} \in A$ with S_∞ dividing $\text{Aut}(\mathcal{N})$.

From the proof of Sami's lemma we obtain that through all generic extensions A has uncountably many orbits but not perfectly many. Thus in particular if we force to make the continuum of the ground model countable then there must be some new equivalence class in A.

So let $\mathbb{P} = \text{Coll}(\omega, 2^{\aleph_0})$ be the forcing to collapse the continuum of V. By the completeness lemma for forcing we may find some $\mathbb{P}$-term σ such that

$$\mathbb{P} \Vdash \sigma[\dot{G}] \in A \wedge \forall \mathcal{N} \in A \ (\mathcal{N} \not\cong \sigma[\dot{G}]).$$

Step 1: There exists $p \in \mathbb{P}$ such that

$$(p, p) \Vdash_{\mathbb{P} \times \mathbb{P}} \sigma[\dot{G}_l] \cong \sigma[\dot{G}_r],$$

where here $\dot{G}_l \times \dot{G}_r$ names the generic object on $\mathbb{P} \times \mathbb{P}$. In other words, the isomorphism type of the model $\sigma[G]$ does not depend on the choice of the generic filter G.

The proof of the claim goes back to some early arguments by Harrington and Silver. The rough idea is that if it failed then in $V^{\text{Coll}(\omega, |\mathcal{P}(\mathbb{P} \times \mathbb{P})|)}$ we could find a "perfect set" of generic filters, such that any two distinct filters pass through some pair of conditions forcing that their evaluations of the term σ leads to non-isomorphic models.

More precisely, we let (D_n) enumerate in $V^{\text{Coll}(\omega, |\mathcal{P}(\mathbb{P} \times \mathbb{P})|)}$ the dense open subset of $\mathbb{P} \times \mathbb{P}$ found in V and we choose conditions $\{p_s : s \in 2^{<\mathbb{N}}\}$ such that for $t \subset s$ we have $p_s \leq p_t$ and for $s, t \in 2^n$ (some $n \in \mathbb{N}$), $s \neq t$, we have

$$(p_s, p_t) \in \bigcap_{k < n} D_k,$$

$$(p_s, p_t) \Vdash \sigma[\dot{G}_l] \not\cong \sigma[\dot{G}_r].$$

For $w \in 2^{\mathbb{N}}$ we let G_w be the filter generated by $\{p_{w|n} : n \in \mathbb{N}\}$. It is not hard to see that

$$w \mapsto \sigma[G_w]$$

is a continuous injection and that for $w \neq w'$ we have

$$\sigma[G_w] \not\cong \sigma[G_{w'}].$$

Step 2: For $(\mathbb{P}, p, \sigma)$ as above, there exists some $\varphi \in V$ (already in the ground model) such that

$$p \Vdash_{\mathbb{P}} \varphi \text{ is Scott sentence of } \sigma[\dot{G}].$$

This follows from general facts about forcing. Let τ be a term for an object that exists in the generic extension but whose value does not depend on the generic:

$$(p, p) \Vdash_{\mathbb{P} \times \mathbb{P}} \tau[\dot{G}_l] = \tau[\dot{G}_r].$$

Then one can argue by induction on the set theoretical rank of τ that there must exist some $a \in V$ with

$$p \Vdash_{\mathbb{P}} \tau[\dot{G}] = a.$$

Step 3: Iterate step 1 to find some sequence $(\mathbb{P}_\alpha, p_\alpha, \sigma_\alpha)$ such that at each $\alpha \neq \beta$

$$p_\alpha \Vdash_{\mathbb{P}_\alpha} \sigma_\alpha[\dot{G}] \in A,$$

$$(p_\alpha, p_\alpha) \Vdash_{\mathbb{P}_\alpha \times \mathbb{P}_\alpha} \sigma_\alpha[\dot{G}_l] \cong \sigma_\alpha[\dot{G}_r],$$

$$(p_\alpha, p_\beta) \Vdash_{\mathbb{P}_\alpha \times \mathbb{P}_\beta} \sigma_\alpha[\dot{G}_l] \not\cong \sigma_\beta[\dot{G}_r].$$

Step 4: Apply step 2 repeatedly to now obtain a sequence (φ_α) of infinitary formulas[9] such that at each α

$$p_\alpha \Vdash_{\mathbb{P}_\alpha} \varphi_\alpha \text{ is Scott sentence of } \sigma_\alpha[\dot{G}].$$

At each α let $\gamma(\alpha)$ be the Scott height of any model of φ_α. Let

$$(\psi_\beta^\alpha)_{\beta < \delta(\alpha)}$$

enumerate without repetitions the canonical $\gamma(\alpha)$-types appearing in any model of φ_α.

Since $\varphi_\alpha \neq \varphi_\beta$ for all $\alpha \neq \beta$ we have

$$\gamma(\alpha), \delta(\alpha) \to \infty$$

as

$$\alpha \to \infty.$$

Step 5: Choose some $\delta(\alpha) \geq \beth_{\aleph_1}$.

Step 6: Choosing some very large λ, obtain an ω-model

$$(N; \hat{\varphi}, \hat{H}, \hat{X}, \mathcal{M}_0, \hat{A}, (\hat{\mathbb{P}}, \hat{p}, \hat{\sigma}), \hat{\in})$$

for the theory of the structure

$$(V; \varphi_\alpha, H, X, \mathcal{M}_0, A, (\mathbb{P}_\alpha, p_\alpha, \sigma_\alpha), \in),$$

which is in turn generated by the indiscernibles $(\hat{\psi}_q)_{q \in \mathbb{Q}}$, where for each $q \in \mathbb{Q}$ the structure N satisfies that $\hat{\psi}_q'$ is a canonical $\hat{\gamma}$ type for any model (appearing in any generic extension of N) satisfying $\hat{\varphi}$.

We can do this by the proof of $\beth_{\aleph_1}$ being the Hanf number of $\mathcal{L}_{\omega_1, \omega}$.

---

[9]In $\mathcal{L}\infty, \omega$.

Step 7: Use the generating indiscernibles $(\hat{\psi}_q)_{q \in \mathbb{Q}}$ to inject $\text{Aut}(\mathbb{Q}; <)$ into $\text{Aut}(N; \hat{\varphi}_\alpha, \dots)$.

Step 8: Let $G \subset \hat{\mathbb{P}}$ be N-generic below $\hat{p}$ and let $\mathcal{A} = \hat{\sigma}[G]$. Then show that each automorphism from step 7 can in some sense be lifted into the automorphism group of $\mathcal{A}$.

This uses a kind of back and forth argument and the fact that $N[G]$ believes that $\hat{\varphi}$ is the Scott sentence of $\mathcal{A}$.

Step 9: Note that since $N[G]$ is an ω-model we must that $\mathcal{A}$ is an expansion of $\mathcal{M}_0$. Hence we can obtain $\text{Aut}(\mathbb{Q}; <)$ as the continuous image of a closed subgroup of $\text{Aut}(\mathcal{M}_0)$. It is not hard to show that S_∞ divides $\text{Aut}(\mathbb{Q}, <)$, and thus S_∞ appears as the continuous image of a closed subgroup of $\text{Aut}(\mathcal{M}_0)$.

Clearly the above argument depended heavily on the existence of the Scott analysis and our ability to play some kind of combinatorial game with the various "virtual Scott sentences." In general there does not seem to be a perfect analog of the Scott sentence for arbitrary Polish group actions[10]; instead it turns out that the Becker-Kechris theorem 2.3 can be used entirely to supplant the combinatorial analysis above.

Remarks about the general case (See Hjorth [to appear] for details).

Suppose now that we have again that $A \subset X$ provides a $\mathbf{\Sigma}_1^1$ counterexample to $\text{TVC}(H, \mathbf{\Sigma}_1^1)$.

By the same arguments as before find some sequence $(\mathbb{P}_\alpha, p_\alpha, \sigma_\alpha)$ such that at each $\alpha \neq \beta$

$$p_\alpha \Vdash_{\mathbb{P}_\alpha} \sigma_\alpha[\dot{G}] \in A,$$

$$(p_\alpha, p_\alpha) \Vdash_{\mathbb{P}_\alpha \times \mathbb{P}_\alpha} \sigma_\alpha[\dot{G}_l] E_H \sigma_\alpha[\dot{G}_r],$$

$$(p_\alpha, p_\beta) \Vdash_{\mathbb{P}_\alpha \times \mathbb{P}_\beta} \neg \sigma_\alpha[\dot{G}_l] E_H \sigma_\beta[\dot{G}_r].$$

Now one can apply a kind of variation of the Becker-Kechris theorem to obtain some ordinal length sequence (τ_α) of "virtual Polish topologies," such that at each α

$$p_\alpha \Vdash_{\mathbb{P}_\alpha} \sigma_\alpha[\dot{G}] \text{ is } G_\delta \text{ in } (X, \tau_\alpha).$$

In some sense we can do this so that a complete description of τ_α exists in V — this is parallel to φ_α being in the ground model V for the special case above.

Then the back and forth argument from step 8 and its appeal to the basic properties of the Scott sentence is replaced by a kind of generalized back and forth argument (applied to the group H) that uses the Effros lemma on G_δ orbits. $\dashv$

4.3. Extensions and speculations. The method of proof allows some other uses. For instance virtually word for the word the same argument as above yields:

---

[10]But chapter 6 of Hjorth [2000] gives an extended discussion of what we can still hope for.

THEOREM 4.2. *Let $\sigma \in \mathcal{L}_{\omega_1,\omega}$ and suppose $A \subset \mathrm{Mod}(\sigma)$ is analytic and has exactly $\aleph_1$ many models up to isomorphism.*[11] *Then there is a countable model $\mathcal{M}$ of σ for which S_∞ divides $\mathrm{Aut}(\mathcal{M})$.*

Rather less clear is whether we could hope to characterize other properties of the equivalence relations of Polish group actions in terms of the algebraic properties of those groups. For instance one can hope for some criteria of when a Polish group satisfies some version of the Glimm-Effros. For this I suspect there are technical reasons why insisting on Borel functions witnessing smoothness may be too restrictive, so I will ask the question in the following way:

QUESTION 4.7. Is there a Polish group H such that the following two properties are equivalent for an arbitrary Polish G?

1. There is a closed subgroup G_0 of G and a continuous homomorphism $\pi\colon G_0 \twoheadrightarrow H$ onto H.
2. There is a Polish G-space X such that:
   there is no continuous injection
   $$\rho\colon 2^{\mathbb{N}} \to X$$
   with all $\vec{x}, \vec{y} \in 2^{\mathbb{N}}$ having
   $$\vec{x} E_{\mathbb{Z}_2^{<\mathbb{N}}} \vec{y} \Leftrightarrow (\rho(\vec{x}) E_G \rho(\vec{y}));$$
   there is no absolutely $\mathbf{\Delta}_2^1$ function (in the sense of say Hjorth [2000]) $\pi\colon X \to \mathbb{R}$ with
   $$\forall x_1, x_2 \in X \ (x_1 E_G x_2 \Leftrightarrow \pi(x) = \pi(y)).$$

Finally while we have a counterexample for the Glimm-Effros dichotomy for $\mathcal{L}_{\omega_1,\omega}$ given above, and hence a counterexample for continuous actions of S_∞, a strictly first order example seems to be missing.

QUESTION 4.8. For a countable first order theory T we can consider $\mathrm{Mod}(T)$, the space of all models for T with underlying set $\mathbb{N}$, equipped with the topology τ_{FO} given by first order formulas.

Does there exist a countable first oder T and an absolutely $\mathbf{\Delta}_2^1$ in the codes function
$$\pi\colon \mathrm{Mod}(T) \to 2^{<\omega_1}$$
such that the following both occur?

1. π is onto;
2. for all $\mathcal{M}, \mathcal{N} \in \mathrm{Mod}(T)$ we have
   $$\mathcal{M} \cong \mathcal{N} \Leftrightarrow \pi(\mathcal{M}) = \pi(\mathcal{N}).$$

---

[11] As in previous usage, here "exactly" is intended to indicate that there is no perfect set of non-isomorphic models.

If no such example can be found we might ask more optimistically:

QUESTION 4.9. Does first order logic satisfy the Glimm-Effros dichotomy? That is to say, if T is a countable first order theory, is it that case that one of the following must hold?

1. There is a Borel function θ from $\text{Mod}(\mathcal{L})$ to $\mathbb{R}$ such that for $\mathcal{M}, \mathcal{N} \models T$

$$\mathcal{M} \cong \mathcal{N} \Leftrightarrow \theta(\mathcal{M}) = \theta(\mathcal{N}).$$

2. There is a continuous injection

$$\rho: 2^{\mathbb{N}} \to \text{Mod}(\mathcal{L})$$

such that for all $\vec{x}, \vec{y} \in 2^{\mathbb{N}}$

$$\vec{x} E_{\mathbb{Z}_2^{<\mathbb{N}}} \vec{y} \Leftrightarrow (\rho(\vec{x}) \cong \rho(\vec{y})).$$

REFERENCES

H. BECKER [1998], *Polish group actions: dichotomies and generalized elementary embeddings*, **Journal of the American Mathematical Society**, vol. 11, pp. 397–449.

H. BECKER AND A. S. KECHRIS [1996], *The descriptive set theory of Polish group actions*, London Mathematical Society Lecture Notes Series, vol. 232, Cambridge University Press, Cambridge.

J. BURGESS [1978], *Equivalences generated by families of Borel sets*, **Proceedings of the American Mathematical Society**, vol. 69, pp. 323–326.

J. BURGESS [1979], *A selection theorem for group actions*, **Pacific Journal of Mathematics**, vol. 80, pp. 333–336.

E. G. EFFROS [1965], *Transformation groups and C^*-algebras*, **Annals of Mathematics**, vol. 81, pp. 38–55.

E. G. EFFROS [1981], *Polish transformation groups and classification problems*, **General topology and modern analysis (Proc. Conf., Univ. California, Riverside, Calif., 1980)** (New York-London), Academic Press, pp. 217–227.

J. FELDMAN [1974], *Borel structures and invariants for measurable transformations*, **Proceedings of the American Mathematical Society**, vol. 46, pp. 383–394.

M. FOREMAN [to appear], *A descriptive view of ergodic theory*.

H. FRIEDMAN [1973], *Countable models of set theory*, **Cambridge summer school in mathematical logic** (Berlin) (A. R. D. Mathias and H. Rogers, editors), Springer-Verlag, pp. 539–573.

H. FRIEDMAN AND L. STANLEY [1989], *A Borel reducibility theory for classes of countable structures*, **The Journal of Symbolic Logic**, vol. 54, pp. 894–914.

S. GAO [1998], *Automorphism groups of countable structures*, **The Journal of Symbolic Logic**, vol. 63, pp. 891–896.

J. GLIMM [1961], *Locally compact transformation groups*, **Transactions of the American Mathematical Society**, vol. 101, pp. 124–138.

A. GREGORCZYK, A. MOSTOWSKI, AND C. RYLL-NARDZEWSKI [1961], *Definability of sets of models of axiomatic theories*, **Bulletin of the Polish Academy of Sciences**, vol. 9, pp. 163–7.

HARISH-CHANDRA [1953], *Representations of a semisimple Lie group on a Banach space. I*, **Transactions of the American Mathematical Society**, vol. 75, pp. 185–243.

L. A. HARRINGTON, A. S. KECHRIS, AND A. LOUVEAU [1990], *A Glimm-Effros dichotomy for Borel equivalence relations*, **Journal of the American Mathematical Society**, vol. 3, pp. 903–928.

G. HJORTH [1999], *Sharper changes in topologies*, **Proceedings of the American Mathematical Society**, vol. 127, pp. 271–278.

G. HJORTH [2000], *Classification and orbit equivalence relations*, Mathematical Surveys and Monographs, vol. 75, American Mathematical Society, Providence, RI.

G. HJORTH [to appear], *Vaught's conjecture on analytic sets*, in the **Journal of the American Mathematical Society**.

G. HJORTH AND S. SOLECKI [1999], *Vaught's conjecture and the Glimm-Effros property for Polish transformation groups*, **Transactions of the American Mathematical Society**, vol. 351, pp. 2623–2641.

A. S. KECHRIS [1995], *Classical descriptive set theory*, Graduate Texts in Mathematics, Springer-Verlag, Berlin.

E. G. K. LOPEZ-ESCOBAR [1965], *An interpolation theorem for denumerably long formulas*, **Fundamenta Mathematicae**, vol. 57, pp. 253–272.

G. W. MACKEY [1957], *Borel structures in groups and their duals*, **Transactions of the American Mathematical Society**, vol. 85, pp. 134–165.

G. W. MACKEY [1963], *Infinite-dimensional group representations*, **Bulletin of the American Mathematical Society**, vol. 69, pp. 628–686.

D. E. MILLER [1977], *On the measurability of orbits in Borel actions*, **Proceedings of the American Mathematical Society**, vol. 63, pp. 165–170.

D. E. MILLER [1978], *The invariant Π_α^0 separation principle*, **Transactions of the American Mathematical Society**, vol. 242, pp. 185–204.

C. RYLL-NARDZEWSKI [1964], *On Borel measurability of orbits*, **Fundamenta Mathematicae**, vol. 56, pp. 129–130.

R. L. SAMI [1994], *Polish group actions and the Vaught conjecture*, **Transactions of the American Mathematical Society**, vol. 341, pp. 335–353.

J. H. SILVER [1980], *Counting the number of equivalence classes of Borel and co-analytic equivalence relations*, **Annals of Mathematical Logic**, vol. 18, pp. 1–28.

J. R. STEEL [1978], *On Vaught's conjecture*, **Cabal seminar 76–77** (Berlin), Lecture Notes in Mathematics, vol. 689, Springer, (Proceedings Caltech-UCLA Logic Seminar, 1976–77), pp. 193–208.

E. THOMA [1964], *Über unitäre Darstellungen abzählbarer, diskreter Gruppen*, **Mathematische Annalen**, vol. 153, pp. 111–138, (German).

R. L. VAUGHT [1973], *Descriptive set theory in $L_{\omega_1,\omega}$*, **Cambridge summer school in mathematical logic (Cambridge, England, 1971)** (Berlin), Lecture Notes in Mathematics, vol. 337, Springer, pp. 574–598.

R. L. VAUGHT [1974/1975], *Invariant sets in topology and logic*, **Fundamenta Mathematicae**, vol. 82, pp. 269–294, Collection of articles dedicated to Andrzej Mostowski on his sixtieth birthday.

DEPARTMENT OF MATHEMATICS
UNIVERSITY OF CALIFORNIA
LOS ANGELES CA90095-1555, USA
E-mail: greg@math.ucla.edu

ASPECTS OF GEOMETRIC MODEL THEORY

ANAND PILLAY

§1. **Introduction.** In this paper (based on my tutorial in Utrecht) I want to discuss some themes from contemporary model theory, mainly originating in stability theory and classification theory, and point out some mathematical implications. Model theory has become largely the study of definable sets (or the category of definable sets and functions) in given structures, as well as the study of interpretability and bi-interpretability. These can either be specific, such as the field of p-adic numbers (as in applications), or can be arbitrary structures which satisfy some model-theoretic hypotheses (stability, ω_1-categoricity, o-minimality). Among the themes or topics I will touch on are: dimension theory, how a structure is built up from "irreducible bits" (geometries), the fine structure of these "irreducible bits", modularity, orthogonality, equivariant model theory (definable groups and group actions), quotients and Galois theory. This paper is aimed at the non model-theorist logician. I want to explain a little of what is going on in model theory, but at the same time I do not want to simply repeat what has already been said in numerous surveys of this kind.

I assume acquaintance with the basic concepts and results of first order logic and model theory: complete theories, compactness theorem, elementary substructure, saturation, Löwenheim-Skolem theorems. T will usually denote a complete theory in a language L. M will be an L-structure, usually a model of T. We let $x, y, \dots$ denote finite sequences of variables, and $a, b, \dots$ finite sequences of elements of a structure M. If $\phi(x)$ is a formula, possibly with additional parameters from M, then $X = \{a \in M^n : M \models \phi(a)\}$ is called a definable set in M. We also write ϕ^M for X. If the parameters from ϕ belong to $A \subseteq M$ we say that X is A-definable. $tp_M(a/A)$ is the set of all formulas with parameters from A which are satisfied by a in M. A function $f : X \to Y$ is said to be definable (in M) if its graph is. So the category associated to M is the category of sets and functions definable in M. One may also want to (and should) include quotient objects X/E where E is a definable equivalence relation on X, as definable sets. This is unproblematic and will be discussed

Supported by NSF grants DMS 9696268 and 0070179.

Logic Colloquium '99
Edited by J. van Eijck, V. van Oostrom, and A. Visser
Lecture Notes in Logic, 17

later. A rather basic notion is that of algebraic closure: suppose M is a structure $A \subseteq M$ and a a tuple from M. We will say that a is algebraic over A (in M) ($a \in acl_M(A)$) if there is a formula $\phi(x)$ over A (that is, with parameters from A) which is satisfied by a and which has only finitely many realizations in M. By compactness this is equivalent to there being some cardinal bound to the set of realisations of $tp_M(a/A)$ in any elementary extension of M. If we require in addition that a is the unique realization of $\phi(x)$ we say a is in the definable closure of A, $a \in dcl_M(A)$.

§2. Morley's Theorem.

§2. Morley's Theorem. Many of the elements of geometric model theory are either present in or naturally suggested by the Baldwin-Lachlan proof of Morley's Theorem [1], and I will introduce them in this way. Morley's theorem was the beginning of stability theory and classification theory. The context is a countable complete theory T. For κ an infinite cardinal, T is said to be κ-categorical if T has exactly one model of cardinality κ up to isomorphism. Morley's theorem states that if T is κ-categorical for some uncountable κ then it is κ-categorical for all uncountable κ. This is a fundamental result on the expressive power of first order logic.

PROOF of Morley's Theorem. Assume T to be κ-categorical, where κ is an uncountable cardinal.

Step 1. Show that T is ω-stable (or totally transcendental). This means that for any countable model M of T the Stone space $S(M)$ of complete types over M in finitely many (or one) variable is countable. If not then there will be a model M' of T of cardinality κ realising uncountably many types over some countable elementary submodel M_0. On the other hand an Ehrenfeucht-Mostowski argument using Skolem functions and indiscernibles yields a model M' of T of cardinality κ such that over any countable subset A of M' only countably many types are realised in M'. Contradiction to κ-categoricity.

Consequences of ω-stability.

The first consequence is the existence of prime models over all sets: For any model M of T and subset A of M there is an elementary substructure $M(A)$ of M which contains A such that whenever M' is a model of T containing A such that $(M, a)_{a \in A} \equiv (M', a)_{a \in A}$ then there is an A-elementary embedding of $M(A)$ in M'. (Shelah subsequently proved that $M(A)$ is unique up to A-isomorphism.)

Another consequence (coming from the existence of Cantor-Bendixon rank on types) is that any formula $\phi(x)$ (with parameters from a model of T) has ordinal valued Morley rank.

DEFINITION 2.1. *Let $M \models T$. Let $\phi(x)$ be a formula over M. We define $RM(\phi(x)) \geq 0$ if $\phi(x)$ is consistent (has a solution in M), and $RM(\phi(x)) \geq$*

$\alpha + 1$ *if there is an elementary extension* M' *of* M *and formulas* $\psi_i(x)$
over M' *for* $i < \omega$, *pairwise inconsistent* (*in* M'), *each implying* $\phi(x)$, *and*
with $RM(\psi_i(x)) \geq \alpha$ *for all* $i < \omega$.
(*Also for* δ *limit*, $RM(\phi(x)) \geq \delta$ *if* $RM(\phi(x)) \geq \alpha$ *for all* $\alpha < \delta$.)

The definition above makes sense for any theory T. In any case for the
theory T currently under consideration, every formula has ordinal-valued
Morley rank. In fact it was proved later (by Baldwin) that every formula has
finite Morley rank.

If $RM(\phi(x)) = \alpha$ then $\phi(x)$ has an associated Morley degree (or multiplicity), the largest k such that there exist $\psi_i(x)$ for $i < k$ over some elementary
extension M' of M, which imply $\phi(x)$, are pairwise inconsistent, and have
Morley rank α.

In particular there is some formula $\phi(x)$ over a model M of T (where x
can even be chosen to be a single variable) which has Morley rank 1 and
Morley degree 1. Such a formula is also called *strongly minimal*, and has the
alternative characterization (which again makes sense in any theory):

DEFINITION 2.2. *Let* $\phi(x)$ *be a formula over a model* M *of* T. *We call*
$\phi(x)$ *strongly minimal if it has infinitely many realizations in* M, *and for any*
elementary extension M' *of* M *and formula* $\psi(x)$ *over* M', *it is not the case*
that both $\phi(x) \wedge \psi(x)$ *and* $\phi(x) \wedge \neg\psi(x)$ *have infinitely many realizations.*

In the present context we will assume that some strongly minimal $\phi(x)$ can
be found without parameters. (This is a delicate point. In general it is proved
that $\phi(x)$ can be found with parameters in the prime model of T).

The set of realisations of $\phi(x)$ in a model M is what is called a strongly
minimal set. Here are some basic properties and definitions.

(a) Algebraic closure on strongly minimal sets satisfies Steinitz exchange:
if M is a model of T, $A \subset M$ and a, b satisfy $\phi(x)$ in M then, if $b \in$
$acl(A, a) \setminus acl(A)$, then $a \in acl(A, b)$.

(b) A set $\{a_i : i \in I\}$ of realizations of $\phi(x)$ in a model M of T will be called
(algebraically) independent, if $a_i \notin acl\{a_j : j \in I, j \neq i\}$ for each $i \in I$.

(c) If $\{a_i : i < \alpha\}$ is an independent set of realisations of $\phi(x)$ in $M \models T$
and $\{b_i : i < \alpha\}$ is an independent set of realizations of $\phi(x)$ in $M' \models T$, then
the map taking a_i to b_i ($i < \alpha$) is elementary.

(d) A *basis* for $\phi(x)$ in M is by definition a maximal independent subset
of ϕ^M. If $\{a_i : i < \alpha\}$ is such a basis then ϕ^M is contained in $acl_M(\{a_i : i < \alpha\})$.

(e) Any two bases for $\phi(x)$ in M have the same cardinality, which we call
$\dim(\phi, M)$.

Step 2. For any model M of T, M is prime and minimal over ϕ^M. Minimality means that there is no proper elementary substructure M' of M containing ϕ^M.

It is enough to show minimality. Suppose not. So there is a proper elementary substructure M' of M with $\phi^{M'} = \phi^M$, a so-called Vaughtian pair. We may assume that both M and M' are countable. An argument (using ω-stability) enables us to find an elementary extension M'' of M of cardinality κ with $\phi^{M''} = \phi^M$. This contradicts κ-categoricity of T, as there will exist another model N of T of cardinality κ in which ϕ^N has cardinality κ.

End of proof. Let M and N be two models of T of cardinality $\lambda > \omega$. As M is prime over ϕ^M the latter has cardinality $= \lambda$, and thus $\dim(\phi, M) = \lambda$. Similarly $\dim(\phi, N) = \lambda$. Let I be a basis for ϕ^M and J a basis for ϕ^N. So I and J both have cardinality λ and there is an elementary map taking I to J (by (c)). This extends (by passing to the algebraic closures) to an elementary map taking ϕ^M to ϕ^N, and then also to an elementary embedding f of M in N (as M is prime over ϕ^M). As N is minimal over ϕ^N, f must be onto N, so M and N are isomorphic. $\dashv$

The proof shows that the isomorphism type of any model M of T is determined by $\dim(\phi, M)$, and it follows fairly easily that the number of countable models of T must be either 1 or ω (the Baldwin-Lachlan Theorem).

§3. Fine structure of uncountably categorical theories.

Step 1 above gave us in particular a "geometry" (the strongly minimal formula $\phi(x)$), and Step 2 told us that this formula controls every model. This last property is roughly what we call "unidimensionality". It is technically convenient to identify the models of T (T any complete theory) with elementary submodels of a big saturated model $\overline{M}$ (or class if you wish), and work inside $\overline{M}$. We will follow this convention from now on. The geometric model-theoretic point of view (represented historically by Zilber) at this point asks: (I) exactly how does the geometry ϕ control the whole structure, or why and how is every model M prime over ϕ^M, and (II) what exactly are the possibilities for $\phi(x)$? Zilber especially needed answers to these questions in the case where T is also ω-categorical, in order to prove that totally categorical theories are not finitely axiomatizable.

Let us begin by looking at question (I). One possibility of course is when $\phi(x)$ is "$x = x$", namely defines the whole structure. In this situation we say that T itself is strongly minimal, and we even call $M \models T$ a strongly minimal structure. Another possibility is the existence of a formula $\psi(x, y)$ defining a finite-to-one function from $\overline{M}$ onto X, where X is some $\emptyset$-definable subset of $(\phi^{\overline{M}})^n$. In this case, each model M is equal to $acl(\phi^M)$. Are there any other possibilities? Group actions and a kind of analogue of "fibre bundles" (from differential geometry) turn out to give examples (and essentially the only other examples). I will describe this construction in full generality. The data will be a structure P (in some language), a definable subset X of P^n and a definable (in P) family $(G_a : a \in X)$ of definable groups (let's suppose X

and the family of groups to be $\emptyset$-definable). For each $a \in X$, let Y_a be a principal homogeneous space for G_a (namely a set on which G_a acts regularly, or equivalently strictly 1-transitively). From this data we manufacture a new structure M. The universe of M will be the disjoint union of P and the Y_a, where P is equipped with its original structure. The language for M will also contained a function symbol π, interpreted as the canonical surjection $\pi\colon \bigcup_a Y_a \to X$, as well as another function symbol $f(-, -, -)$ such that for each $a \in X$, $f(a, -, -)$ defines the action of G_a on Y_a. We will also allow arbitrary additional relations on M as long as no new structure is induced on P (namely every $\emptyset$-definable subset of P^m definable in the new structure M should be already $\emptyset$-definable in the original structure P). We call M a definable fibre bundle over P, with data $(G_a\colon a \in X)$. It is not hard to see that M is prime and minimal over P (and likewise for any N elementarily equivalent to M). If $Th(P)$ is uncountably categorical, so is $Th(M)$. Note that if each group G_a is finite, then $M = acl_M(P)$. Also if X is a singleton $\{a\}$, and b any element of Y_a then $M = acl_M(P, b)$.

Zilber (see [26]) essentially proved that definable fibre bundles explain minimality over the strongly minimal set $\phi(x)$. (Shelah's semiregular types technology [22] together with an input from Hrushovski give a general account for superstable theories.)

PROPOSITION 3.1. *If $Th(M)$ is uncountably categorical and $P \subseteq M$ is a strongly minimal set, then there are structures $P = P_0, P_1, \ldots, P_k = M$ such that P_{i+1} is a definable fibre bundle over P_i. Moreover the relevant groups G_a can be taken to be living in P and to be elementary abelian, finite simple nonabelian, or infinite without definable proper infinite normal subgroups.*

One consequence of the proposition is that T has finite Morley rank. In any case we see in this result the strong relationship between the internal structure of a given model M and the problem of the classification of all models of $Th(M)$.

A kind of restatement of the above proposition, which has a Galois-theoretic flavour is:

REMARK 3.2. *Suppose T is uncountably categorical, $\phi(x)$ is strongly minimal (or actually any formula with infinitely many realizations), and $M \models T$. Then $M = \phi^M \cup \{a_i\colon i < \alpha\}$ where for each $\beta < \alpha$ either (i) $a_\beta \in acl(\phi^M \cup \{a_i\colon i < \beta\})$ or (ii) $tp(a_\beta / \phi^M \cup \{a_i\colon i < \beta\})$ is isolated by a formula $\chi(y)$ which defines a principal homogeneous space for a definable group G contained in ϕ^M. Moreover G can be chosen to be infinite, connected, and simple.*

The Galois-theoretic content of say case (ii) is that the group of elementary permutations of $dcl(\phi^M \cup \{a_i\colon i \leq \beta\})$ which fix $\phi^M \cup \{a_i\colon i < \beta\}$ pointwise is isomorphic to G.

The notion of "almost strongly minimality" and variants, will be important later.

DEFINITION 3.3. *T is said to be almost strongly minimal if there is some strongly minimal set P in $\overline{M}$ such that, after naming finitely many elements of $\overline{M}$ (namely adding finitely many constants to the language), $\overline{M} = acl_{\overline{M}}(P)$.*

Given a strongly minimal structure P containing an infinite definable group G, the basic fibre bundle constructed from P from the data P and $G_a = G$ for all $a \in P$, will be uncountably categorical and not strongly minimal. The reader may find it worthwhile looking in detail at the theory of $((Z/4Z)^\omega, +)$. This is a definable fibre bundle (with additional structure given by the group operation) over $(Z/2Z)^\omega$, and is not almost strongly minimal. Given a structure P and data $(G_a : a \in X)$, the issue of what the possible bundles over P with this data can be is essentially a cohomological question.

Now we pass to (II). The issue is to classify strongly minimal sets. What does this mean? The strongly minimal subset $P = \phi^{\overline{M}}$ of the model $\overline{M}$ can be viewed as a structure in its own right: for each formula $\psi(x_1, \ldots, x_n)$ of L let R_ψ be the set of realizations of ψ each of whose coordinates is in P. The resulting structure $(P, R_\psi)_\psi$ has quantifier-elimination. Zilber's hope was to classify such structures. Classification up to bi-interpretability is probably the most reasonable notion here, although the finer notion of identifying two structures if there is a bijection between their universes taking definable sets to definable sets is also useful.

There are three important examples of strongly minimal structures:

(i) An infinite set in the empty language (that is, equality the only relation).

(ii) An (infinite-dimensional) vector space V over a field F, in the language containing just $+$, 0 and f_a for each $a \in F$, representing scalar multiplication.

(iii) An algebraically closed field in the field language $(+, -, \cdot, 0, 1)$.

Let us fix a strongly minimal structure P, possibly living as above in $\overline{M}$. One can study P by studying the behaviour of algebraic closure. Coming out of Steinitz exchange mentioned in the previous section, we assign a dimension or rank to any finite tuple from P: for any set A of parameters (maybe from $\overline{M}$) and finite tuple b from P, $\dim(b/A)$ is defined to be the cardinality of any maximal A-algebraically independent subtuple of b (which is well-defined). (In fact $\dim(a/A)$ coincides with the Morley rank of $tp(a/A)$ which is the smallest Morley rank of a formula in this type.)

DEFINITION 3.4. *Let b, c be tuples from P and A and set of parameters. We say that b is independent from c over A, if $\dim(b, c/A) = \dim(b/A) + \dim(c/A)$.*

DEFINITION 3.5. (i) *We call P degenerate if for any subset B of P, $acl(B) \cap P = \bigcup_{b \in B}(acl(b) \cap P)$.*

(ii) *We call P modular if after naming a small set of parameters, we have, for all tuples b, c from P, b is independent from c over $acl(b) \cap acl(c)$.*

Degenerate implies modular. Example (i) above is degenerate, example (ii) modular and nondegenerate and example (iii) nonmodular.

Zilber conjectured, very early on:

Zilber conjecture. If P is nonmodular, then an infinite (necessarily algebraically closed) field is definable in the structure P.

In fact he tentatively conjectured that in the nonmodular case the structure P itself is essentially an algebraically closed field and all sets definable in P are defined in the field language.

The full conjecture was disproved by Hrushovski [8]. However the conjecture was proved by Hrushovski and Zilber [13] under certain additional assumptions on P of a topological-geometric nature, namely assuming P to be a "Zariski geometry". I will not go into the definition. Zariski geometries are treated in detail in Marker's tutorial in Haifa [16]. Also in the concrete examples we'll be considering there are alternative proofs.

There are some rather more intuitive or geometric ways of seeing the above notions (modularity, ...). P has Morley rank 1. P^2 (which one should think of as 2-space over P) has Morley rank 2, and contains various Morley rank 1 definable subsets, such as the diagonal $\{(x,x): x \in P\}$, or for any $a \in P$, $\{(x,a): x \in P\}$. Think of strongly minimal subsets of P^2 as curves over P. Modularity essentially says that if $X \subseteq P^m$ is definable and $(C_a: a \in X)$ is a definable family of curves (strongly minimal subsets of P^2) with pairwise intersection finite, then $RM(X) \leq 1$: there are no 2 or higher dimensional families of curves. Degeneracy says that X is finite: there is *no* infinite family of curves.

Modularity (in this and more general contexts) is a central concept of geometric model theory. There is a well-developed general theory, and somewhat surprisingly, the degenerate/modular/nonmodular trichotomy has fundamental meaning in many mathematical contexts. In the next result, I summarise the structural consequences (due to Zilber [26]) of these notions in the uncountably categorical context. T will be uncountably categorical, $\overline{M}$ a (big) model of T and $P = \phi^{\overline{M}}$ a strongly minimal subset of $\overline{M}$.

PROPOSITION 3.6. (i) *If P is degenerate, then there are no infinite definable groups in $\overline{M}$. In particular $\overline{M} = acl(P)$.*

(ii) *If P is modular and nondegenerate then there is an infinite strongly minimal group definable in $\overline{M}$ (in fact essentially on P itself).*

(iii) *If P is modular, then for any definable group G in $\overline{M}$, G is abelian-by-finite, G has no infinite definable family of connected subgroups, and every definable (in $\overline{M}$) subset of G is a finite Boolean combination of cosets (translates of subgroups).*

§4. Quotients. I fudged over certain issues of definability in the last section, because certain of the objects mentioned (definable homogeneous spaces and definable groups for example) may only exists as quotients of definable sets by definable equivalence relations. So in this section I will discuss the status and model-theoretic treatment of such quotient objects as well as more general quotients (by type-definable equivalence relations). I will also discuss various Galois groups attached to theories. T is an arbitrary complete theory and we work in $\overline{M}$ a big saturated model of T. (I ignore set-theoretic complications.) A quotient-definable set is something of the form X/E where X is a definable set and E a definable equivalence relation on X. It makes perfect sense to speak of a definable map from X/E to Y/F: it will be a map f induced by a definable relation R between X and Y. In various kinds of geometry the issue of quotient objects (for example the space of orbits of a manifold under the action of a Lie group), is a very delicate matter, because one wants the quotient object to exist as a geometric object of the same kind. However, there is absolutely nothing at the general model-theoretic level which is problematic about such quotient objects (quotients of definable sets by definable equivalence relations), in the sense that they remain entirely within the framework of first order model theory. One formalism for seeing this is Shelah's formation of the many-sorted structure $\overline{M}^{eq}$. This is obtained by adding a new sort S_E for each $\emptyset$-definable equivalence relation E on $\overline{M}^n$, together with a function f_E from $\overline{M}^n \to S_E$. Then any Y/F (where Y, F may be defined with parameters) can be naturally identified with a definable subset of some sort S_E in $\overline{M}^{eq}$. Precisely for this reason, we should really understand the category of definable sets and functions in $\overline{M}^{eq}$. The structures P_i given in Lemma 3.1 are really definable in M^{eq}. This all suggests that we should from the start consider many-sorted structures, working in many-sorted logic. In fact there is no harm in even allowing a sort S_ϕ for each formula $\phi(x)$ of L (without parameters). To "understand" such a structure then amounts to identifying a certain family $(S_i)_i$ of sorts, classifying the definable subsets of these sorts (quantifier-elimination), and showing that for any sort S and $\emptyset$-definable equivalence relation E on S, there is a $\emptyset$-definable bijection between S/E and a definable subset of one of the S_i (elimination of imaginaries). Nevertheless, back in the one-sorted situation we make:

DEFINITION 4.1. *T has elimination of imaginaries if for any $\emptyset$-definable set $X \subseteq \overline{M}^n$ and $\emptyset$-definable equivalence relation E on X there is a $\emptyset$-definable bijection f between X/E and some definable $Y \subseteq \overline{M}^m$.*

REMARK 4.2. (i) *The definition above has a very minor discrepancy with the usual notion of elimination of imaginaries, but agrees if there are two distinct constants in T.*

(ii) *If T has elimination of imaginaries, then the right hand side of Definition 3.1 also holds where we no longer demand $\emptyset$-definability of X, E and f.*

Maybe the only general model-theoretic result regarding elimination of imaginaries is:

FACT 4.3. *Suppose T is strongly minimal. Then, up to naming finitely many parameters T has weak elimination of imaginaries: for any X/E as in Definition 4.1, there is a $\emptyset$-definable set Y and a $\emptyset$-definable surjective function $f : Y \to X/E$ with finite fibres.*

FACT 4.4. *The theory of algebraically closed fields of a given characteristic, the theory of real closed fields and the theory of differentially closed fields of characteristic 0 all eliminate imaginaries.*

In each of the theories above definable groups can be definably equipped with unique "geometric" structure: algebraic groups, Nash groups and differential algebraic groups respectively. Similarly for definable homogeneous spaces. So Fact 4.3 proves that quotients of such groups by definable subgroups exist as geometric objects.

Elements of the form a/E (a a finite tuple from $\overline{M}$ and E a $\emptyset$-definable equivalence relation) are usually called imaginaries. As we have explained there is no intrinsic model-theoretic problem with these objects. Suppose now that E is an equivalence relation on tuples of fixed, but possibly infinite (although small, relative to the saturation of $\overline{M}$) length, which is defined by a possibly infinite set of formulas (without parameters say). An equivalence class a/E is what we have recently called a *hyperimaginary*. A type-definable set of such objects: X/E where X is a set of tuples of the right length defined by a possibly infinite set of formulas, is a *hyperdefinable set*. The status of such objects (hyperimaginaries, and hyperdefinable sets) *is* a model-theoretic problem. If E happens to be an intersection of *definable* equivalence relations E_i (at least when restricted to $tp(a)$), then a/E can be identified naturally with the sequence $(a/E_i)_i$ of imaginaries, namely with a "pro-imaginary". We will say that T eliminates hyperimaginaries if this always happens: any type-definable equivalence relation is equivalent, on the set of realizations of any complete type, to the intersection of definable equivalence relations (any hyperimaginary is a pro-imaginary). Any stable theory (for example, an uncountably categorical theory) has elimination of hyperimaginaries. However, even if T does *not* eliminate hyperimaginaries, these objects still remain within first order model theory (namely subject to the compactness theorem). Although we cannot in a sensible way add new sorts S_E for each such E or even talk about definable sets of hyperimaginaries, we can make sense out of the complete type of a hyperimaginary (as a certain partial type), and this is enough to do model theory. Hyperdefinable, non pro-definable sets arise naturally in nonstandard analysis (where E might be the relation of being infinitely close). For example the so-called nonstandard hull $\hat{B}$ of a Banach space B is a hyperdefinable set in a saturated model $^*B$ of B. Henson and Iovino have developed some stability theory for Banach spaces using Henson's Banach space logic. An alternative treatment would be to use standard stability theory directly on the hyperdefinable set $\hat{B}$.

By $acl^{eq}(A)$ we mean the set of elements of $\overline{M}^{eq}$ in the algebraic closure of A. For T the theory of algebraically closed fields of characteristic 0, $acl^{eq}(\emptyset)$ is precisely $\overline{Q}$. The automorphism group of $\overline{M}$, $Aut(\overline{M})$, acts naturally on $acl^{eq}(\emptyset)$, and the corresponding group, $Aut(\overline{M})$ quotiented by the normal subgroup consisting of those σ which fix $acl^{eq}(\emptyset)$ pointwise, has the natural structure of a profinite group (which is precisely the absolute Galois group of Q in the characteristic 0 algebraically closed fields case). We denote this group by $Gal_{pf}(T)$. It is an invariant of the bi-interpretability type of T. The structure of $Gal_{pf}(T)$ has various implications. For example, if T is ω-categorical and $Gal_{pf}(T)$ is finite (even after naming any finite set of parameters), then Lascar [14] proved that the bi-interpretability type of T can be recovered from the abstract group $Aut(\overline{M})$. Also Hrushovski [9] remarks that if T is uncountably categorical and finitely axiomatizable, then only finitely many finite simple groups occur as quotients of $Gal_{pf}(T)$ (even after naming finitely many parameters). Also there is a Galois theory. For any T there is a Galois correspondence between closed subgroups of $Gal_{pf}(T)$ and definably closed subsets of $acl^{eq}(\emptyset)$.

The analogous construction can be made for hyperimaginaries. Let $bdd^{heq}(A)$ be the set of hyperimaginaries which have small orbit under $Aut_A(\overline{M})$. Then $Aut(\overline{M})$ acts on $bdd^{heq}(\emptyset)$. As above we obtain a group which we denote $Gal_c(T)$, the closed Galois group. This has naturally the structure of a compact (Hausdorff) topological group, and is again an invariant of the bi-interpretability type of T. $Gal_{pf}(T)$ is naturally a quotient of $Gal_c(T)$. The two groups are equal just if there is no hyperdefinable (over an element $c \in acl^{eq}(\emptyset)$) connected compact Lie group acting as automorphisms of some hyperdefinable (over c) set X. Again there is a Galois correspondence between closed subgroups of $Gal_c(T)$ and definably closed subsets of $bdd^{heq}(\emptyset)$. (See [15].)

There is a third, natural but rather mysterious group $Gal(T)$, which lives rather in the "descriptive set theory" of $\overline{M}$. Consider equivalence relations E on possibly infinite tuples, which are invariant under $Aut(\overline{M})$. Let $bdd^{inveq}(A)$ be the set of classes of such E which have small orbit under $Aut(\overline{M})$. Again $Aut(\overline{M})$ acts on $bdd^{inveq}(\emptyset)$ and $Gal(T)$ is the resulting group. $Gal_c(T)$ is a quotient of $Gal(T)$. Ziegler recently gave an example where $Gal(T) \neq Gal_c(T)$. We know very little about $Gal(T)$. For example what can be the cardinality of the quotient $Gal(T)/Gal_c(T)$?

§5. Finite rank structures. The basic ingredients in the proof of Morley's theorem were generalized by Shelah [22] to build the enormous machinery of stability theory which he used to solve the spectrum problem for countable theories. Analogously, the geometric ideas from section 3 have much wider applicability. The key points are the identification of certain "geometries", the fine structure of these geometries, and structural consequences for parts of the ambient structure controlled by these geometries. A new feature, absent in the ω_1-categorical context, will be orthogonality. For example, there may be several strongly minimal formulas ϕ_i which together control the whole structure (every model M of T is prime over $\bigcup_i \phi_i^M$) but the ϕ_i may have no mutual interaction. Also in the general superstable case, the geometries that have to be considered may have infinite Morley rank. Here we will restrict our attention to finite rank structures where it is the strongly minimal geometries that are relevant. This is a small generalization of the situation considered in section 3. The possible presence of orthogonal strongly minimal sets is the new feature. In subsequent applications, these finite rank structures may arise as definable sets in some ambient (infinite rank or even unranked) structures. Also at this point, the reader may wish to think of our theories and structures being many sorted.

DEFINITION 5.1. *We will say that a structure M (equivalently $Th(M)$) has finite Morley rank, if in $Th(M)$ every formula has finite Morley rank. We will call M finite-dimensional, if in addition, there are a finite set $\{\phi_i : i = 1, \ldots, k\}$ of strongly minimal formulas in T^{eq} such that every model N of $Th(M)$ is prime and minimal over $\bigcup_i \phi_i^N$.*

We assume for the remainder of this section that T is a theory of finite Morley rank, $M \models T$ and $\overline{M}$ a saturated model of T. We will work in $\overline{M}$ unless we say otherwise. Recall that the Morley rank of a complete type $p(x)$ over a set A is the minimum of the Morley ranks of the formulas in p. We write $RM(a/A)$ for $RM(tp(a/A))$. If a happens to be a tuple from a strongly minimal set, this coincides with $\dim(a/A)$ as defined in section 3.

DEFINITION 5.2. *Let a be a finite tuple, $A \subset B$. We say a is independent from B over A if $RM(a/B) = RM(a/A)$.*

REMARK 5.3. *This coincides with independence in the sense of nonforking. We have the basic properties: a is independent from B over A if a is independent from $A \cup \{b\}$ over A for all finite tuples b from B iff b is independent from $A \cup \{a\}$ over A for all finite tuples b from B. Given a and $A \subseteq B$, there is a' realising $tp(a/A)$ such that a' is independent from B over A. $tp(a/A)$ is stationary if for all $B \supset A$, and a', a'' realising $tp(a/A)$ independent from B over A, $tp(a'/B) = tp(a''/B)$. $tp(a/A)$ is stationary iff it has Morley degree 1.*

DEFINITION 5.4. *T is modular, if for all finite tuples a, b, a is independent from b over $acl^{eq}(a) \cap acl^{eq}(b)$.*

This agrees with Definition 3.5 (ii) if T happens to be strongly minimal. Here are some generalizations of Proposition 3.6.

FACT 5.5. (i) *T is modular iff each strongly minimal formula in $\overline{M}^{eq}$ is modular.*

(ii) *If T is modular and G is a definable group in $\overline{M}^{eq}$ then G is abelian-by-finite, G has no infinite definable family of infinite connected subgroups, and all definable subsets of G are Boolean combinations of cosets.*

DEFINITION 5.6. *Let X, Y be definable sets in $\overline{M}^{eq}$, both defined over A say. We say that X and Y are fully orthogonal if whenever b is a finite tuple from X and c a finite tuple from Y then b is independent from c over A.*

REMARK 5.7. (i) *The content of full orthogonality is that any definable subset of $X^n \times Y^m$ is a finite union of products of definable subsets of X^n and definable subsets of Y^m.*

(ii) *Restricted to strongly minimal sets, full orthogonality is usually just called orthogonality. Nonorthogonality is an equivalence relation on strongly minimal sets.*

If T is also finite-dimensional, then Proposition 3.1 holds, in the sense that there are (given M) $P_0, \ldots, P_n = M$ such that $P_0 = \phi_1^M \cup \ldots \cup \phi_k^M$ and P_{i+1} is a definable fibre bundle over P_i.

Groups of finite Morley rank. We mean a group G with possibly additional structure such that $Th(G)$ has finite Morley rank. In many applications such groups will arise as definable groups in some ambient structure. A result of Lascar (see [20]) says that any finite Morley rank group is finite-dimensional (in the sense of 5.1). From Proposition 3.1, we see that the structure of simple groups of finite Morley rank is relevant to the fine structure of uncountably categorical theories. An underlying and fundamental result due to Macintyre is that any infinite field of finite Morley rank is algebraically closed. G is said to be connected if G has no proper definable subgroup of finite index. All the above notions have various equivariant implications and interpretations. For example

FACT 5.8. *If H_1, H_2 are definable connected fully orthogonal subgroups of G, then H_1 and H_2 commute. Moreover any definable subset X of $H_1 \cdot H_2$ is essentially the product of a definable subset X_1 of H_1 with a definable subset X_2 of H_2.*

Let X be a definable subset of G of Morley multiplicity 1. The *stabilizer* of X in G, $Stab_G(X)$, is by definition the set of $g \in G$ such that $RM(X) = RM(X \cap g \cdot X)$. This is a definable subgroup of G. A kind of stability-theoretic analogue of the *socle* of a group (subgroup generated by minimal normal subgroups) is the largest connected definable subgroup of G contained in the algebraic closure of some finite set $(\phi_i(x))_i$ of strongly minimal formulas. (This should be read in a saturated model.) We will call this object $s(G)$, hopefully without ambiguity. If G happened to be uncountably categorical, this is precisely the maximal connected almost strongly minimal subgroup of G. As remarked in section 3, if G is $(Z/4Z)^\omega$, this is $2G$. A useful and relatively elementary result relating the structure of definable sets in a commutative group G to $s(G)$ is the following [10]:

LEMMA 5.9. *Let G be a commutative connected group of finite Morley rank, defined in an ambient structure $\overline{M}$ over a set A. Assume that every connected definable subgroup of G is $acl^{eq}(A)$-definable. Let $H = s(G)$. Let X be any definable set of Morley degree 1 which has finite stabilizer. Then, up to a set of smaller Morley rank, X is contained in a single translate of H.*

§6. **Examples.** *Algebraically closed fields.* The theory of algebraically closed fields of a fixed characteristic say 0 (ACF_0) is strongly minimal, with quantifier-elimination in the language $(+, -, \cdot, 0, 1)$. The dichotomies and results from sections 3 and 5 are largely vacuous here. All definable sets are mutually nonorthogonal, all definable groups are almost strongly minimal, nothing is modular. The basic objects of algebraic geometry are certain definable sets, varieties, defined by finite systems of polynomial equations. Morley rank and algebraic-geometric dimension coincide for such objects. Any definable set is a finite Boolean combination of such things, even a finite Boolean combination of smooth projective varieties. Definable functions are piece-wise rational. One of the aims of algebraic geometry is the classification of varieties up to birational isomorphism. This is pretty close to classifying definable sets up to definable isomorphism. The general model theory of ACF_0 says very little about this problem. We will see later however that the model theory of certain enriched structures (such as differentially closed fields) is meaningful for issues such as the deformation theory of algebraic varieties.

A definable group can be uniquely equipped with the geometric structure of a variety (pieced together from finitely many affine varieties with rational transition maps) such that multiplication becomes a morphism. A class of such groups which is very important for geometry and arithmetic is the class of abelian varieties, connected algebraic groups whose underlying variety is a closed subvariety of some projective space P^n. These are commutative groups, and any smooth projective curve which is not isomorphic to P^1 will embed in a unique smallest such abelian variety (its Jacobian variety). Hence their importance for the study of curves, at least. Any strongly minimal definable set X in $(K, +, \cdot)$ is, up to finite, a smooth projective curve C. Assume $K = C$. A smooth projective curve C is a compact Riemann surface (so a compact 2-dimensional manifold), and has a finite number of "handles" g, the genus of the curve. (g is the dimension of its Jacobian variety.) This is a fundamental invariant of the curve, and also of the strongly minimal set X. We will use below the following fact: the strongly minimal set (or curve if you wish) has genus ≥ 2 if and only if there is no definable group structure on X, even after adding or subtracting finitely many points. See [21] for more background.

Compact complex manifolds. A complex manifold M is a topological space with a covering by open sets homeomorphic to open subsets of some C^n such that the transition maps are holomorphic (complex analytic). If M is such, so are $M \times M$, $M \times M \times M$ etc. By an analytic subset or subvariety) of M we mean a subset X of M such that for every $a \in M$, there is an open neighbourhood U of a in M and holomorphic functions $f_1, \ldots, f_n$ on U such that $U \cap X = \{x \in U : f_1(x) = \ldots = f_n(x) = 0\}$. Let M be a compact complex manifold and consider M as a relational structure by adding a predicate R_X for each analytic subvariety X of M^n. Zilber pointed out (using a theorem of Remmert) that M is a structure of finite Morley rank with quantifier-elimination (but clearly not saturated as every element of M is essentially named by a constant), in fact even a "Zariski structure" in a generalized sense. We can actually consider the whole category of compact complex manifolds as a many sorted structure; the relations on $M_1 \times \ldots \times M_n$ being again the analytic subvarieties. This category is again a structure of finite Morley rank (every sort has finite Morley rank). It turns out that the machinery developed earlier *is* meaningful for complex compact manifolds (either one at a time, or the whole category). We let $\mathcal{A}$ denote the many-sorted structure. Among the sorts is $P^1(C)$ (we will just say P^1), which is, by Chow's theorem, essentially just the structure $(C, +, \cdot)$ considered above. Basically all the general theory we have discussed has meaning in the structure $\mathcal{A}$. A complex torus T is a complex Lie group of the form C^n / Λ where Λ is a lattice of real rank $2n$. It is precisely a compact complex Lie group so is already a sort in $\mathcal{A}$ (and the group operation is definable). For suitably general Λ and for $n > 1$, T will be fully orthogonal to P^1. The complex analytic literature ([23]) already contains the result that any complex torus which is fully orthogonal to P^1 must be modular. This implies:

PROPOSITION 6.1. *Suppose T is a complex torus with no proper subtori. Then ether T is modular or T is (definably) isomorphic to an abelian variety.*

The Zilber conjecture is true in $\mathcal{A}$ (via Zariski geometries). This may have interesting implications for the classification of compact complex manifolds up to bimeromorphic equivalence. Some recent results ([19]) are:

PROPOSITION 6.2. *Let G be a strongly minimal modular group definable in $\mathcal{A}$. Then G is definably isomorphic to a complex torus.*

PROPOSITION 6.3. *Suppose X is a strongly minimal set definable in $\mathcal{A}$. Suppose $\overline{X}$ is a connected compact complex manifold such that X is a definable open subset of $\overline{X}$. Then X is degenerate if and only if (i) there are no nonconstant meromorphic functions (to P^1) on $\overline{X}$, and (ii) there is no generically surjective meromorphic map from $\overline{X}$ to any space of the form T/G where T is a complex torus and G a finite group of (holomorphic) automorphisms of T.*

Definable groups and homogeneous spaces come into the picture much as in Proposition 3.1. For example, it follows from the general theory that if M is a compact Kähler manifold, and $f : M \to X$ is the algebraic reduction of M (f a definable map onto an algebraic variety X of maximal possible dimension) and the general fibre of f is isomorphic to an algebraic variety Y, then Y is a homogeneous space for a complex algebraic group.

Differential equations. One algebraic route to the study of (algebraic) differential equations, is via differential rings and fields. A differential field is a field F equipped with a derivation D. The theory of differential fields of characteristic 0 has a model companion, the theory of differentially closed fields DCF_0. This theory is ω-stable, but of infinite Morley rank. The definable sets of finite rank turn out to witness very nicely the geometric-model-theoretic themes discussed above. In fact, via model theory we see quite amazing analogies between the category of definable sets of finite Morley rank in a model of DCF_0 and the category $\mathcal{A}$ discussed in the previous section. (See [18] for more on this).

We fix a large model $(K, +, \cdot, D)$ of DCF_0. k denotes the field of constants (which is a strongly minimal set). Again the Zilber conjecture is true for strongly minimal sets definable in this structure, via the Zariski geometries theorem. There exists a direct differential algebraic-geometric proof in the case where the strongly minimal set is already a subset of a finite Morley rank group [2], yielding Proposition 6.4 below (first proved in [12]).

Suppose A to be an abelian variety (defined in the algebraically closed field $(K, +, \cdot)$). By $A^\sharp$ we mean the smallest definable (in $(K, +, \cdot, D)$) subgroup of A containing the group of torsion points of A.

PROPOSITION 6.4. (i) $A^\sharp$ *has finite Morley rank, and for any finite Morley rank G with $A^\sharp < G < A$, $A^\sharp = s(G)$.*

(ii) *If A is defined over the field k of constants, then $A^\sharp$ is precisely $A(k)$, the group of points of A with coefficients in k.*

(iii) $A^\sharp$ *is modular if and only if A has no abelian subvariety isomorphic (as an algebraic group) to an abelian variety defined over k (A has k-trace 0).*

(iv) *Let A_1, A_2 be simple abelian varieties, each nonisomorphic to any abelian variety defined over k. Then there is a rational isogeny from A_1 to A_2 if and only if $A_1^\sharp$ and $A_2^\sharp$ are nonorthogonal.*

The analogue of 6.3 is:

PROPOSITION 6.5. *Let X be a strongly minimal set in $(K, +, \cdot, D)$. Then exactly one of the following holds*:

(i) X *is degenerate,*

(ii) X *is definably isomorphic to a definable subset of k^n (for some n),*

(iii) *there is a simple abelian variety A with k-trace 0 and a generically surjective definable map from X to $A^\sharp/G$ for some finite group G of definable automorphisms of $A^\sharp$.*

Fundamental questions remain concerning degenerate strongly minimal sets. An important invariant of a definable set of finite Morley rank is its "order" (more or less the order of the differential polynomials defining it). In the order 1 case, the situation is rather clear (by a finiteness theorem of Jouanolou). But for higher orders nothing is known.

Again the definable homogeneous space technology from section 3 is relevant. It both explains and generalizes the classical Picard-Vessiot Galois theory of linear differential equations.

Finiteness theorems. Faltings proved [4] that a curve X of genus ≥ 2 defined over $\mathbf{Q}$ has only finitely many points with coordinates in $\mathbf{Q}$. A version over function fields (conjectured by Lang) was proved earlier by Manin:

PROPOSITION 6.6. *Let X be a curve defined over F where F is a function field over an algebraically closed k (characteristic 0). Then either* (i) *X is not isomorphic to a curve defined over k and $X(F)$ is finite, or* (ii) *X is isomorphic to a curve X_0 defined over k and all but finitely many points of $X(F)$ come from points of $X_0(k)$ via this isomorphism.*

As is well-known, model theoretic methods give a proof of the first part (using especially 6.4 (iii)). Rather easier aspects of the theory (a small part of (i) and (ii) of 6.4) yield the second part, which I will outline. This second part is called the Theorem of di Franchis and can be restated:

PROPOSITION 6.7 (characteristic 0). *Let X be a curve of genus ≥ 2 and W be any variety. Then there are only finitely many generically surjective rational maps from W to X.*

We sketch a proof, coming essentially from [11]. Let k be an algebraically closed field over which X and W are defined. There is no harm in assuming k to have infinite transcendence degree. Let F be the function field of W. A generically surjective rational map from W to X corresponds to a point of $X(F) \setminus X(k)$, so we must show there are finitely many such things. Let D be a derivation on F whose field of constants is k. Extend to a derivation D on a differentially closed field K containing F (still with constants k). Let A be an abelian variety containing X and defined over k (the Jacobian variety of X). The only use of $genus(X) \geq 2$ will be that the algebraic-geometric stabilizer $Stab(X) = \{a \in A : a + X = X\}$ is finite (otherwise X would be a translate of its stabilizer). Now there is a certain definable (in the differentially closed field) homomorphism, the logarithmic derivative, from $A(K)$ to some K^n, whose kernel is $A(k)$. As $A(F)/A(k)$ is finitely generated (by the Lang-Néron theorem) there is a finite Morley rank definable subgroup G of A containing $A(F)$. Let $Y = (X(K) \setminus X(k)) \cap G$. We must show Y to be finite. Suppose otherwise, and let Z be a definable subset of Y of the same Morley rank as Y and of Morley degree 1. As X does not contain any infinite coset of a subgroup of A and Z is infinite, $Stab_G(Z)$ is finite, so by 6.4 (i), (ii), and 5.9, we may assume that Z is contained in a single coset of $A(k)$. Let

$a \in Z$ (so $a \notin A(k)$). As Z is infinite, there are infinitely many $b \in A(k)$ such that $b + a \in Z$. Let $k_0 < k$ be a finitely generated field such that the curve X is defined over k_0. As the induced structure on k is the pure field structure, we can find $b \notin acl(k_0)$ such that $b + a \in Z$. In particular, working in the algebraically closed field K, a is a generic point of X over k_0, a is independent from b over k_0 and $b + a$ is also a generic point of X over k_0. So $b \in Stab(X)$ and as $b \notin acl(k_0)$ this shows $Stab(X)$ to be infinite, a contradiction.

It should be remarked that proofs like the above (using auxiliary model-theoretic structures) automatically yield good bounds, in the above case doubly exponential bounds, as a function of the shape of the equations defining X and the rank of the finitely generated group $A(F)/A(k)$. (See [11].)

Faltings [5] subsequently proved a generalization of Mordell's conjecture, again conjectured by Lang: If A is a complex abelian variety, X a subvariety and Γ a finitely generated subgroup of A then $X \cap \Gamma$ is a finite union of cosets. The same holds for semiabelian varieties, algebraic groups which are extensions of an abelian variety by $(C^*)^n$. (As is well-known, methods using the above machinery yield a proof of this in the function field case, also in positive characteristic [10].)

One can ask whether an analogous (absolute) statement holds in the category $\mathcal{A}$. Here the analogue of a semiabelian variety is a extension A of a complex torus by some $(C^*)^n$. Such a group A is commutative and has a compactification $\overline{A}$ living in $\mathcal{A}$ such that A and its group structure are definable (in $\mathcal{A}$). The analogue of a subvariety is an analytic subvariety of A definable in $\mathcal{A}$. We then have:

PROPOSITION 6.8. *Suppose A is an extension of a complex torus by $(C^*)^n$. Suppose X is a "subvariety" of A and Γ a finitely generated subgroup of A. Then $X \cap \Gamma$ is a finite union of cosets.*

Sketch of proof. This is a straightforward reduction to the semiabelian variety case using methods from [10]. We write the group operation additively. We may assume that X is irreducible, and $X \cap \Gamma$ is "Zariski dense" in X, namely X is the smallest "analytic" subvariety of A containing $X \cap \Gamma$. We want to show that X is a translate of a definable subgroup of A. Let $S = \{a \in A : a + X = X\}$. Let $\pi : A \to A/S$ be the natural homomorphism. Then $\pi(\Gamma)$ is finitely generated and $\pi(\Gamma) \cap \pi(X)$ is Zariski-dense in $\pi(X)$. Also X is a union of translates of S. So (replacing A, X, Γ, by $\pi(A)$, $\pi(X)$, $\pi(\Gamma)$) we may assume that S is finite (or even trivial). Let $s(A)$ be the definable socle of A. By 5.9 we may assume that X is contained in $s(A)$. By the results in section 6, $s(A)$ is the almost direct sum of A_1 a semiabelian variety, and A_2 a modular complex torus. A_1 and A_2 are fully orthogonal, so by 5.7, $X = X \cap A_1 + X \cap A_2$. As A_2 is modular $X \cap A_2$ is a translate of a subgroup A_3 of A_2. But then A_3 is contained in S, so A_3 is a point. So up to translation X is contained in A_1, a semiabelian variety. By [25] (the generalization of

Faltings theorem to semiabelian varieties), X is a translate of a semiabelian subvariety. We obtain both a proof of the proposition and a contradiction.

§7. Variants. The main problem is the development of geometric model theory outside the finite Morley rank context as well as finding applications. The theory is fully in place ([22], [7]) for superstable theories (T is superstable if for all sufficiently large models M of T there are at most $|M|$-many complete types over M). There is a rank on types, the U-rank, which is ordinal valued but may be infinite. Strongly minimal formulas are replaced by regular types, types of U-rank ω^α. There is as yet no formulation of a "Zariski geometry" on regular types, and also no general theorems regarding the Zilber conjecture in this context.

In fact the general theory is in place for stable theories, except that regular types no longer coordinatize the structure. (T is stable if for any model M of T there are at most $|M|^{|T|}$ complete types over M.) See [17] for a reasonably comprehensive account.

There are not many stable theories. In the past five years there has been an enormous amount of work done generalizing the machinery of stability theory to a larger class of theories, the *simple* theories. The model companion of fields equipped with an automorphism, $ACFA$, is simple but unstable. The validity of the Zilber conjecture here ([3]) has led to more model-theoretic applications to diophantine geometry [6].

The study of o-minimal structures (neither stable, nor simple) is another thriving area [24]. However a lot of the general geometric theory is vacuous here, as 1-dimensionality is built into the situation. One would like to see a theory of finite-dimensional structures which are coordinatized by o-minimal structures, in which the principal homogeneous spaces from section 3 play a role.

In some sense the various kinds of theories for which there is a developed model theory correspond to structures surrounding number theory: the complex field (strongly minimal), the real field (o-minimal), ultraproducts of finite fields (simple of SU-rank 1). There is however no really "general" theory in place corresponding to the field of p-adics, and one would hope to see some progress here.

REFERENCES

[1] J. BALDWIN and A. H. LACHLAN, *On strongly minimal sets*, **The Journal of Symbolic Logic**, vol. 36 (1972), pp. 79–96.

[2] A. BUIUM and A. PILLAY, *A gap theorem for abelian varieties*, **Mathematical Research Letters**, vol. 4 (1997), pp. 211–219.

[3] Z. CHATZIDAKIS and E. HRUSHOVSKI, *Model theory of difference fields*, **Transactions of the American Mathematical Society**, vol. 351 (1999), pp. 2997–3071.

[4] G. Faltings, *Endlichkeitssätze für abelsche Varietäten über Zahlköpern*, **Inventiones Mathematicae**, vol. 73 (1983), pp. 349–366.

[5] ———, *The general case of S. Lang's conjecture*, **Barsotti symposium in algebraic geometry**, Perspectives in Math., vol. 15, Academic Press, 1994, pp. 175–182.

[6] E. Hrushovski, *Difference fields and the Manin-Mumford conjecture over number fields*, to appear in **Annals of Pure and Applied Logic**.

[7] ———, *Contributions to stable model theory*, **Ph.D. thesis**, Berkeley, 1986.

[8] ———, *A new strongly minimal set*, **Annals of Pure and Applied Logic**, vol. 62 (1993), pp. 147–166.

[9] ———, *Finitely axiomatizable ω_1-categorical theories*, **The Journal of Symbolic Logic**, vol. 59 (1994), pp. 838–844.

[10] ———, *Mordell-Lang conjecture for function fields*, **Journal of the American Mathematical Society**, vol. 9 (1996), pp. 667–690.

[11] E. Hrushovski and A. Pillay, *Effective bounds for transcendental points on subvarieties of semiabelian varieties*, **American Journal of Mathematics**, vol. 122 (2000), pp. 439–450.

[12] E. Hrushovski and Z. Sokolovic, *Minimal sets in differentially closed fields*, to appear in **Transactions of the American Mathematical Society**.

[13] E. Hrushovski and B. Zilber, *Zariski geometries*, **Journal of the American Mathematical Society**, vol. 9 (1996), pp. 1–56.

[14] D. Lascar, *Category of models of a complete theory*, **The Journal of Symbolic Logic**, vol. 82 (1982), pp. 249–266.

[15] D. Lascar and A. Pillay, *Hyperimaginaries and automorphism groups*, to appear in **The Journal of Symbolic Logic**.

[16] D. Marker, *Strongly minimal sets and geometries*, **Proceedings of logic colloquium'95** (J. Makovsky, editor), Springer, 1998.

[17] A. Pillay, **Geometric stability theory**, Oxford University Press, 1996.

[18] ———, *Some model theory of compact complex manifolds*, **Hilbert's 10th problem: relations with arithmetic and algebraic geometry** (Jan Denef et al., editors), vol. 270, AMS, 2000, pp. 323–338.

[19] A. Pillay and T. Scanlon, *Meromorphic groups*, preprint, 2000.

[20] B. Poizat, **Groupes stables**, Nur al-Mantiq Wal-mar'ifah, Villeurbanne, 1987.

[21] I. Shafarevich, **Basic algebraic geometry**, Springer, 1994.

[22] S. Shelah, **Classification theory**, North-Holland, 1990.

[23] K. Ueno, **Classification theory of algebraic varieties and coompact complex spaces**, vol. 439, Springer, 1975.

[24] L. van den Dries, **Tame topology and o-minimality**, Cambridge University Press, 1998.

[25] P. Vojta, *Integral points on subvarieties of semi-abelian varieties*, **Inventiones Mathematicae**, vol. 126 (1996), pp. 133–181.

[26] B. Zilber, **Uncountably categorical theories**, Translations of Mathematical Monographs, vol. 117, AMS, 1993.

DEPARTMENT OF MATHEMATICS
1409 W. GREEN STREET
UNIVERSITY OF ILLINOIS AT URBANA-CHAMPAIGN
URBANA, IL 61801, USA
E-mail: pillay@math.uiuc.edu

RESEARCH ARTICLES

THE INTUITIONISTIC ARITHMETICAL HIERARCHY

WOLFGANG BURR

Abstract. Due to the lack of a theorem on prenex normal forms for intuitionistic logic, the arithmetical hierarchy is not suitable for intuitionistic arithmetic. The main purpose of this paper is to fill this obvious gap: we introduce intuitionistic counterparts of the classes in the arithmetical hierarchy. These are shown to have nice properties and lead to a normal form theorem for intuitionistic arithmetic and moreover to suitable fragments of Heyting arithmetic. The present paper extends and refines the results of Burr [1].

§1. **Intuitionistic counterparts of Π_n.** By classical predicate logic, every formula is equivalent to some prenex one. This so-called normal form theorem gives rise to the well-known arithmetical hierarchy: the complexity of an (arithmetical) formula is measured in terms of the number of alternating quantifiers of its prenex normal form. A formula is in Π_n (Σ_n) if it is equivalent to some formula with n alternating quantifiers starting with an universal (existential) one and a matrix without any unrestricted quantifiers. However, the theorem on prenex normal forms does not hold in the intuitionistic framework. More precise: there are formulae which are not equivalent to any prenex formula within intuitionistic predicate logic. Thus, the arithmetical hierarchy is not sufficient for the intuitionistic setting. The main purpose of this paper is to introduce classes Φ_n which are shown to be reasonable counterparts of the classes Π_n and hence form an intuitionistic arithmetical hierarchy. Before we give the definition, we should ask which conditions we would like to have satisfied by Φ_n?

First of all, we are looking for normal forms, hence each formula should be equivalent to a formula in some Φ_n (in an intuitionistic basic theory). Moreover, classically Φ_n should neither be more nor less than Π_n, that is, classically Π_n and Φ_n should be equivalent[1]. Natural closure conditions would increase the usability of our classes and hence are desirable. Finally, truth-predicates (in other word, universal or complete formulae) for Φ_n which are itself in Φ_n should exist.

---

[1] Two formula classes are equivalent over some theory T if each formula of the one class is (provably in T) equivalent to some formula of the other class and vice versa.

Logic Colloquium '99
Edited by J. van Eijck, V. van Oostrom, and A. Visser
Lecture Notes in Logic, 17

We are now ready to give our main definition and to show that our classes fulfill all requirements listed above. Our definition is an improved version of that given in [1]. As an advantage, the new version is much more compact and hence easier to handle in practice. Another difference is the definition $\Phi_1 := \Pi_1$. For a discussion of this definition see section 3. This is our main definition:

DEFINITION 1 (The formula classes Φ_n).

  (i) $\Phi_0 := \Delta_0$, $\Phi_1 := \Pi_1$
  (ii) Suppose Φ_{n-2}, Φ_{n-1} are defined for $n \geqslant 2$. Then Φ_n consists of all formulae

$$\forall x(\psi \to \exists z \chi)$$

where $\psi \in \Phi_{n-1}$, $\chi \in \Phi_{n-2}$ and x, z are finite tuples of variables and x may occur both in ψ and χ.[2]

Notation. Before we start with the technical part of this section we should fix some notation. $\mathcal{L}_A = \{0, 1, +, \cdot, <\}$ denotes the language of arithmetic. By Δ_0 ($=: \Pi_0 =: \Sigma_0$) we denote the class of formulae that do not contain any unrestricted quantifiers. $i\Delta_0$ is the intuitionistic $\mathcal{L}_A$-theory with defining axioms for 0, 1, $+$, $\cdot$, $<$, intuitionistic first order predicate logic with identity and the induction-schema $I\varphi \equiv \varphi(0) \to \forall x(\varphi(x) \to \varphi(x + 1)) \to \forall x\varphi(x)$ for $\varphi \in \Delta_0$. Suppose Γ is a class of $\mathcal{L}_A$-formulae. Then $i\Gamma$ denotes $i\Delta_0$ plus induction-schema for all $\varphi \in \Gamma$. HA (Heyting Arithmetic) denotes $i\Delta_0$ augmented by induction for arbitrary formulae. $I\Delta_0$, $I\Gamma$ and PA (Peano Arithmetic) denote the classical counterparts of $i\Delta_0$, $i\Gamma$ and HA respectively. HA$^\omega$ denotes Heyting arithmetic in all finite types (see e.g., [1]).

To start with, we show that classically Π_n and Φ_n coincide.

LEMMA 2 ($\Pi_n = \Phi_n$ in $I\Delta_0$). *The classes Π_n and Φ_n are equivalent in $I\Delta_0$.*

PROOF. By induction on n. The cases $n = 0, 1$ are obvious. Now suppose $n \geqslant 2$ and $\varphi \equiv \forall x(\psi \to \exists z \chi)$. By induction hypothesis $\psi \in \Pi_{n-1}$ and $\chi \in \Pi_{n-2}$. Then we have by means of classical logic

$$\forall x(\psi \to \exists z \chi) \leftrightarrow \forall x(\underbrace{\neg \psi}_{\Sigma_{n-1}} \vee \underbrace{\exists z \chi}_{\Sigma_{n-1}})$$

and this is in Π_n ($I\Delta_0$ is needed to merge blocks of identical quantifiers).

For the converse direction, let $\varphi \in \Pi_n$. Then $\varphi \equiv \forall x \exists y \varphi_0$, $\varphi_0 \in \Pi_{n-2}$. By induction hypothesis there is $\psi_0 \in \Phi_{n-2}$ equivalent to φ_0. Hence φ is equivalent to $\forall x(0 = 0 \to \exists y \psi_0)$ which is Φ_n.   ⊣

---

[2]Note that we allow sequences of quantifiers since Φ_n is in general not closed under restricted quantifiers (even not in HA). We can merge blocks of identical quantifiers if we add function symbols for pairing and projection functions to our language.

LEMMA 3 (Closure conditions). *Suppose $n \geqslant 1$. Then we have within $i\Delta_0$
(by $\varphi \in \Phi_n$ we mean: φ is in $i\Delta_0$ provably equivalent to a formula in Φ_n):*

(i) $\varphi \in \Phi_n \Rightarrow \forall x \varphi \in \Phi_n$

(ii) $\varphi \in \Phi_{n-2} \Rightarrow \exists x \varphi \in \Phi_n$

(iii) $\varphi_0, \varphi_1 \in \Phi_{n-2} \Rightarrow \varphi_0 \vee \varphi_1 \in \Phi_n$

(iv) $\varphi_0, \varphi_1 \in \Phi_n \Rightarrow \varphi_0 \wedge \varphi_1 \in \Phi_n$

(v) $\varphi_0 \in \Phi_{n-1}, \varphi_1 \in \Phi_n \Rightarrow \varphi_0 \to \varphi_1 \in \Phi_n$

PROOF. By induction on n. The case $n = 1$ is obvious. Suppose $n \geqslant 2$. Then (i) and (ii) hold by definition.

For (iii) note that $\varphi_0 \vee \varphi_1 \leftrightarrow \exists z(z = 0 \to \varphi_0 \wedge z \neq 0 \to \varphi_1)$. By induction hypothesis and (ii) this yields the claim.

(iv) Suppose $\varphi_i \equiv \forall x(\psi_i \to \exists z \chi_i)$ and $\psi_i \in \Phi_{n-1}$, $\chi_i \in \Phi_{n-2}$ for $i = 0, 1$. Then the induction hypothesis yields

$(\varphi_0 \wedge \varphi_1)$

$$\leftrightarrow \forall v x[(v = 0 \to \psi_0 \wedge v \neq 0 \to \psi_1) \to (v = 0 \to \exists z \chi_0 \wedge v \neq 0 \to \exists z \chi_1)].$$

This is equivalent to

$$\forall v x[\underbrace{(v = 0 \to \psi_0 \wedge v \neq 0 \to \psi_1)}_{\Phi_{n-1}} \to \exists z \underbrace{(v = 0 \to \chi_0 \wedge v \neq 0 \to \chi_1)}_{\Phi_{n-2}}].$$

Hence this is equivalent to a formula in Φ_n.

(v) Suppose $\varphi_0 \in \Phi_{n-1}$ and $\varphi_1 \equiv \forall x(\psi \to \exists z \chi) \in \Phi_n$ with $\psi \in \Phi_{n-1}$, $\chi \in \Phi_{n-2}$. Whence we have

$$(\varphi_0 \to \varphi_1) \leftrightarrow (\varphi_0 \to \forall x(\psi \to \exists z \chi)) \leftrightarrow \forall x(\underbrace{(\varphi_0 \wedge \psi)}_{\Phi_{n-1}} \to \exists z \chi).$$

This concludes the proof of the lemma. $\dashv$

Due to the closure conditions the classes Φ_n provide normal forms for intuitionistic arithmetic:

COROLLARY 4 (Normal form theorem). *Every $\varphi \in \mathcal{L}_A$ is equivalent in $i\Delta_0$ to a formula $\varphi \in \Phi_n$ for some n. Hence the formulae in $\bigcup_{n \in \omega} \Phi_n$ are normal forms for $\mathcal{L}_A$.*

Since each Δ_0-formula is equivalent to some $\vee, \wedge$-free formula, we obtain as an interesting observation that (in $i\Delta_0$) $\vee$ and $\wedge$ are definable. Hence

COROLLARY 5. $\{\bot, \to, \forall, \exists\}$ *is complete for intuitionistic arithmetic.*

DEFINITION 6 (Truth-predicates for Φ_n). We define truth-predicates T_n for each natural number n. Let $\text{Sat}_{\Delta_0}(x)$ be a truth-definition for Δ_0-formulae, cf. e.g., [2, I.1.d]. It is clear that all properties for Sat_{Δ_0} are provable already in $i\Sigma_1$, in particular for every $\varphi(x_1, \ldots, x_n) \in \Delta_0$ we have $i\Sigma_1 \vdash \text{Sat}_{\Delta_0}(\ulcorner \varphi(\dot{x}_1, \ldots, \dot{x}_n) \urcorner) \leftrightarrow \varphi(x_1, \ldots, x_n)$. Now we define:

1. $T_0 := \text{Sat}_{\Delta_0}$.
2. Put T_1 such that for $\varphi(x) \in \Delta_0$ we have

$$T_1(\ulcorner \forall x \varphi(x) \urcorner) \leftrightarrow \forall x \text{Sat}_{\Delta_0}(\ulcorner \varphi(\dot{x}) \urcorner).$$

3. Now suppose T_{n-1} and T_{n-2} are already defined for $n \geqslant 2$. We define T_n such that for all $\psi \in \Phi_{n-1}$ and $\chi \in \Phi_{n-2}$ we have

$$T_n(\ulcorner \forall x(\psi \to \chi) \urcorner) \leftrightarrow \forall x[T_{n-1}(\ulcorner \psi(\dot{x}) \urcorner) \to \exists z T_{n-2}(\ulcorner \chi(\dot{z}, \dot{x}) \urcorner)].$$

This (and similar 2.) is possible, since it is primitive recursive to decide for a given z whether it is the Gödelnumber of a formula $\forall x(\psi \to \exists z \chi) \in \Phi_{n+1}$ and if this is the case, to recover the Gödelnumbers of the sub-formulae ψ, χ from x.

These truth-predicates fulfill our requirements:

THEOREM 7. $T_n \in \Phi_n$. $i\Sigma_1 \vdash T_n(\ulcorner \varphi(\dot{x}_1, \ldots, \dot{x}_n) \urcorner) \leftrightarrow \varphi(x_1, \ldots, x_n)$, $\varphi \in \Phi_n$.

PROOF. Straightforward. $\dashv$

1.1. Intuitionistic counterparts of $I\Pi_n$. One major motivation to search for intuitionistic counterparts of Π_n was the lack of suitable intuitionistic versions of the well-known fragments $I\Pi_n$ of Peano arithmetic. This is due to the fact the hierarchy $(i\Pi_n)_{n \in \mathbb{N}}$ collapses as shown by Visser and Wehmeier [7]: Let $PNF = \{\varphi \in \mathcal{L}_A \mid \varphi \text{ is prenex}\}$ then iPNF proves the same Π_2-sentences as $i\Pi_2$ and thus is dramatically weak.

It turns out that for $n \geqslant 2$, $i\Phi_n$ are suitable intuitionistic counterparts of $I\Pi_n$. Before we give arguments in favor of this claim, we again ask for requirements which should be met by the counterparts of $I\Pi_n$. Fore sure, when adding classical logic we again want to have neither more nor less than $I\Pi_n$. The second requirement is that both theories should prove the same Π_2-sentences. The reason for putting the second requirement is as follows: The totality of a recursive functions is expressed by a Π_2-theorem $\forall x \exists y T(e, x, y)$ (where T is Kleene's T predicate and e a Kleene-index of some recursive function) and hence the class of Π_2-theorems of a theory characterizes its computational content. And it is well-known that in many cases a classical theory and its intuitionistic version coincide in terms of their computational content. In particular, for PA and HA this is the case and hence it is natural to demand this for their fragments as well. Indeed, we can show

THEOREM 8. $i\Phi_n$ and $I\Pi_n$ prove the same Π_2-sentences for all $n \neq 1$.

PROOF. See [1], corollary 2.5 and note, that $\Phi'_n \subseteq \Phi_n$ holds[3] for the refined version of Φ_n as well. $\dashv$

Due to the presence of truth predicates, we easily deduce:

COROLLARY 9. The fragments $i\Phi_n$ are all finitely axiomatized.

---

[3] Φ'_n is introduced in [1] and somehow the almost negative fragment of Φ_n.

§2. Intuitionistic counterparts of Σ_n. In the classical setting, we have dually to the classes Π_n the Σ_n-hierarchy. This situation is usually visualized by the picture in Fig. 1. This gives rise to the question if there are natural intuitionistic counterparts Ψ_n of Σ_n? Classically, we have $\Sigma_n = \neg\Pi_n$, where $\neg\Gamma = \{\neg\varphi \mid \varphi \in \Gamma\}$, however, due to the peculiarities of intuitionistic logic, we cannot expect $\neg\Phi_n$ to be a good candidate for an intuitionistic counterpart of Σ_n. Indeed, $i\neg\Phi_n$ is dramatically weak, (cf. the proof of theorem 12 and note that all formulae in $\neg\Phi_n$ are Harrop). To start with, let us ask for criteria that should be met by Ψ_n. Obviously, we would like to have classically

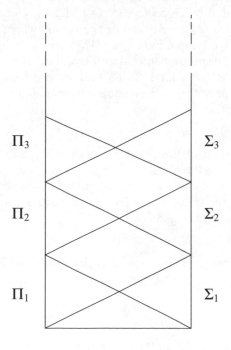

FIGURE 1

$\Psi_n = \Sigma_n$ and furthermore that $I\Sigma_n$ and $i\Psi_n$ prove the same Π_2-sentences. Moreover, every formula should be intuitionistically equivalent to a formula in some Ψ_n and we would like to be able to define suitable truth-predicates. This is the definition we suggest:

DEFINITION 10 (The classes Ψ_n).

(i) $\Psi_0 := \Delta_0$, $\Psi_1 := \Sigma_1$
(ii) For $n \geqslant 2$, let Ψ_n consists of all formulae $\varphi \to \psi$ where $\varphi \in \Phi_n$ and $\psi \in \Phi_{n-1}$.

It is easy to verify that classically we have $\Psi_n = \Sigma_n$. Since $\bigcup_n \Phi_n = \mathcal{L}_A(i\Delta_0)$ we easily find any $\varphi \in \mathcal{L}_A$ in some Ψ_n (provably in $i\Delta_0$). Similar to the preceding section we easily obtain suitable truth-predicates for Ψ_n. Moreover we have:

THEOREM 11. $I\Sigma_n$ and $i\Psi_n$ prove the same Π_2-sentences for each $n \in \mathbb{N}$.

PROOF. We use a technique due to Leivant [4]. For $n = 0, 1$ the assertion is well-known, so let $n \geqslant 2$. Since $i\Psi_n$ contains $i\Sigma_1$ we may freely introduce primitive recursive function and hence each Σ_1 formula is equivalent to a formula $\exists x \, t(x) = 0$ where t is a primitive recursive term. Let φ^- be the negative translation introduced by Gödel (see [5, Def. 2.3.4.]). Since prime formulae of $\mathcal{L}_A$ are decidable in $i\Delta_0$ we may choose the variant where $(s = t)^- :\equiv s = t$ and $\perp^- :\equiv \perp$. Let $\vdash_M$ (respectively $\vdash_I$ and $\vdash_C$) denote derivability in minimal intuitionistic logic (respectively intuitionistic and classical logic), see [4]. We write $\Gamma' := \{\varphi' \mid \varphi \in \Gamma\}$ for $\Gamma \subseteq \mathcal{L}_A$ and $'$ a translation. Note that the decidability of prime formulae is provable over minimal logic as well. Hence theorem 2.3.5 of [5] is

$$\Gamma \vdash_C \varphi \Rightarrow \Gamma^- \vdash_M \varphi^-. \tag{1}$$

Let φ^ψ be the variant of the Friedman (Dragalin) translation introduced in [4]: $(a = b)^\psi \equiv a = b$; $(\perp)^\psi \equiv \psi$ and $()^\psi$ commutes with $\wedge, \vee, \rightarrow, \forall, \exists$. The difference to the usual version is that atomic formulae besides $\perp$ are unchanged by the translation. This modified translation still works for minimal logic and 1.1 of [4] yields

$$\Gamma \vdash_M \varphi \Rightarrow \Gamma^\psi \vdash_M \varphi^\psi. \tag{2}$$

Now suppose $I\Sigma_n$ proves a Π_2-sentences. That is without loss of generality a sentence $\forall x \exists y \, t(x, y) = 0$. This means that for $\Gamma := \{I\varphi \mid \varphi \in \Sigma_n\}$ we have

$$\Gamma \vdash_C \exists y \, t(x, y) = 0. \tag{3}$$

(1) and (2) yield for $\psi :\equiv \exists y \, t(x, y) = 0$:

$$((\Gamma)^-)^\psi \vdash_M (\neg\neg\psi)^\psi \text{ i. e., } ((\Gamma)^-)^\psi \vdash_M \psi \tag{4}$$

since $(\neg\neg\psi)^\psi \leftrightarrow [(\psi \rightarrow \psi) \rightarrow \psi] \leftrightarrow \psi$. For an instance $I\varphi$ of the induction schema we have (provably in $i\Delta_0$)

$$((I\varphi)^-)^\psi \leftrightarrow I(\varphi^-)^\psi, \tag{5}$$

hence, to finish the proof we need to show that for each $\varphi \in \Sigma_n$ there is some φ' such that $\vdash_C \varphi \leftrightarrow \varphi'$ and for $\psi \in \Sigma_1$

$$(\varphi^-)^\psi \in \Psi_n. \tag{6}$$

(4), (5) and (6) imply the assertion, thus it remains to show (6). Let $\psi \in \Sigma_1$. By induction on n we show: For $\varphi \in \Pi_n$ there is φ' with $\vdash_C \varphi \leftrightarrow \varphi'$ and

$$((\varphi')^-)^\psi \in \Phi_n \text{ (provably in } i\Delta_0). \tag{7}$$

1. $n = 2$. For $\varphi \equiv \forall x \exists y\, s(x, y) = 0$ let $\varphi' \equiv \forall x \neg \forall y\, s(x, y) > 0$. Then $((\varphi')^-)^\psi \equiv \forall x(\forall y\, s(x, y) > 0 \to \psi)$ which is in Φ_2.

2. $n > 2$. Let $\varphi \in \Pi_n$. Then there is $\varphi_0 \in \Pi_{n-1}$ in a way such that $\vdash_C \varphi \leftrightarrow \forall x \neg \varphi_0$ and by induction hypothesis $((\varphi_0)^-)^\psi$ is in Φ_{n-1}. Thus

$$((\forall x \neg \varphi_0)^-)^\psi \equiv \forall x(((\varphi_0)^-)^\psi \to \psi).$$

The later formula is in Φ_n and (7) follows. Now we are ready to prove (6): Let $\varphi \in \Sigma_n$. Then there is $\varphi_0 \in \Pi_n$ such that $\vdash_C \varphi \leftrightarrow \neg \varphi_0$. By (7) we find φ_0' with $\vdash_C \varphi \leftrightarrow \neg \varphi_0'$ and $((\varphi_0')^-)^\psi$ is equivalent in $i\Delta_0$ to a formula in Φ_n and

$$((\neg \varphi_0')^-)^\psi \leftrightarrow [((\varphi_0')^-)^\psi \to \psi].$$

By definition, the later formula is in Ψ_n.   ⊣

REMARK. The definition of Ψ_n is not fully satisfying. This is because Ψ_n is not really dual to Φ_n. We already mentioned that the dual definition $\Psi_n := \neg \Phi_n \equiv \Phi_n \to \bot$ is not sufficient. However, we may read $\Phi_n \to \Sigma_1$ as a generalized version of negation, or better, as the Friedman translation of negation. Checking the preceding proof yields that this definition is already sufficient to obtain the full Π_2-strength of the associated fragments of HA. However, by this definition we do not get a normal form theorem. Therefore, we put $\Psi_n :\equiv \Phi_n \to \Phi_{n-1}$ to include all of $\mathcal{L}_A$ by definition. Another disadvantage is the lack of closure conditions. Indeed, we might have defined Ψ_n in a way such that it is closed under existential quantification, e.g., by setting $\Psi_{n+2} := \exists x(\Phi_{n+1} \to \Phi_n)$, but still, for $\wedge, \vee, \to$ we would not obtain adequate conditions. Our choice has been motivated by its compactness and the fact that in practice, closure under existential quantification does not seem to be really useful.

In general, Φ_n seem to be much more natural classes than Ψ_n: In mathematical practice (e.g., in well-ordering-proofs, in proofs of the totality of some recursive function etc.), those formulas for which induction is applied are naturally located in the Φ_n hierarchy. By definition, they occur within the Ψ_n hierarchy as well, however at a later stage.

§3. **The rôle of Π_1.** In Fig. 2 we represent the fragments $i\Pi_n$, $i\Psi_n$ and $I\Pi_n (= I\Sigma_n)$ in terms of their Π_2-strength. We notice one irregular point, namely $i\Phi_1$ which is by definition $i\Pi_1$. This is the only theory which is not Π_2 equivalent to its classical counterpart $I\Pi_1 (= I\Sigma_1)$ as shown by Wehmeier [7]: $i\Pi_1$ does not prove the totality of 2^x, since all its provably recursive functions are bounded by polynomials. Wehmeier's result gives rise to the question, whether there is at all an intuitionistic counterpart of $I\Pi_1$, which may fill the gap in Fig. 2. More precise, we are looking for a class $\Gamma \subseteq \mathcal{L}_A$ in a way such that classically $\Gamma = \Pi_1$ and $i\Gamma$ and $I\Pi_1$ prove the same Π_2-sentences. I

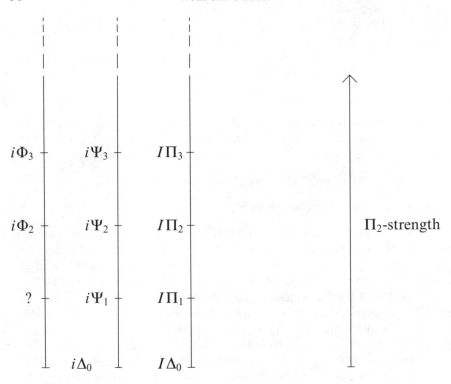

FIGURE 2

conjecture that there is no such class. The following is a partial answer to the question. We define classes Π and Σ which are generalizations of Π_1 and Σ_1:

(i) $\Delta_0 \subseteq \Pi \cap \Sigma$.
(ii) Let $\varphi, \varphi' \in \Pi$ and $\psi, \psi' \in \Sigma$. Then $\varphi \wedge \varphi', \psi \to \varphi, \forall x \varphi \in \Pi$ and $\psi \vee \psi', \psi \wedge \psi', \varphi \to \psi, \exists x \psi \in \Sigma$.

Π (respectively Σ) are those formulae which are syntactically recognizable (i.e., by applying the de Morgan rules) being classically in Π_1 (Σ_1). An argument due to Kohlenbach [3] yields the following theorem

THEOREM 12. *All provably recursive functions of $i\Pi$ are bounded by polynomials. Hence $i\Pi$ proves strictly less Π_2-sentences than $I\Pi_1$.*

PROOF. We easily verify that all formulae in Π are Harrop, where Harrop formulae are defined inductively: 1) atomic formulae are Harrop; 2) if φ, ψ are Harrop and χ is arbitrary, then $\varphi \wedge \psi$, $\forall x \varphi$ and $\chi \to \varphi$ are Harrop. Now we consider mq-realizability defined on page 214 of [6]. An easy induction on Harrop formulae proves that for φ Harrop, x in $x \, mq \, \varphi$ is the empty sequence. An inspection of the soundness theorem 3.4.5 of [6] now yields that in order to interpret the induction schema for such φ we do not need any recursors R_σ at all. Also, in the interpretation of PL10 we may replace R_σ by case distinction functionals D_σ. Hence, $i\Pi$ is mq-interpretable in a theory we call $HA^{\omega-}$: this is HA^ω without recursors R_σ but with additional functionals $+, \cdot, D_\sigma$.[4] The soundness theorem applied to $i\Pi \vdash \exists y \varphi(x, y)$ now provides us with a term $t[x] \in HA^{\omega-}$ such that $HA^{\omega-} \vdash \varphi(x, t[x])$. Kohlenbach's results (see Theorem 4.3 of [1]) yield a polynomial $p(x)$ that majorizes $t[x]$ and hence $HA^{\omega-} \vdash \forall x \exists y \leqslant p(x) \varphi(x, y)$. Since HA^ω is conservative over HA ([6, 3.6.3]) the theorem is proved. ⊣

The theorem says that a weak intuitionistic basic theory augmented by induction for all formulae which are Π_1 by (classical) rules of de Morgan does not prove the totality of exponentiation. This gives a hint which may affirm my conjecture: It is rather unlikely that induction for some formula that is — although not syntactically — logically equivalent to some Π_1-formula yields an intuitionistic proof of the totality of $2^{()}$.

REFERENCES

[1] WOLFGANG BURR, *Fragments of Heyting arithmetic*, **The Journal of Symbolic Logic**, vol. 65 (2000), pp. 1223–1240.

[2] Petr Hájek and Pavel Pudlák (editors), **Metamathematics of first-order arithmetic**, Springer, 1993.

[3] ULRICH KOHLENBACH, *Relative constructivity*, **The Journal of Symbolic Logic**, vol. 63 (1998), no. 4, pp. 1218–1238.

[4] DANIEL LEIVANT, *Syntactic translations and provably recursive functions*, **The Journal of Symbolic Logic**, vol. 50 (1985), no. 3, pp. 682–688.

[5] A. S. TROELSTRA and D. VAN DALEN, **Constructivism in mathematics. Vol. I**, North-Holland Publishing Co., Amsterdam, 1988.

[6] Anne S. Troelstra (editor), **Metamathematical investigation of intuitionistic arithmetic and analysis**, Lecture Notes in Mathematics, no. 344, Springer, Heidelberg/New York, 1973.

[7] KAI F. WEHMEIER, *Fragments of HA based on Σ_1-induction*, **Archive for Mathematical Logic**, vol. 37 (1997), no. 1, pp. 37–49.

INSTITUT FÜR MATHEMATISCHE LOGIK UND GRUNDLAGENFORSCHUNG
DER WESTFÄLISCHEN WILHELMS-UNIVERSITÄT MÜNSTER
EINSTEINSTR. 62, D-48149 MÜNSTER, GERMANY
E-mail: Wolfgang.Burr@math.uni-muenster.de

---

[4]In [6] HA^ω is called $N\text{-}HA^\omega$.

ON SOLVABLE GROUPS AND RINGS DEFINABLE IN O-MINIMAL STRUCTURES

MÁRIO J. EDMUNDO

§1. **Introduction.** Let $\mathcal{N} = (N, <, \dots)$ be an o-minimal structure. Below definable will mean $\mathcal{N}$-definable with parameters.

Pillay in [16] adapts Hrushovski's proof of Weil's Theorem that an algebraic group can be recovered from birational data to show that a definable group G can be equipped with a unique definable space structure making the group into a topological group, and definable homomorphisms between definable groups are topological homomorphisms. In fact, as remarked in [11], if $\mathcal{N}$ is an o-minimal expansion of a real closed field G equipped with the above unique definable space structure is a C^p group for all $p \in \mathbb{N}$; and definable homomorphisms between definable groups are C^p homomorphisms for all $p \in \mathbb{N}$. Moreover, again by [11], the definable space structure on a definable subgroup is the sub-space structure.

By [16] definable groups satisfies the descending chain condition (DCC) on definable subgroups. This is used to show that the definably connected component of the identity G^0 of a definable group G is the smallest definable subgroup of G of finite index. Also infinite such groups have infinite definable abelian subgroups; a definable subgroup H of G is closed and the following are equivalent (i) H has finite index in G, (ii) $\dim H = \dim G$, (iii) H contains an open neighbourhood of the identity element of G and (iv) H is open in G. Finally, by [18] an infinite abelian definable group G has unbounded exponent and the subgroup $\text{Tor}(G)$ of torsion points of G is countable. In particular, if $\mathcal{N}$ is $\aleph_0$-saturated then G has an element of infinite order.

One dimensional definable manifolds are classified in [17] and the following is deduced. Suppose that G is one-dimensional definably connected definable group. Then by [16] G is abelian, and G is torsion-free or for each prime p the set of p-torsion points of G has p elements. In the former case G is an ordered abelian divisible definably simple group.

Note that if I is a one-dimensional definably connected ordered definable group, then the structure $\mathcal{I}$ induced by $\mathcal{N}$ on I is o-minimal. In particular, we have the following results from [7]. Suppose that $(I, 0, 1, +, <)$ is a

Logic Colloquium '99
Edited by J. van Eijck, V. van Oostrom, and A. Visser
Lecture Notes in Logic, 17

one-dimensional definably connected torsion-free definable group, where 1 is a fixed positive element. Let $\Lambda(\mathcal{I})$ be the division ring of all $\mathcal{I}$-definable endomorphisms of $(I, 0, +)$. Then exactly one of the following holds: (1) $\mathcal{I}$ is *linearly bounded* with respect to $+$ (i.e. for every $\mathcal{I}$-definable function $f : I \to I$ there is $r \in \Lambda(\mathcal{I})$ such that $\lim_{x \to +\infty}[f(x) - rx] \in I$), or (2) there is a $\mathcal{I}$-definable binary operation $\cdot$ such that $(I, 0, 1, +, \cdot, <)$ is a real closed field. Also, up to $\mathcal{I}$-definable isomorphism there is at most one $\mathcal{I}$-definable group $(I, 0, *)$ such that $\mathcal{I}$ is linearly bounded with respect to $*$ and at most one $\mathcal{I}$-definable (real closed) field $(I, 0, 1, \oplus, \otimes)$.

Moreover, the following are equivalent: (i) $\mathcal{I}$ is linearly bounded with respect to $+$, (ii) for every $\mathcal{I}$-definable function $f : A \times I \to I$, where $A \subseteq I^n$, there are $r_1, \ldots, r_l \in \Lambda(\mathcal{I})$ such that for every $a \in A$ there is $i \in \{1, \ldots, l\}$ with $\lim_{x \to +\infty}[f(a, x) - r_i x] \in I$ and (iii) there is no infinite definable subset of $\Lambda(\mathcal{I})$.

Let $(I, 0, 1, +, <)$ be as above and let $\Lambda := \Lambda(\mathcal{I})$. Then $\mathcal{I}$ is called *semibounded* if every $\mathcal{I}$-definable set is already definable in the reduct $(I, 0, 1, +, <, (B_k)_{k \in K}, (\lambda)_{\lambda \in \Lambda})$ of $\mathcal{I}$ where $(B_k)_{k \in K}$ is the collection of all bounded $\mathcal{I}$-definable sets. According to [2], the following are equivalent: (i) $\mathcal{I}$ is semibounded, (ii) there is no $\mathcal{I}$-definable function between a bounded and an unbounded subinterval of I, (iii) there is no $\mathcal{I}$-definable (real closed) field with domain an unbounded subinterval of I, (iv) for every $\mathcal{I}$-definable function $f : I \to I$ there are $r \in \Lambda$, $x_0 \in I$ and $c \in I$ such that for all $x > x_0$, $f(x) = rx + c$ and (v) $\mathcal{I}$ satisfies the "structure theorem".

Let $(I, 0, 1, +, \cdot, <)$ be a real closed field definable in $\mathcal{N}$. Let $\mathcal{K}(\mathcal{I})$ be the ordered field of all $\mathcal{I}$-definable endomorphisms of the multiplicative group $(I^{>0}, \cdot, 1)$. Note that $\mathcal{K}(\mathcal{I}) \to I$, $\alpha \to \alpha'(1)$ is an embedding of ordered fields. The elements of $\mathcal{K}(\mathcal{I})$ are called *power functions* and for $\alpha \in \mathcal{K}(\mathcal{I})$ with $\alpha'(1) = r$ we write $\alpha(x) = x^r$. By [6] exactly one of the following holds: (1) $\mathcal{I}$ is *power bounded* (i.e., for every $\mathcal{I}$-definable function $f : I \to I$ there is $r \in \mathcal{K}(\mathcal{I})$ such that ultimately $|f(x)| < x^r$) or (2) $\mathcal{I}$ is *exponential* (i.e., there is an $\mathcal{I}$-definable ordered group isomorphism $e : (I, 0, +, <) \to (I^{>0}, 1, \cdot, <)$). Moreover, the following are equivalent: (i) $\mathcal{I}$ is power bounded, (ii) for every $\mathcal{I}$-definable function $f : A \times I \to I$, where $A \subseteq I^n$, there are $r_1, \ldots, r_l \in \mathcal{K}(\mathcal{I})$ such that for every $a \in A$, if the function $x \to f(a, x)$ is ultimately nonzero then, there is $i \in \{1, \ldots, l\}$ with $\lim_{x \to +\infty}[f(a, x)/x^{r_i}] \in I$ and (iii) there is no infinite definable subset of $\mathcal{K}(\mathcal{I})$.

If $\mathcal{I}$ is power bounded, then we know that $(I, 0, +, <)$ and $(I^{>0}, 1, \cdot, <)$ are the only (up to $\mathcal{I}$-definable isomorphism) $\mathcal{I}$-definable one-dimensional torsion-free ordered groups. The *Miller-Starchenko conjecture* says that in an o-minimal expansion $\mathcal{I}$ of an ordered field every $\mathcal{I}$-definable one-dimensional torsion-free ordered group is $\mathcal{I}$-definable isomorphic to either $(I, 0, +, <)$ or $(I^{>0}, 1, \cdot, <)$. (In the general case we only know (see [7]) that up to

$\mathcal{I}$-definable isomorphisms there are at most two $\mathcal{I}$-definably connected, $\mathcal{I}$-definable one-dimensional torsion-free ordered groups). Suppose that the Miller-Starchenko conjecture does not hold for $\mathcal{I}$, then we call the unique $\mathcal{I}$-definable group $(I, 0, \oplus, <)$ which is not $\mathcal{I}$-definably isomorphic to $(I, 0, +, <)$ or $(I^{>0}, 1, \cdot, <)$ the *Miller-Starchenko group of* $\mathcal{I}$. Note the following: if G is an $\mathcal{I}$-definable one-dimensional torsion-free ordered group, then we can assume that $G = (I, 0, \oplus, <)$, and $\alpha \colon G \to (I, 0, +)$ is an abstract C^1 isomorphism iff $\forall s \in G, \ \alpha'(s)\frac{\partial\oplus}{\partial x}(0, s) = \alpha'(0)$ where for all $t, s \in G, \oplus(t, s) := t \oplus s$ i.e., α is Pfaffian over $(I, 0, 1, +, \cdot, \oplus, <)$ (note that, by associativity of $\oplus$, for all $s \in G$, $\frac{\partial\oplus}{\partial x}(0, s) \neq 0$).

The notion of definably compact groups was introduced in [15]. Let G be a definable group. We say that G is *definably compact* if for every definable continuous embedding $\sigma \colon (a, b) \subseteq N \to G$, where $-\infty \leq a < b \leq +\infty$, there are $c, d \in G$ such that $\lim_{x \to a^+} \sigma(x) = c$ and $\lim_{x \to b^-} \sigma(x) = d$, where the limits are taken with respect to the topology on G. In [15] the following result is established. Let G be a definable group which is not definably compact. Then G has a one-dimensional definably connected torsion-free (ordered) definable subgroup.

The trichotomy theorem [13] and the theory of non orthogonality from [11] are used to prove the following.

THEOREM 1.1 ([1]). *Let U be a definable group and let A be a definable normal subgroup of U. Then we have a definable extension $1 \to A \to U \xrightarrow{j} G \to 1$ with a definable section $s \colon G \to U$.*

If we take in Theorem 1.1 A to be the radical of U i.e., the maximal definably connected definable solvable normal subgroup of U we get that G is either finite or definably semisimple i.e., it has no infinite proper abelian definable normal subgroup. Definable definably semisimple groups are classified in [11] (see also [12] and [10]). Below, $\mathbf{G}$ is the structure $(G, \cdot)$ where $\cdot$ is the group operation of G.

FACT 1.2 ([11] and [10]). *Let G be a definably semisimple $\mathbf{G}$-definably connected definable group. Then $G = G_1 \times \cdots \times G_l$ and for each $i \in \{1, \ldots, l\}$ there is an o-minimal expansion $\mathcal{I}_i$ of a real closed field definable in $\mathcal{N}$ such that there is no definable bijection between a distinct pair among the I_i's, G_i is $\mathcal{I}_i$-definably isomorphic to a I_i-semialgebraic subgroup of $GL(n_i, I_i)$ which is a direct product of I_i-semialgebraically simple, I_i-semialgebraic subgroups of $GL(n_i, I_i)$.*

§2. Solvable definable groups.
Theorem 1.1 allows one to develop the o-minimal analogue of the classical group extension theory from [3] and [4]. The connection between definable extensions and the second cohomology group associated to that extension is main tool used in the following results from

[1] which give a classification of definable solvable groups module definably compact definable abelian groups.

THEOREM 2.1 ([1]). *Let U be a definable solvable group. Then U has a definable normal subgroup V such that U/V is a definably compact definable solvable group and $V = K \times W_1 \times \cdots \times W_s \times V_1' \times V_1 \times \cdots \times V_k' \times V_k$ where K is the definably connected definably compact normal subgroup of U of maximal dimension and for each $j \in \{1, \ldots, s\}$ (resp., $i \in \{1, \ldots, k\}$) there is a semi-bounded o-minimal expansion $\mathcal{J}_j$ of a group (resp., an o-minimal expansion $\mathcal{I}_j$ of a real closed field) definable in $\mathcal{N}$ such that there is no definable bijection between a distinct pair among the $\mathcal{J}_j$'s and $\mathcal{I}_i$'s, W_j is a direct product of copies of the additive group of $\mathcal{J}_j$, V_i' is a direct product of copies of the linearly bounded one-dimensional torsion-free $\mathcal{I}_i$-definable group and V_i is an $\mathcal{I}_i$-definable group such that $Z(V_i)$ has an $\mathcal{I}_i$-definable subgroup Z_i such that $Z(V_i)/Z_i$ is a direct product of copies of the linearly bounded one-dimensional torsion-free $\mathcal{I}_i$-definable group and there are $\mathcal{I}_i$-definable subgroups $1 = Z_i^0 < Z_i^1 < \cdots < Z_i^{m_i} = Z_i$ such that for each $l \in \{1, \ldots, m_i\}$, Z_i^l/Z_i^{l-1} is the additive group of $\mathcal{I}_i$, and $V_i/Z(V_i)$ $\mathcal{I}_i$-definably embeds into $GL(n_i, I_i)$.*

We also give a different proof the following result about definably compact definable groups. This result was proved in [14, Corollary 5.2] under the assumption that $\mathcal{N}$ has definable Skolem function. With Theorem 1.1, this assumption on $\mathcal{N}$ is unnecessary.

THEOREM 2.2. *Let U be a definably compact, definably connected definable group. Then U is either abelian or $U/Z(U)$ is a definable semi-simple group. In particular, if U is solvable then it is abelian.*

Theorem 2.1 gives a partial solution to the Peterzil-Steinhorn splitting problem for solvable definable group with no definably compact parts (see [15]). We say that a definable abelian group U *has no definably compact parts* if there are definable subgroups $1 = U_0 < U_1 < \cdots < U_n = U$ such that for each $j \in \{1, \ldots, n\}$, U_j/U_{j-1} is a one-dimensional definably connected torsion-free definable group. We say that a definable solvable group U *has no definably compact parts* if U has definable subgroups $1 = U_0 \trianglelefteq U_1 \trianglelefteq \cdots \trianglelefteq U_n = U$ such that for each $i \in \{1, \ldots, n\}$, U_i/U_{i-1} is a definable abelian group with no definable compact parts. Peterzil and Steinhorn ask in [15] if a definable abelian group U of dimension two and with no definably compact parts is a direct product of one-dimensional definably connected torsion-free definable groups. Theorem 2.1 above reduces this problem to the case where U is a group definable in a definable o-minimal expansion $\mathcal{I}$ of a real closed field $(I, 0, 1, +, \cdot, <)$ and we have an $\mathcal{I}$-definable extension $1 \to A \to U \to G \to 1$ where $A = (I, 0, +, <)$ and $G = (I, 0, *, <)$ is a one-dimensional torsion-free $\mathcal{I}$-definable group. We prove that in this case, there is an $\mathcal{I}$-definable 2-cocycle $c \in Z^2_{\mathcal{I}}(G, A)$ for U such that U is $\mathcal{I}$-definably isomorphic to $A \times G$ iff there is an $\mathcal{I}$-definable function $\alpha \colon G \to A$ such that $\forall s \in G$, $\alpha'(s)\frac{\partial *}{\partial x}(0, s) = \alpha'(0) + \frac{\partial c}{\partial x}(0, s)$.

Let $\mathcal{I}$ be an o-minimal expansion of a real closed field $(I, 0, 1, +, \cdot, <)$ and suppose that we have an abelian $\mathcal{I}$-definable extension $1 \to A \to U \to G \to 1$ where $A = (I, 0, +, <)$ and $G = (I, 0, *, <)$ is a one-dimensional torsion-free $\mathcal{I}$-definable group. We shall say that U is a *Peterzil-Steinhorn $\mathcal{I}$-definable group* if U is not $\mathcal{I}$-definably isomorphic to $A \times G$. A corollary of our main result is the following fact.

THEOREM 2.3 ([1]). *Let $\mathcal{I} = (I, 0, 1, +, \cdot, <, \dots)$ be an o-minimal expansion of a real closed field with no Peterzil-Steinhorn $\mathcal{I}$-definable groups. Then every $\mathcal{I}$-definable solvable group U with no $\mathcal{I}$-definably compact parts is $\mathcal{I}$-definably isomorphic to a group definable of the form $U' \times G_1 \dots G_k \cdot G_{k+1} \dots G_l$ where U' is a direct product of copies of linearly bounded one-dimensional torsion-free $\mathcal{I}$-definable groups, for $i = 1, \dots, k$, $G_i = (I, 0, +)$ and for $i = k + 1, \dots, l$, $G_i = (I^{>0}, 1, \cdot)$. In particular, $G := G_1 \dots G_k \cdot G_{k+1} \dots G_l$ $\mathcal{I}$-definably embeds into some $GL(n, I)$ and U is $\mathcal{I}$-definably isomorphic to a group definable in one of the following reducts $(I, 0, 1, +, \cdot, \oplus)$, $(I, 0, 1, +, \cdot, \oplus, e^t)$ or $(I, 0, 1, +, \cdot, \oplus, t^{b_1}, \dots, t^{b_r})$ of $\mathcal{I}$ where $(I, 0, \oplus)$ is the Miller-Starchenko group of $\mathcal{I}$, e^t is the $\mathcal{I}$-definable exponential map (if it exists), and the t^{b_j}'s are $\mathcal{I}$-definable power functions. Moreover, if U is nilpotent then U is $\mathcal{I}$-definably isomorphic to a group definable in the reduct $(I, 0, 1, +, \cdot, \oplus)$ of $\mathcal{I}$.*

Let G be a definable group and X a subset of G. By DCC on definable subgroups, the intersection of all definable subgroups of G containing X is a definable subgroup of G. This is the smallest definable subgroup of G containing X and we denote it by $d(X)$ and call it the *definable subgroup of G generated by X*.

We prove the o-minimal version of the Lie-Kolchin-Mal'cev theorem. The proof is a modification of that in [8] for the finite Morley rank case and uses the classification given in [1] of all definable G-modules and this in turn uses Theorem 2.1 and the classification of definable faithful and irreducible G-modules given in [5].

THEOREM 2.4 ([1]). *If U is a definably connected definable solvable group, then $U^{(1)}$ is a $\bigvee$-definable nilpotent normal subgroup and $d(U^{(1)})$ is a definable nilpotent normal subgroup.*

The next two results are about the structure induced by $\mathcal{N}$ on a definable group. They are proved using methods developed in [1].

THEOREM 2.5 ([1]). *Let U be a definable group and let $\{T(x) \colon x \in X\}$ be a definable family of non empty definable subsets of U. Then there is a definable function $t \colon X \to U$ such that for all $x, y \in X$ we have $t(x) \in T(x)$ and if $T(x) = T(y)$ then $t(x) = t(y)$.*

Peterzil and Starchenko proved in [14], using the theory of $\bigvee$-definable groups and assuming that $\mathcal{N}$ has definable Skolem functions, that if $U = (U, \cdot)$ is a definable group which is not abelian-by-finite, then a real closed field is interpretable in U. In [1] we get the result below, again avoiding the theory of $\bigvee$-definable groups and without assuming that $\mathcal{N}$ has definable Skolem functions. (In fact, by Theorem 2.5 all the results from [14] can now be proved without the assumption that $\mathcal{N}$ has definable Skolem functions).

THEOREM 2.6. *Let U be a definable group which is not abelian-by-finite. Then a real closed field is definable in $(N, <, U, \cdot)$.*

§3. **Definable rings.** We start by recalling some facts from [16] and [9] about definable rings. Let U be a definable ring. Then U can be equipped with a unique definable space structure making the ring into a topological ring, and definable homomorphisms between definable rings are topological homomorphisms. In fact, it follows from the results in [11], that if $\mathcal{N}$ is an o-minimal expansion of a real closed field then, U equipped with the above unique definable space structure is a C^p ring for all $p \in \mathbb{N}$ and definable homomorphisms between definable rings are C^p homomorphisms for all $p \in \mathbb{N}$.

The following fact follows from the analogue result for definable groups. Let U be a definable ring. Then U satisfies the descending chain condition (DCC) on definable left (resp., right and bi-) ideals. Let U^0 be the definable connected component of zero in the additive group of U. Then U^0 is the smallest definable ideal of U of finite index. We say that U is definably connected if $U^0 = U$.

In [15] the following result is established. Let U be an infinite definable associative ring without zero divisors. Then U is a division ring and there is a one-dimensional definable subring I of U which is a real closed field such that U is either I, $I(\sqrt{-1})$, or the ring of quaternions over I.

Theorem 2.1, the result from [9] about rings definable in o-minimal expansions of real closed fields and Wedderburn theory are used to prove the following.

THEOREM 3.1 ([1]). *Let U be a definable ring. Then there is a definable left ideal $V = K \oplus \bigoplus_{j=1}^{m} W_j \oplus \bigoplus_{i=1}^{n} V_i' \oplus \bigoplus_{i=1}^{n} V_i$ of U such that K is the definably compact, definably connected definable left ideal of U of maximal dimension and for each $j = 1, \ldots, m$ (resp., $i = 1, \ldots, n$) there is a semi-bounded o-minimal expansion $\mathcal{J}_j$ of a group (resp., an o-minimal expansion $\mathcal{I}_i$ of a real closed field) definable in $\mathcal{N}$ such that there is no definable bijection between a distinct pair among the J_j's and the I_i's, W_j is a direct product of copies of the additive group of $\mathcal{J}_j$ and has zero multiplication, V_i' is a direct product of copies of the linearly bounded one-dimensional torsion-free $\mathcal{I}_i$-definable group and has zero multiplication, each V_i is an $\mathcal{I}_i$-definable ring such that if $\overline{V}_i := V_i / ann_{V_i} V_i$ is non-trivial then $\overline{V}_i$ is a finitely generated I_i-algebra (and therefore I_i-definable)* *and if it is associative then it is $\mathcal{I}_i$-definably isomorphic to a finitely generated I_i-subalgebra of some $M_{n_i}(I_i)$ and has a nilpotent finitely generated ideal Z_i such that $\overline{V}_i/Z_i$ is $\mathcal{I}_i$-definably isomorphic to $\bigoplus_{j=1}^{m_i} M_{k_{i,j}}(D_{i,j})$ where for each $j = 1, \ldots, m_i$, $D_{i,j}$ is either I_i, $I_i(\sqrt{-1})$, or the ring of quaternions over I_i. Moreover, U/V is a definably compact definable ring.*

Similarly to Theorem 2.2 we get the following result, which is also a consequence of Lemma 5.6 of [14] and Theorem 1.1.

THEOREM 3.2. *A definably compact, definably connected definable ring has zero multiplication.*

Acknowledgements. Part of the work presented here is contained in the authors DPhil Thesis which was financially supported by JNICT grant PRAXIS XXI/BD/5915/95. I would like to thank my thesis adviser Professor Alex Wilkie and Kobi Peterzil for their constant support. I would also like to thank the EPSRC for current financial support.

REFERENCES

[1] M. EDMUNDO, *Solvable groups definable in o-minimal structures*, Preprint (submitted), 2000.

[2] ———, *Structure theorems for o-minimal expansions of groups*, **Annals of Pure and Applied Logic**, vol. 102 (2000), pp. 159–181.

[3] S. EILENBERG and S. MACLANE, *Cohomology theory in abstract groups I*, **Annals of Mathematics**, vol. 48 (1947), pp. 51–78.

[4] ———, *Cohomology theory in abstract groups II*, **Annals of Mathematics**, vol. 48 (1947), pp. 326–341.

[5] D. MACPHERSON, A. MOSLEY, and K. TENT, *Permutation groups in o-minimal structures*, Preprint, 1999.

[6] C. MILLER, *A growth dichotomy for o-minimal expansions of ordered fields*, Preprint, 1996.

[7] C. MILLER and S. STARCHENKO, *A growth dichotomy for o-minimal expansions of ordered groups*, **Logic colloquium 93** (W. Hodges et al., editors), Oxford University Press, 1994, Preprint 1995.

[8] A. NESIN, *Solvable groups of finite morley rank*, **Journal of Algebra**, vol. 121 (1989), pp. 26–39.

[9] O. OTERO, Y. PETERZIL, and A. PILLAY, *On groups and rings definable in o-minimal expansions of real closed fields*, **The Bulletin of the London Mathematical Society**, vol. 28 (1996), pp. 7–14.

[10] Y. PETERZIL, A. PILLAY, and S. STARCHENKO, *Linear groups definable in o-minimal structures*, Preprint, 1999.

[11] ———, *Definably simple groups in o-minimal structures*, **Transactions of the American Mathematical Society**, vol. 352 (2000), no. 10, pp. 4397–4419.

[12] ———, *Simple algebraic groups over real closed fields*, **Transactions of the American Mathematical Society**, vol. 352 (2000), no. 10, pp. 4421–4450.

[13] Y. PETERZIL and S. STARCHENKO, *A trichotomy theorem for o-minimal structures*, **Proceedings of the London Mathematical Society**, vol. 77 (1998), pp. 481–523.

[14] ———, *Definable homomorphisms of abelian groups in o-minimal structures*, **Annals of Pure and Applied Logic**, vol. 101 (1999), no. 1, pp. 1–27.

[15] Y. PETERZIL and C. STEINHORN, *Definable compacteness and definable subgroups of o-minimal groups*, **The Journal of the London Mathematical Society**, vol. 59 (1999), no. 2, pp. 769–786.

[16] A. PILLAY, *On groups and fields definable in o-minimal structures*, **Journal of Pure and Applied Algebra**, vol. 53 (1988), pp. 239–255.

[17] V. RAZENJ, *One-dimensional groups over an o-minimal structure*, **Journal of Pure and Applied Algebra**, vol. 53 (1991), pp. 269–277.

[18] A. STRZEBONSKI, *Euler characteristic in semialgebraic and other o-minimal groups*, **Journal of Pure and Applied Algebra**, vol. 96 (1994), pp. 173–201.

[19] L. VAN DEN DRIES, *Tame topology and o-minimal structures*, Cambridge University Press, 1998.

THE MATHEMATICAL INSTITUTE
24-29 ST GILES, OX1 3LB OXFORD
UNITED KINGDOM
E-mail: edmundo@maths.ox.ac.uk

LOGICAL TOPOLOGIES AND SEMANTIC COMPLETENESS

VALENTIN GORANKO

To Johan van Benthem, on the occasion of his 50th birthday, with high respect.

Abstract. We study the generic problem of proving semantic completeness of a logical system with respect to a class of "standard models", provided a weaker completeness result with respect to a larger class of "general models" has been obtained. We propose a natural topological approach to this problem based on the notion of *logical topology* and the related concept of *logical approximation*. After some general results regarding these concepts we discuss them in the framework of first-order logic. The paper ends with an example of a particular application of ideas developed here.

§1. Introduction.

1.1. The relative completeness problem. The present study is motivated by the following, often arising in logical studies, generic problem. Suppose a deductive system L (of any nature) in a certain logical language is intended to axiomatize a class of *standard models* SM, and a completeness theorem has been established with respect to a larger class of *general models* GM, i.e., it has been proved that

$$L \vdash \phi \text{ iff } L \models_{GM} \phi.$$

The goal is **to prove completeness of L with respect to the standard models**, i.e.,

$$L \vdash \phi \text{ iff } L \models_{SM} \phi.$$

Here are three illustrative cases:

1. *Finite model property* in classical, modal, etc. logics. The "general models" are all models for the logic L, and the "standard models" are the *finite* models. While completeness with respect to general models

Key words and phrases. Logical topologies, Stone topology, logical approximation, semantic completeness.

This work was supported by a research grant GUN 2034353 of the National Research Foundation of South Africa and by the SASOL research fund of the Faculty of Natural Sciences at Rand Afrikaans University.

Logic Colloquium '99
Edited by J. van Eijck, V. van Oostrom, and A. Visser
Lecture Notes in Logic, 17
© 2004, ASSOCIATION FOR SYMBOLIC LOGIC

is a uniform result in classical logic, due to Gödel's completeness theorem, completeness with respect to "standard" (i.e., finite) models is an essentially nontrivial property, as Trakhtenbrot's theorem testifies.

2. *Kripke-frame completeness* in modal logic. The "general models" are all Kripke models for the logic L, and the "standard models" are those Kripke models based on frames for L. Again, the completeness with respect to the class of general models is a general result in modal logic (based on the standard canonical construction) but the completeness with respect to the standard models, i.e., Kripke completeness, is the non-trivial and important one. For more details and sample results, see [Benthem, 1984].

3. *First-order approximation of Π_1^1-theories.* Often a structure, or a class of structures, can be characterized by means of a Π_1^1-sentence, but not in first-order logic. Typical examples are: the ordering of natural numbers (with the induction axiom), the ordering of the reals (with the continuity axiom), the class of all well-orderings etc. On the other hand, every Π_1^1-sentence Φ can be 'approximated' by the first-order schema Φ_1 obtained from Φ by restricting the universal second-order quantification to all instances of parametrically first-order definable relations and functions. Now, the models of Φ are the "standard models" and the models of Φ_1 are the "general models". The question '*does the scheme Φ_1 axiomatize the first-order fragment of Φ?*' is essentially the question of relative completeness we discuss here. A number of elaborated positive completeness results of this type have been obtained in [Doets, 1987] and, using his techniques, in [Backofen, Rogers, and Vijay-Shanker, 1995] and [Venema, 1993].

There is no general method for solving the problem described above, but usually some specific model-theoretic constructions are applied which transform general into standard models while preserving satisfiability.

Here we do not offer a general solution to that problem either, but rather a *methodology* based on a natural topological approach which can be applied in various particular cases. Because of space limitations we only outline one example of a non-trivial application at the end of the paper.

1.2. A topological approach. The idea in a nutshell is to **find an appropriate topology** $\mathcal{T}$ **on the class[1] of general models** GM **such that:** (i) **The class of standard models** SM **is dense in** GM **with respect to** $\mathcal{T}$, i.e., **the closure in** $\mathcal{T}$ **of** SM **is** GM. (ii) **Validity is a closed property with respect to** $\mathcal{T}$.

"Closedness" here is used in the standard topological sense: a property is closed if the set (class) of points which satisfy it is closed. Since every

---

[1] Here we talk about topologies on (proper) classes, rather than on sets. This foundational issue will not affect what follows, as we will see later that every such topology is essentially equivalent to a "small", i.e., set-based, one.

closed and dense set in a topological space coincides with the whole space, the following observation is immediate.

> If the conditions above hold for some topology $\mathcal{T}$ on **GM** then completeness with respect to **GM** implies completeness with respect to **SM**.

Alternatively, one can associate with every general model its *theory*, i.e., the set of its valid formulae in the language under consideration, and look for an appropriate topology *on the set of all theories of general models*, for which an analogous result can be stated. This approach has some technical advantages, but the two approaches are essentially equivalent, as it will be shown further.

Although some intimate connections between logic and topology have been established and studied (see e.g., [Barwise and Feferman, 1985]), it seems that, except in abstract model theory, topological methods and results have so far been under-utilized for solving purely logical problems, and there are few publications which more explicitly pursue that direction.

In this paper we suggest a more systematic exploration of the idea of using basic topological techniques and results to obtain relative completeness results in logic.

The preliminary section 1 contains some background from logic and topology. In section 2 we introduce the notion of *logical topology* and the related concept of *logical approximation*, and study their basic properties. In particular, as a direct consequence of the Baire's category theorem, we obtain a general relative completeness result (theorem 3.8, and theorem 4.9 as a particular case in first-order logic) which seemingly has so far not been explicitly noted, despite the well-known relationship of the Baire's theorem to logic (see [Rasiowa and Sikorski, 1963] and [Goldblatt, 1985]). In section 3 we discuss logical topologies and logical approximation in classical logic. We show that, not surprisingly, a topology on the set of all complete theories in a first-order language is logical iff it contains the Stone topology (proposition 4.1) and briefly study a simple and natural extension of the Stone topology in languages with infinite signature. In section 4 we mention a specific application to the first-order theory of trees, and outline a proof of completeness based on ideas and results from the paper. The last section 5 discusses a research agenda arising from this study.

§2. **Preliminaries.** Here we summarize some basic topological facts that will be used further. For details on definitions and related results, [Ebbinghaus, Flum, and Thomas, 1994] and [Hodges, 1993] are general references on the necessary logical background, and e.g., [Engelking, 1985] — on topology.

Let L be a first-order language, $\mathrm{SEN}(L)$ be the set of sentences of L, and $\mathcal{C}(L)$ be the set of complete theories in L. The **Stone topology** $\mathcal{S}(L)$ is defined on the set $\mathcal{C}(L)$ by a base of clopen sets $\{[\phi] \mid \phi \in \mathrm{SEN}(L)\}$ where

$[\phi] = \{T \in \mathcal{C}(L) \mid \phi \in T\}$. It is easy to see that the closed sets in $\mathcal{C}(L)$ are precisely the sets $\{T \in \mathcal{C}(L) \mid \Gamma \subseteq T\}$ where Γ is a closed theory in L and that $\mathcal{S}(L)$ is a totally disconnected compact Hausdorff space.

The topology $\mathcal{S}(L)$ determines a topology $\mathcal{S}_{\mathrm{STR}}(L)$ on the class of all L-structures $\mathrm{STR}(L)$, called by Tarski the **elementary topology**, where the closed subclasses are precisely the first-order axiomatizable classes. This topology is pseudo-metrizable when the language is countable and every first-order axiomatizable class, considered as a subspace of $\mathcal{S}_{\mathrm{STR}}(L)$ then becomes a complete pseudo-metric space.

Given a topology $\mathcal{T}$ and a set A in $\mathcal{T}$, $\mathrm{Cl}_{\mathcal{T}}(A)$ is the closure of A in $\mathcal{T}$.

DEFINITION 2.1. A subset A of a topological space $\mathcal{T}$ on a set X is **dense** if $\mathrm{Cl}_{\mathcal{T}}(A) = X$. A subset A of a set B in a topological space $\mathcal{T}$ over a set X is **dense in** B if $\mathrm{Cl}_{\mathcal{T}|B}(A) = B$, where $\mathcal{T}|B$ is the topology on B induced by $\mathcal{T}$.

DEFINITION 2.2. A topological space $\mathcal{T}$ has the **Baire's property** if every countable intersection of dense open sets in $\mathcal{T}$ is dense in $\mathcal{T}$.

Two well known versions of the **Baire category theorem** state that *every complete pseudo-metric space, as well as every compact Hausdorff space has the Baire's property.*

A topology is first-countable if every point has a countable base of open neighbourhoods. It is easy to see that the Stone topology is first countable iff the language is at most countable. In first-countable topologies closed sets can be characterized in terms of closure under limits of convergent sequences, while in general, they are characterized in terms of convergent nets or clustering filters.

§3. **Logical topologies on theories and structures.** We fix an arbitrary logical language L with specified semantics, i.e., a class of L-structures and a relation $\models$ of validity of L-formulae in L-structures.

Let $\mathcal{T}$ be a topology on the class of L-structures.

DEFINITION 3.1. The topology $\mathcal{T}$ is **logical on a class of L-structures M** if validity is a closed property with respect to the topology on M induced by $\mathcal{T}$. $\mathcal{T}$ is **logical** if it is logical on the class of all L-structures.

For every L-structure A, we denote by $\mathrm{TH}(A)$ the *theory of A*, i.e., the set of L-formulae valid in A.

Now, let $\mathcal{T}$ be a topology on a set $\boldsymbol{TH}$ of theories of L-structures and for every subset $S \subseteq \boldsymbol{TH}$, $\mathrm{Cl}_{\mathcal{T}}(S)$ be the closure of S with respect to $\mathcal{T}$.

The following definition, though it may look somewhat unintuitive, will turn out to match the one of a logical topology on a class of structures given above.

DEFINITION 3.2. The topology $\mathcal{T}$ is **logical on** TH if for every subset $S \subseteq TH$, $\bigcap S = \bigcap \mathrm{Cl}_{\mathcal{T}}(S)$; $\mathcal{T}$ is **logical** if it is logical on the set of all theories of L-structures.

There is a natural duality between the two notions of logical topologies. For every topology $\mathcal{T}$ on a class of structures M we can associate a topology $\mathcal{T}_{\mathrm{TH}}$ on the set of their theories, where the closed sets are of the type $\{\mathrm{TH}(A) \mid A \in C\}$ for each closed set C in $\mathcal{T}$. Conversely, for every topology $\mathcal{T}$ on a set of theories T we can associate a topology $\mathcal{T}_{\mathrm{STR}}$ on the class of all models of theories from T, with closed sets of the type $\{A \mid \mathrm{TH}(A) \in C\}$ for each closed set C in $\mathcal{T}$.

PROPOSITION 3.3.

1. *If $\mathcal{T}$ is a logical topology on a class M of L-structures, then $\mathcal{T}_{\mathrm{TH}}$ is a logical topology on the set T of their theories.*
2. *If $\mathcal{T}$ is a logical topology on a set T of theories of L-structures, then $\mathcal{T}_{\mathrm{STR}}$ is a logical topology on the class M of their models.*

PROOF. 1. It is sufficient to note that for every $S \subseteq T_{\mathrm{TH}}$, the closure of S in $\mathcal{T}_{\mathrm{TH}}$ consists of all theories of structures which are in the closure of $\{A \mid \mathrm{TH}(A) \in S\}$ in $\mathcal{T}$.

2. Likewise.   ⊣

Thus, both notions are essentially equivalent. While most of the ideas and concepts discussed here look more natural when formulated in terms of structures, it is technically more convenient and elegant to state and prove many of the results in terms of theories, so we shall use interchangeably the two frameworks.

PROPOSITION 3.4.   *If $\mathcal{T}$, $\mathcal{R}$ are topologies on a set of theories TH, $\mathcal{T} \subseteq \mathcal{R}$, and $\mathcal{T}$ is logical, then $\mathcal{R}$ is logical, too.*

PROOF. $\mathcal{T} \subseteq \mathcal{R}$ implies $\mathrm{Cl}_R(S) \subseteq \mathrm{Cl}_T(S)$, so $\bigcap S \subseteq \bigcap \mathrm{Cl}_{\mathcal{T}}(S) \subseteq \bigcap \mathrm{Cl}_{\mathcal{R}}(S)$ for every $S \subseteq TH$.   ⊣

DEFINITION 3.5. Let $\mathcal{T}$ be a logical topology on the class of L-structures. A structure A is **logically approximated (with respect to** $\mathcal{T}$**)** in a class of structures M if A belongs to the closure of M (with respect to $\mathcal{T}$). A class of structures K is **logically approximated (with respect to** $\mathcal{T}$**)** by M if every structure from K is logically approximated in M. The closure $\mathrm{Cl}_{\mathcal{T}}(K)$ of K, i.e., the class of all structures logically approximated in the class K, will be called the **logical closure of** K (with respect to $\mathcal{T}$).

Note that if $M \subseteq K$ then K is logically approximated by M with respect to a topology $\mathcal{T}$ iff M is dense in K with respect to $\mathcal{T}$. Thus, the following statement formalizes the idea outlined in the introduction and provides formal grounds for applications of our approach to solving the relative completeness problem.

THEOREM 3.6. *Let $\mathcal{L}$ be a deductive system in the language L, complete for a class of structures K, $\mathcal{T}$ be logical on K, and M be a subclass of K which approximates logically K with respect to $\mathcal{T}$. Then $\mathcal{L}$ is complete for M.*

A direct application of the Baire category theorem yields:

LEMMA 3.7. *Let K be a class of L-structures, $\mathcal{T}$ be a logical topology on K with the Baire's property, and $\{M_k\}_{k \in N}$ be a family of open subclasses of K such that K is logically approximated by each M_k. Then K is logically approximated by $M = \bigcap_{k \in N} M_k$.*

The following theorem is a combination of the previous two statements.

THEOREM 3.8. *Let $\mathcal{L}$ be a deductive system in L, complete with respect to a class of L-structures K, $\mathcal{T}$ be a logical topology on K with the Baire's property and $\{M_k\}_{k \in N}$ be a family of open and dense subclasses of K. Then L is complete with respect to $M = \bigcap_{k \in N} M_k$.*

§4. Logical topologies in first-order logic and elementary approximations of structures.

We now fix an arbitrary first-order language L. With no risk of confusion we shall denote both the Stone topology on $\mathcal{C}(L)$ and the elementary topology $\mathcal{S}_{\mathrm{STR}}(L)$ by $\mathcal{S}$, and the closure operator in both topologies by $\mathrm{Cl}_{\mathcal{S}}$.

Note that for every $S \subseteq \mathcal{C}(L)$, $\mathrm{Cl}_{\mathcal{S}}(S) = \{T \in \mathcal{C}(L) \mid \bigcap S \subseteq T\}$. On the other hand, for every class of L-structures K, $\mathrm{Cl}_{\mathcal{S}}(K)$ is the **elementary closure of K**, i.e., the smallest elementary class which contains K. Thus, $\mathrm{Cl}_{\mathcal{S}}(K)$ is the class $\mathrm{MOD}(\mathrm{TH}(K))$, of all models of the first-order theory of K. Therefore, a theory T is complete for a class K iff $\mathrm{TH}(K) = T$ i.e., K is dense in $\mathrm{MOD}(T)$.

PROPOSITION 4.1. *A topology $\mathcal{T}$ on $\mathcal{C}(L)$ is logical iff it contains the Stone topology.*

PROOF. First, suppose $\mathcal{T}$ is logical and let $S \subseteq \mathcal{C}(L)$ be closed in $\mathcal{S}(L)$, i.e., $S = \{T \in \mathcal{C}(L) \mid \bigcap S \subseteq T\}$. Then $\bigcap S \subseteq \bigcap \mathrm{Cl}_{\mathcal{T}}(S)$, so $\mathrm{Cl}_{\mathcal{T}}(S) \subseteq \mathcal{F}$, i.e., $\mathrm{Cl}_{\mathcal{T}}(S) = S$. For the converse, by proposition 3.4, it suffices to show that the Stone topology is logical. Indeed, for any $S \subseteq \mathcal{C}(L)$, if $T \in \mathrm{Cl}_{\mathcal{S}}(S)$ then $\bigcap S \subseteq T$, hence $\bigcap \mathcal{F} \subseteq \bigcap \mathrm{Cl}_{\mathcal{S}}(S)$. ⊣

Thus, the Stone topology is the weakest logical topology on the class of all L-structures, but there can be even weaker logical topologies suitable on some subclasses.

Sometimes it may easier to deal with logical topologies stronger than the Stone topology. A natural example of such a topology in first-order logic can be introduced by using an appropriate metric (which need not be inducing the Stone topology) on $\mathcal{C}(L)$.

The notion of *quantifier rank of a formula* is introduced as usual in languages with relational signatures, and appropriately modified for languages including constant and functional symbols, as in [Ebbinghaus, Flum, and Thomas, 1994].

Let $\mathrm{SEN}^{(n)}(L)$ be the set of L-sentences of (modified) rank $\leq n$ and for every $\Gamma \subseteq \mathrm{SEN}$, $\Gamma^{(n)} = \Gamma \cap \mathrm{SEN}^{(n)}(L)$.

First, we define *distance* in $\mathcal{C}(L)$ as follows:

$$d(T_1, T_2) = \begin{cases} cl0 & \text{if } T_1 = T_2, \\ \frac{1}{n+1} & \text{if } n \text{ is the least integer such that } T_1^{(n)} \neq T_2^{(n)}. \end{cases}$$

PROPOSITION 4.2.

1. $\langle \mathcal{C}(L), d \rangle$ *is a bounded and complete metric space.*
2. *The topology* $\mathcal{C}_d(L)$ *on* $\mathcal{C}(L)$ *induced by* d *is logical.*

PROOF. 1. To see that d is a metric it is sufficient to note that $d(T_1, T_3) \leq \max(d(T_1, T_2), d(T_2, T_3))$ for any $T_1, T_2, T_3 \in \mathcal{T}(L)$. Boundedness is obvious. For completeness,[2] let $T_1, T_2, \ldots, T_n, \ldots$ be a Cauchy sequence in $\mathcal{T}(L)$. Then, for each $n \in N$ there is $N \in N$ such that for all $p, q > N$, $T_p^{(n)} = T_q^{(n)}$. Let us denote the latter by Γ_n. Thus we obtain a chain of theories $\Gamma_0 \subseteq \Gamma_1 \subseteq \ldots$. Let $\Gamma = \bigcup_{n=0}^{\infty} \Gamma_n$. Γ is a complete theory. Indeed, Γ is closed. For, let $\Gamma \models \phi$ where $\phi \in \mathrm{SEN}^{(m)}(L)$ for some m. Then $\Gamma_k \models \phi$ for some k. Hence $\phi \in \Gamma_{\max(k,m)}$, so $\phi \in \Gamma$. Furthermore, for every $\phi \in \mathrm{SEN}^{(m)}(L)$ for some m, either ϕ or $\neg\phi$ is in Γ_k for every $k \geq m$. Finally, it is clear that $\lim_{n \to \infty} T_n = \Gamma$.

2. We shall prove that $\mathcal{C}_d(L)$ contains the Stone topology. Let S be a closed set in $\mathcal{S}(L)$. Then $S = \{ T \in \mathcal{C}(L) \mid \bigcap S \subseteq T \}$. Since every metric space is first-countable, it is sufficient to show that the limit T in $\mathcal{C}_d(L)$ of any sequence $T_0, T_1, \ldots$ from S is in S. Indeed, $\bigcap S \subseteq T$ since every sentence from $\bigcap S$ with a rank n will belong to all complete theories in the open $\frac{1}{n+1}$-neighbourhood of each theory from S. Thus, S is closed in $\mathcal{C}(L)$. $\dashv$

PROPOSITION 4.3. *For every first-order language L the following are equivalent*:

1. *The language L has finitely many non-logical symbols.*
2. *The topology $\mathcal{C}_d(L)$ coincides with the Stone topology.*
3. *The space $\mathcal{C}_d(L)$ is compact.*
4. *$\mathcal{C}_d(L)$ is totally bounded.*
5. *For every $n \in N$, the set $\{ T^{(n)} \mid T \in \mathcal{C}(L) \}$ is finite.*
6. *For every $n \in N$ there is no infinite independent subset of $\mathrm{SEN}^{(n)}(L)$.*

PROOF. $(1) \Rightarrow (2)$: Let $S \subseteq \mathcal{C}(L)$ be closed in $\mathcal{C}_d(L)$ and $\Gamma = \bigcap S$. We shall prove that $S = \{ T \in \mathcal{C}(L) \mid \Gamma \subseteq T \}$. We only need to show that $T \in S$ whenever $\Gamma \subseteq T$. Indeed, for every $n \in N$, $T^{(n)}$ is finite, so it is included in some $T_n \in S$, otherwise $\neg \bigwedge T^{(n)} \in \Gamma$, so T would be inconsistent. Thus, T is the limit in $\mathcal{C}_d(L)$ of a sequence $T_1, T_2, \ldots$ in S, hence $T \in S$.

---

[2]A stronger result has been proved in [Cifuentes, Sette, and Mundici, 1996]: for any first-order language L, the elementary topology $\mathcal{S}_{\mathrm{STR}}(L)$ is Cauchy complete, i.e., every Cauchy net converges.

(2)⇒(3): Follows from compactness of the Stone topology.

(3)⇔(4): Every complete metric space is compact iff it is totally bounded.

(4)⇔(5): Since the complete theories in every $\frac{1}{n+1}$-neighbourhood of $\mathcal{T}(L)$ share the same $\text{SEN}^{(n)}(L)$-fragment, $\mathcal{T}(L)$ is covered by finitely many open balls of radius $\frac{1}{n+1}$ iff there are finitely many $\text{SEN}^{(n)}(L)$-fragments of theories from $\mathcal{T}(L)$.

(5)⇒(6): Suppose Γ is an infinite independent subset of $\text{SEN}^{(n)}(L)$ for some natural n. For each $\delta \in \Gamma$ we consider the consistent theory $\Gamma_\delta = \Gamma - \{\delta\} \cup \{\neg\delta\}$. All these theories have different $\text{SEN}^{(n)}(L)$-fragments.

(6)⇒(1): If the language has infinitely many non-logical symbols, then there are infinitely many atomic formulae of (at most) one variable and rank not greater than 1, no two of which share non-logical symbols, hence there is an infinite independent set of sentences in $\text{SEN}^{(2)}(L)$.   ⊣

Thus, we see that, $\mathcal{C}_d(L)$ is simpler and easier to deal with in case of infinite languages, where it is stronger than the Stone topology, and especially in uncountable languages where the latter is not first-countable.

Logical approximation with respect to $\mathcal{S}(L)$ will be called *elementary approximation*, and the approximation with respect to $\mathcal{C}_d(L)$ — *strong elementary approximation*. Note that a structure A is *strongly elementarily approximated in a class K* iff for every $n \in N$ there is $A_n \in K$ such that $A \equiv_n A_n$. The class of all structures which are strongly elementarily approximated in K will be called the *strong elementary closure* of K. Thus, every structure, strongly elementarily approximated in a class K, is elementarily approximated in K, but the converse need not hold in a language with an infinite signature. Respectively, every elementary closure is a strong elementary closure, but not conversely, and the elementary closure of any class K contains its strong elementary closure.

Elementary approximation and closure are already well-understood from various classical model-theoretics results, and we shall only mention just two characterizations of elementary approximation. The first one, in $\mathcal{S}(L)$ is essentially equivalent to the compactness theorem (see [Hodges, 1993]): *a theory $T \in \mathcal{C}(L)$ is elementarily approximated by a set $S \subseteq \mathcal{C}(L)$ iff every finite subset of T is included in some complete theory from S.* The second one, in $\mathcal{S}_{\text{STR}}(L)$, is a well-known preservation result: *a structure A is elementarily approximated by a class of structures K iff A elementarily equivalent to an ultraproduct of structures from K.*

Here are two easy characterizations of strong elementary approximations.

DEFINITION 4.4. A net of L-structures $\langle A_i \rangle_{i \in D}$, where D is a directed indexing family, is **strongly convergent** if it is convergent in $\mathcal{C}_d(L)$.

THEOREM 4.5. *A class K of L-structures is a strong elementary closure iff it is closed under elementary equivalence and ultraproducts of strongly convergent nets.*

PROOF. If K satisfies the closure conditions, then every structure strongly elementarily approximated in K belongs to K since the ultraproduct of $\langle A_i \rangle_{i \in D}$ over any free ultrafilter on D is elementarily equivalent to the limit of that net. Conversely, every strong elementary closure is closed under elementary equivalence and therefore, under ultraproducts of strongly convergent nets.

$\dashv$

In the case of a countable language, the result above can be stated in terms of converging sequences, rather than nets.

A simple game-theoretic characterization of strong elementary approximations exists, too.

DEFINITION 4.6. **Ehrenfeucht game with choice of a companion:** Given a structure A, and a class of structures K, the game goes between two players as follows. In his first move Player I selects a natural number n. Then Player II selects a structure B from K. Then the game continues as the usual Ehrenfeucht game for A and B and ends after n more moves. The winning conditions are the same as for the usual Ehrenfeucht games.

PROPOSITION 4.7. *A structure A is strongly elementarily approximated in a class K iff Player II as a winning strategy for every game with choice of a companion.*

Finally, we state a useful result on relative completeness, which follows from theorem 3.8. Recall that a first-order theory T is complete with respect to a class K iff K is dense in $\mathrm{MOD}(T)$, and that every elementary class is itself a compact and Hausdorff space with the induced elementary topology.

DEFINITION 4.8. A class of first-order structures $\mathcal{M}$ is *co-elementary* (in a class of structures $\mathcal{K}$) if its complement (in $\mathcal{K}$) is elementary (in $\mathcal{K}$).

THEOREM 4.9. *Let $\{M_k\}_{k \in N}$ be a family of classes of L-structures, each of them co-elementary in an elementary class $\mathcal{K}$, and let $T = \mathrm{TH}(\mathcal{K})$ be complete with respect to each M_k. Then T is complete with respect to $M = \cap_{k \in N} M_k$.*

§5. An application: a relative completeness result of the first-order theory of coloured ω-trees. In this section we outline a sample completeness result obtained using ideas and results presented here. We give this result just as an illustration, rather than for its own sake, as the idea of the proof can be used to establish a more general fact. For a detailed proof and related results see [Goranko, 1999].

First, we need some definitions. By a *tree* we mean any (strictly) partially ordered set with a least element (root), in which every element has a linearly ordered set of predecessors. The elements of a tree are called *nodes*. A *path* in a tree is any maximal linearly ordered subset. A tree in which every path has the order type of ω will be called an ω-*tree*. The set of predecessors of a node a will be called the *stem of* a. A *sibling* of a node a in a tree T is any node in T with the same stem as a. The *level* k of a tree consists of the nodes which have k-element stems. (Thus, the 0-level consists of the root of the tree). The *finite levels* in a tree are all k-levels for $k \in \omega$. A tree is *finitely branching* (*on a level* k) if every node (on a level k) has finitely many siblings.

Moreover, we can consider trees enriched with finitely many additional unary predicates which will be called *colours*, and the resulting structures *coloured trees*.

THEOREM 5.1. *The first-order theory* CT_ω *of all* (*coloured*) ω-*trees is complete with respect to the class of finitely branching* (*coloured*) ω-*trees.*

PROOF. (Sketch:)
Let $\mathcal{M}_\omega$ be the class of all models of CT_ω and $\mathcal{M}^f$ consist of all trees from $\mathcal{M}_\omega$ which are finitely branching on all finite levels and in which every satisfiable formula of one variable is satisfiable on a finite level.

The proof consists of two major steps. The first step is to prove that CT_ω is complete with respect to $\mathcal{M}^f$. For this we show that $\mathcal{M}^f$ is dense in $\mathcal{M}_\omega$ with respect to the elementary topology. Indeed, $\mathcal{M}^f$ can be represented as an intersection of the family of classes $\{\mathcal{M}_k\}_{k \in N}$, where $\mathcal{M}_k$ consists of the models M of CT_ω finitely branching at the first k levels and satisfying on finite levels the first k formulae of some fixed enumeration of the formulae satisfiable in M. Note that each $\mathcal{M}_k$ is co-elementary in $\mathcal{M}_\omega$. Furthermore, it can be proved, using Ehrenfeucht's theorem, that each $\mathcal{M}_k$ is dense in $\mathcal{M}_\omega$ because every tree from $\mathcal{M}_\omega$ is n-equivalent to a tree from $\mathcal{M}_k$. Now the claim follows by theorem 4.9.

The second step then is to prove completeness of CT_ω with respect to the class of finitely branching ω-trees. For that it suffices to show that every tree T from $\mathcal{M}^f$ is elementarily equivalent to a finitely branching ω-tree. We shall use the omitting types theorem. Consider the 1-type

$$\tau(x) = \{\neg l_k(x) \mid k \in N\}$$

where $l_k(x)$ says that x is on a level k. Note that τ is not principal in T since any generator of that type would be satisfiable in T by a node which belongs to some finite level. Hence, τ is omitted in a countable model T' such that $T' \equiv T$, and hence $T' \models CT_\omega$. Then T' is an ω-tree. Furthermore, T' is finitely branching at every level because it satisfies the same formulae $\chi_{k,m}$ saying that every node on level k has no more that m siblings, as T does. Thus, T' is a finitely branching ω-tree, whence the completeness. ⊣

§6. **Concluding remarks.** In this paper we have only outlined some basic ideas of using topological methods to prove relative completeness and have discussed some rather immediate results regarding logical topologies. This approach can be further developed both from logical and topological perspectives.

From the topological perspective, there is much more to be done, as there is a number of non-trivial topological results which can be usefully reformulated in logical terms and applied for solving relative completeness (and other) problems. For instance, it is known (see [Fraïssé, 1967]) that $S_{\text{STR}}(L)$ are *uniform spaces*, which brings additional useful properties, little explored and used in logic so far.

A major logical perspective is to study logical topologies in second-order, infinitary, modal, etc. logics and to apply them to non-trivial completeness problems in these logics. Some of the results in first order logic easily generalize to a wide variety of other logical languages and systems. For instance, an analogue of the Stone topology can be introduced in every logical language with a disjunction, over the class of theories which are consistent and *prime* in sense that $\alpha \vee \beta \in T$ iff $\alpha \in T$ or $\beta \in T$. Then it is easy to check that $\text{Cl}(S) = \{T \in TH \mid \bigcap S \subseteq T\}$ defines a topological closure, and the logical topologies in that language are precisely the extensions of the resulting topology. Still, one can search for other useful constructions of topologies, logical *on a class of structures*.

One of the problems mentioned in the introduction can be re-phrased more generally as *elementary approximation of second-order properties*: *given a second-order theory T, and a first-order fragment T_1 of T, is the class of models of T_1 elementarily approximated in the class of models of T?* In other words: *is T_1 complete for the class of models of T?* Equivalently: *is T_1 the full first-order fragment of T?* It seems natural to explore this problem using logical topologies.

Finally, there is a number of basic model-theoretic constructions used in modal logic to transform Kripke models into 'standard' ones for the logic under consideration, such as *filtration* and *bisimulation*, (introduced in modal logic by van Benthem (see [Benthem, 1983]) under the name of *zig-zag relation*). We hope that these constructions can be linked with the topological framework discussed here and thus the toolkit for proving completeness in modal logic can be strengthened and expanded.

REFERENCES

[1995] R. BACKOFEN, J. ROGERS, and K. VIJAY-SHANKER, *A first-order axiomatization of the theory of finite trees*, **Journal of Logic, Language and Information**, vol. 4 (1995), pp. 5–39.

[1985] J. Barwise and S. Feferman (editors), **Model-theoretic logics,**, Springer-Verlag, New York, 1985.

[1996] J. CIFUENTES, A. SETTE, and D. MUNDICI, *Cauchy completeness in elementary logic*, **The Journal of Symbolic Logic**, vol. 61 (1996), no. 4, pp. 1153–1157.

[1987] H. C. DOETS, *Completeness and definability: Applications of the Ehrenfeucht game in intensional and second-order logic*, **Ph.D. thesis**, Department of Mathematics and Computer

Science, University of Amsterdam, 1987.

[1994] H.-D. EBBINGHAUS, J. FLUM, and W. THOMAS, *Mathematical logic*, 2nd ed., Springer-Verlag, Berlin, 1994.

[1985] R. ENGELKING, *General topology*, 2nd ed., PWN, Warsaw, 1985.

[1967] R. FRAÏSSÉ, *Course de logique mathématique*, Gauthier-Villars, Paris, 1967.

[1985] R. GOLDBLATT, *On the role of the Baire category theorem and dependent choice in the foundations of logic*, The Journal of Symbolic Logic, vol. 50 (1985), pp. 412–422.

[1999] V. GORANKO, *Trees and finite branching*, **Proceedings of the second panhellenic logic symposium** (P. Kolaitis and G. Koletsos, editors), 1999, pp. 107–111.

[1993] W. HODGES, *Model theory*, Cambridge University Press, Cambridge, 1993.

[1963] H. RASIOWA and R. SIKORSKI, *The mathematics of metamathematics*, PWN, Warsaw, 1963.

[1983] J. VAN BENTHEM, *Modal logic and classical logic*, Bibliopolis, Naples, 1983.

[1984] ——— , *Correspondence theory*, **Handbook of philosophical logic** (D. M. Gabbay and F. Guenthner, editors), vol. 2, Reidel, Dordrecht, 1984, pp. 167–247.

[1993] Y. VENEMA, *Completeness via completeness: Since and Until*, **Diamonds and defaults** (M. de Rijke, editor), Kluwer Academic Publishers, Dordrecht, 1993, pp. 279–286.

DEPARTMENT OF MATHEMATICS
RAND AFRIKAANS UNIVERSITY
PO BOX 524, AUCKLAND PARK 2006, JOHANNESBURG
SOUTH AFRICA
E-mail: vfg@na.rau.ac.za

VALUED FIELDS AND ELIMINATION OF IMAGINARIES

DEIRDRE HASKELL

When model theory is applied to study a particular mathematical structure, care must be taken in the choice of language so that the definable sets are both tractable and mathematically interesting. This is seen in the classical result of Tarski-Seidenberg, that in the theory of real closed fields in the language of rings with an additional relation symbol for the ordering, the theory has elimination of quantifiers and the definable sets are the semialgebraic sets. Mathematically, in order to fully understand a structure, it is natural to study its quotients by appropriate equivalence relations. Unfortunately, even if we restrict to definable equivalence relations, the study of the quotient structure is in general beyond the realm of model theory, as it is not definable in the original language. We thus try to characterize for what theories the quotient structures are definable in a given language, or to choose an appropriate language so that they become definable.

The equivalence classes of a $\emptyset$-definable equivalence relation are called imaginaries. Individually, any imaginary is definable with parameters, but the collection of all the equivalence classes for a given relation is not. If there were a formula which, for each equivalence class, chose a unique representative of the class, then this formula would define a copy of the quotient structure. More generally, such a formula does not have to choose a representative; it will suffice to have some unique tuple which identifies each equivalence class. This leads to the following definition.

DEFINITION 1. A complete theory T in a language $\mathcal{L}$ has *elimination of imaginaries* if, for every model $\mathcal{M}$ of T and for any formula $\varphi(\bar{x}, \bar{y})$ defining an equivalence relation on M^n for some n, there is a formula $\theta(\bar{x}, \bar{z})$ such that for every $\bar{a} \in M^n$ there is a unique $\bar{b} \in M^m$ such that $\forall \bar{x}(\varphi(\bar{x}, \bar{a}) \leftrightarrow \theta(\bar{x}, \bar{b}))$.

If T eliminates imaginaries then any automorphism σ of $\mathcal{M}$ which fixes the tuple $\bar{b}$ will fix setwise the equivalence class of φ characterized by $\bar{b}$. Vice versa, this property can be used as an equivalent definition (provided the language has at least two constant symbols). We can also change the focus of the definition to all definable sets. For an automorphism will fix a set D, defined by the formula $\varphi(\bar{x}, \bar{d})$, if and only if it fixes the equivalence

Logic Colloquium '99
Edited by J. van Eijck, V. van Oostrom, and A. Visser
Lecture Notes in Logic, 17

class containing $\bar{d}$ of the $\emptyset$-definable equivalence relation $E(\bar{y}, \bar{y}')$ defined by $\forall \bar{x}(\varphi(\bar{x}, \bar{y}) \leftrightarrow \varphi(\bar{x}, \bar{y}'))$. We thus have the following equivalent definition.

DEFINITION 2. A complete theory T in a language $\mathcal{L}$ has *elimination of imaginaries* if for every model $\mathcal{M}$ of T and for every definable set $D \subset M^n$ for any n, there is a finite tuple $\bar{c} \in M^m$ (called a *code* for D) such that, for all automorphisms σ of $\mathcal{M}$, σ fixes $\bar{c}$ pointwise if and only if σ fixes D setwise.

Shelah [9] introduced the imaginary elements of a structure $\mathcal{M}$ in the construction of the associated structure $\mathcal{M}^{\mathrm{eq}}$, which contains all of its imaginaries. Given a $\emptyset$-definable equivalence relation E on M^n, the language $\mathcal{L}^{\mathrm{eq}}$ is formed by adding a sort for the set of equivalence classes of E, and a function from M^n to the sort. If A is an equivalence class, and $\bar{a} \in A$ we write $\ulcorner A \urcorner$ for the image of $\bar{a}$ under this function. The theory T^{eq} is formed by adding axioms which say that the elements of each new sort are the equivalence classes of the appropriate equivalence relation, and that the function takes any tuple in M^n to the correct equivalence class. (For a more careful exposition of the definition of $\mathcal{M}^{\mathrm{eq}}$ and elimination of imaginaries, see the books by Buechler [1] or Hodges [3].) With this terminology, we say that an imaginary i in $\mathcal{M}^{\mathrm{eq}}$ is *coded in* $\mathcal{M}$ if there is a tuple $\bar{c} \in \mathrm{dcl}_{\mathcal{M}}(i)$ with $i \in \mathrm{dcl}_{\mathrm{eq}}(\bar{c})$. If finite sets can be coded in the structure, this can be weakened to requiring the code $\bar{c} \in \mathrm{acl}_{\mathcal{M}}(i)$. Notice that this condition that the finite sets be coded is nontrivial (in fact, elimination of imaginaries can founder on precisely this point); the tuple (a_1, a_2) is not in general a code for the set $\{a_1, a_2\}$. If a_1, a_2 are elements of a field, the set is coded rather by the tuple of coefficients of the polynomial $(X - a_1)(X - a_2)$.

In the above definitions, if the language is the full sorted language $\mathcal{L}^{\mathrm{eq}}$ then the models are the sorted structures $\mathcal{M}^{\mathrm{eq}}$. Since, in the way described above, every definable set D is an equivalence class of a $\emptyset$-definable relation, there is an element $\ulcorner D \urcorner$ in $\mathcal{M}^{\mathrm{eq}}$ (unique up to interdefinability) which codes D. So $\mathcal{M}^{\mathrm{eq}}$ automatically eliminates imaginaries (once it is shown that the process of going to $\mathcal{M}^{\mathrm{eq}}$ does not need to be iterated). For the purposes of pure model theory, it is natural to work in $\mathcal{M}^{\mathrm{eq}}$ and take advantage of the elimination of imaginaries. But for applications of model theory, we generally want to use our knowledge of the definable sets in the structure itself to draw other conclusions. Thus it is useful to know if a theory eliminates imaginaries in the language for which we know something about the definable sets. For example, real closed fields, and more generally, o-minimal expansions of real closed fields, have elimination of imaginaries. (By the o-minimality, a definable set in one variable has a definable representative; use some uniformity and definable Skolem functions to lift this to sets in more than one variable.) Pillay [6] uses the o-minimality to study groups definable in an o-minimal expansion of a real closed field, and because such theories eliminate imaginaries, his results apply also to groups which are definable in a quotient of the structure.

By contrast, valued fields do not have elimination of imaginaries in an obvious language. Recall that a valued field is a field K together with a valuation function $|\ |$ which is a map from K to an ordered group $\Gamma \cup \{0\}$ satisfying

$$|xy| = |x||y|$$
$$|x + y| \leq \max\{|x|, |y|\}.$$

(Contrary to standard practice, I am writing the group Γ multiplicatively.) The p-adic numbers provide an example. Fix a prime p, and define

$$|\ |: \mathbb{Q} \to \{p^n : n \in \mathbb{Z}\} \cup \{0\} \text{ by } \left|p^r \frac{q}{s}\right| = p^{-r}, \text{ where } p \nmid qs, \text{ and } r \in \mathbb{Z};$$
$$|0| = 0.$$

The valuation function gives rise to a metric on $\mathbb{Q}$, and the field of p-adic numbers $\mathbb{Q}_p$ is the completion of $\mathbb{Q}$ with respect to this metric. The valuation function extends naturally to $\mathbb{Q}_p$ and to its algebraic closure. The field $\mathbb{Q}_p$ is a *p-adically closed field*, and the theory of p-adically closed fields can be recursively axiomatised in a natural language in which we have quantifier elimination (see, for example [7]). The algebraic closure of $\mathbb{Q}_p$ is an example of an *algebraically closed valued field*, and this theory can also be axiomatised and has quantifier elimination in the language of fields with a predicate Div which defines the valuation [8]. In both cases, the definable sets have lots of good properties; for example a definable set in one variable is either finite or has interior in the valuation topology.

Neither of these theories has elimination of imaginaries; for example, the relations $|x| = |y|$ and $|x - y| \leq 1$ cannot be coded. Macintyre and Scowcroft [5] showed that if representatives for the equivalence classes of these relations are added to the language, then the statement that the theory of p-adically closed fields eliminates imaginaries is independent of ZFC; a rather surprising result. Instead, to try to eliminate imaginaries, one can try to add a small part of $\mathcal{M}^{\text{eq}}$. The idea is to add enough sorts to eliminate imaginaries, but not so many that we cannot control the definable sets. In the work of Haskell, Hrushovski and Macpherson [2] that I describe here, we work in a language which has sorts for the *torsors* as I explain below. We show that the theory of algebraically closed valued fields in this language has elimination of imaginaries. Whether the analogous language will suffice to eliminate imaginaries in a p-adically closed field is still a question.

Before going into details, let me review some basic facts about valued fields, and present a picture which illustrates why some relations cannot be eliminated. A valued field K has a subring called the valuation ring; $R = \{x \in K : |x| \leq 1\}$. The valuation ring has a unique maximal ideal $\mathcal{M} = \{x \in K : |x| < 1\}$. The quotient $R/\mathcal{M}$ is called the residue field k. For $\mathbb{Q}_p$, the valuation ring is $\{x = \sum_{i=M}^{\infty} a_i p^i : M \geq 0\}$, the maximal ideal is

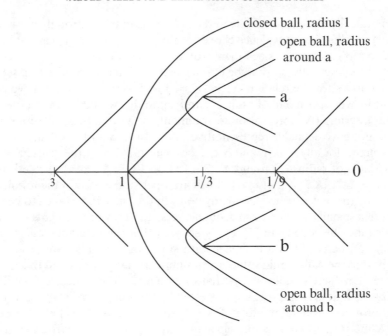

$\{x = \sum_{i=M}^{\infty} a_i p^i : M > 0\}$ and the residue field is isomorphic to the finite field with p elements, $\mathbb{F}_p$. For the algebraic closure of $\mathbb{Q}_p$, the residue field is the algebraic closure of $\mathbb{F}_p$.

It is useful to think of a valued field as a set of infinite branches in a tree. The value group can be thought of as the set of nodes on the branch which is the 0 of the field. By translation, the set of nodes on any other branch can also be identified with the value group. For any $a, b \in K$, $|a|$ is the node at which a branches off from 0, and $|a - b|$ is the node at which a branches off from b. Suppose $|a| = |b| = 1$. Then a and b are representatives of elements of the residue field k, and they branch off from 0 at the node which is the 1 of the value group. If $|a - b| = 1$ then they branch off from each other at the same place, and represent different elements of k. If $|a - b| < 1$, they branch off from each other further up, and represent the same element of k. The "open ball" $B_{<1}(a) := \{x \in K : |x - a| < 1\}$ can be identified with an element of k. If $b \notin B_{<1}(a)$ then $B_{<1}(b)$ is a disjoint open ball and is identified with a different element of k. The "closed ball" $B_{\leq 1}(a) := \{x \in K : |x - a| \leq 1\}$ is the entire valuation ring and the set of open subballs of $B_{\leq 1}(a)$ of the same radius is isomorphic to the residue field (though, in general, not canonically). This picture is repeated at every node of the tree. The diagram shown is of the 3-adics.

For an algebraically closed valued field, the value group will be divisible, hence densely ordered and can be shown to be o-minimal. The residue field will be algebraically closed, so infinite and can be shown to be strongly minimal. We thus see that the equivalence classes of the relation $|x| = |y|$ cannot have definable representatives, since there would be one at each node on 0. Thus the set of representatives would be infinite without interior. Similarly, a set of representatives for the relation $|x| = |y| = 1 \,\&\, |x - y| < 1$ would have one element in every open ball of radius 1, and again this would be an infinite set without interior. (A more elaborate argument shows that these relations not only do not have definable representatives, but also cannot be coded.)

In general, for any $a \in K$ and $\alpha \in \Gamma$, we can talk about the "closed" and "open" balls of radius α around a: $B_{\leq \alpha}(a) := \{x \in K : |x - a| \leq \alpha\}$ and $B_{<\alpha}(a) := \{x \in K : |x - a| < \alpha\}$. Both are open and closed in the topology induced by the metric from $|\ |$, and any point in the ball can be taken to be the center. These balls of course are definable sets, and they are elements of $\mathcal{M}^{\mathrm{eq}}$ which cannot be coded. In [4], Holly showed that the definable sets in one variable can be coded in a language with sorts for the balls. We originally thought that the balls would suffice to eliminate all imaginaries, but this is not the case. One needs additionally what one might call 'n-dimensional balls'. The objects that play this role are suggested by thinking of balls algebraically. A ball around 0 is a submodule of K; $B_{\leq \gamma}(0) = \{x \in K : |x| \leq \gamma\} = cR$ is an R-module, where c is any element with $|c| = \gamma$. Similarly, the open ball $B_{<\gamma}(0) = c\mathcal{M}$. A ball around a different point is a coset of cR or $c\mathcal{M}$: $B_{\leq \gamma}(a) = a + cR$. From this point of view, a generalization to n dimensions of a ball around 0 is a definable R-submodule of K^n, and the generalization of a general ball is a coset of a definable R-module, that is, a *torsor*. In [2], we include sorts in the language for a collection of definable torsors, and eliminate imaginaries with respect to them.

From now on, I will take $\mathcal{K}$ to be a very large, saturated, homogeneous model of an algebraically closed valued field, with field K, value group Γ, valuation ring R and residue field k. For each natural number n, let S_n be the set of free R-submodules of K^n on n generators and $\mathcal{S} = \bigcup_{n=1}^{\infty} S_n$. Then S_1 is the set of modules of the form γR for some γ, that is, the closed balls containing 0. For each $s \in S_n$, let $\mathrm{red}(s) = s/\mathcal{M}s$ be the reduction of s modulo the maximal ideal $\mathcal{M}$. As for the $n = 1$ case, $\mathrm{red}(s)$ is isomorphic to k^n, and is an n-dimensional vector space over k. Each element of $\mathrm{red}(s)$ is a coset of $\mathcal{M}s$, so is a torsor. We define $T_n = \bigcup\{\mathrm{red}(s) : s \in S_n\}$ and $\mathcal{T} = \bigcup_{n=1}^{\infty} T_n$. In particular, T_1 is the set of open balls of the form $B_{<|a|}(a)$. The sorts $\mathcal{S}$ and $\mathcal{T}$ are included in the language, as are K, k and Γ. Collectively, we refer to these as the geometric sorts G. I will omit the details of the language we put on the sorts, and simply note that the theory of algebraically closed fields admits quantifier elimination in the sorted language.

It is relatively straightforward to prove that the sorts $\mathcal{S} \cup \mathcal{T}$ suffice to code all the torsors. Furthermore, we can prove that the geometric sorts suffice to code all the definable sets, that is, the theory eliminates imaginaries in our language.

THEOREM 1. *The theory of algebraically closed valued fields eliminates imaginaries with respect to the sorts G.*

The following proposition is key to our proof of elimination of imaginaries.

PROPOSITION 1. *Let M be a sufficiently saturated homogeneous structure (in a sorted language, with at least one $\emptyset$-definable symbol which for convenience we will write ∞), and suppose that M has an $\mathrm{Aut}(M)$-invariant family $\mathcal{U}$ of definable sets with the following property. For every $a \in M^n$ there is a sequence $(a_1, \ldots, a_m)$ from M^{eq} such that $\mathrm{dcl}(a) = \mathrm{dcl}(a_1, \ldots, a_m)$ and for each $i \leq m$, there is $U \in \mathcal{U}$ such that $a_i \in U$ and U is $a_1 \ldots a_{i-1}$-definable. Suppose also that whenever g is a definable function from a set in $\mathcal{U}$ to M then g is coded in M. Then all definable subsets of M^n are coded in M, so $\mathrm{Th}(M)$ has elimination of imaginaries.*

PROOF. Let $A \subset M^n$ be a definable set which needs to be coded in M (over an arbitrary base set C of parameters). By assumption, for each element a of A, there is a tuple $(a_1, \ldots, a_m)$ from M^{eq} such that $\mathrm{dcl}(a) = \mathrm{dcl}(a_1, \ldots, a_m)$ and for each $i \leq m$, there is $U \in \mathcal{U}$ such that $a_i \in U$ and U is $a_1 \ldots a_{i-1}$-definable. By compactness, we can assume m is independent of a and the 'coding' is uniform. Let A' be the definable set of such tuples. Then $\mathrm{dcl}(\ulcorner A \urcorner) = \mathrm{dcl}(\ulcorner A' \urcorner)$, so $\mathrm{dcl}(C \ulcorner A \urcorner) = \mathrm{dcl}(C \ulcorner A' \urcorner)$ and so it suffices to code A' over C. We argue by induction on m, so we assume that over any base set F of parameters, if $\ell < m$ and A^* is a definable set of ℓ-tuples $(b_1, \ldots, b_\ell)$ from M with each b_i in an element of $\mathcal{U}$ defined over $F \cup \{b_j : j < i\}$, then A^* is coded in M over F.

To start the induction, observe that if $m = 1$ then by compactness there are finitely many C-definable sets $U_1, \ldots, U_r \in \mathcal{U}$ such that $A' \subset U_1 \cup \ldots \cup U_r$. Let $g_i : U_i \to U_i \cup \{\infty\}$ be given by $g_i(x) = x$ if $x \in A' \cap U_i$, and $g_i(x) = \infty$ otherwise. By assumption, each function g_i is coded over C by some sequence e_i from M, and $(e_1, \ldots, e_r)$ codes A' over C.

Now assume the result for $m - 1$. Let B be the set of first co-ordinates of tuples from A'. Each such a_1 lies in a set from $\mathcal{U}$, and again by compactness, we can assume they all lie in the same set U. For each $a \in B$, let $A'(a) := \{b : (a, b) \in A'\}$. By induction, each $A'(a)$ is coded in M over Ca by a sequence $c_a = (c_a^1, \ldots, c_a^\ell) \in M^\ell$. By compactness, we may suppose ℓ is fixed. By assumption, each coordinate function $a \mapsto c_a^i$ is coded over C, and a tuple listing these codes is a code for A' over C (in M).   ⊣

The family $\mathcal{U}$ from the proposition is given in our case by a subfamily of the *unary sets* (it will follow from elimination of imaginaries that the subfamily is in fact the whole family).

DEFINITION 3. A *definable 1-module* is an R-module (living in $\mathcal{K}^{\mathrm{eq}}$) which is definably isomorphic to one of $\gamma R/\delta R$, $\gamma R/\delta M$, $\gamma M/\delta R$ or $\gamma M/\delta M$, for $0 \le \delta \le \gamma \in \Gamma$. A *1-torsor* is a definable torsor of a definable 1-module or an intersection of definable torsors of definable 1-modules. A *unary set* is a 1-torsor or an interval $[0, \alpha]$ or $[0, \alpha)$ in Γ.

For example, the unary set $0 + \gamma R/\delta R$ is precisely the set of distinct balls $B_{\le \delta}(a)$, where $\gamma \ge |a| \ge \delta$. However, the ball $B_{\le \delta}(a)$ is also itself the unary set $a + \delta R/0R$. Thus we can think of a ball either as a unary set or as an element of a unary set, and this alternation in point of view is often useful.

We define an element a to be *generic* in a unary set U over an algebraically closed set of parameters C if a does not lie in any proper unary subset of U defined over C. The generic type of U is the type of a generic element of U. It is easy to show that an element of U realises the generic type of a unique unary subset of U.

To prove Theorem 1 we use Proposition 1, where we take the family $\mathcal{U}$ to be the following collection of unary sets:

  (i)   intervals in Γ,
  (ii) the definable 1-torsors U which are coded in G, and whose subtorsors
        are coded in G over $\ulcorner U \urcorner$,
  (iii) the ∞-definable 1-torsors which are intersections of 1-torsors in (ii).

We need to know that this family satisfies the hypotheses of the proposition. Since the sorts in the structure, in this case, are G, the issue is whether elements of $\mathcal{S} \cup \mathcal{T}$ have unary codes in the sense of the proposition. I explain how this works in the special case of an R-module A in S_2. Let $A_0 = \{x : (x, 0) \in A\}$, $A_1 = \{y : (0, y) \in A\}$, $B_0 = \{x : \exists y \, (x, y) \in A\}$ and $B_1 = \{y : \exists x \, (x, y) \in A\}$. Then A_0, A_1, B_0, B_1 are all R-submodules of K, hence isomorphic (though not canonically so) to R. In particular, A_0, A_1, B_0 and B_1 are simultaneously unary sets, elements of unary sets and elements of S_1. Also, B_0/A_0 and B_1/A_1 are R-modules, isomorphic to each other and to $R/\alpha R$ for some $\alpha < 1$. I claim that (A_0, A_1, B_0, B_1, A) is a unary code for A. To show this, we need to verify that A is an element of an (A_0, A_1, B_0, B_1)-definable unary set. Now A can be identified with a homomorphism in $\mathrm{Hom}(B_0/A_0, B_1/A_1)$. As B_0/A_0 and B_1/A_1 are R-modules, $\mathrm{Hom}(B_0/A_0, B_1/A_1)$ is also an R-module and isomorphic to $R/\alpha R$, hence is a unary set as required.

We thus need to prove that definable sets from the unary sets in $\mathcal{U}$ to G are coded, that is, to prove the following proposition. Below, we use the notation $\ulcorner f \urcorner$ for the imaginary element in $\mathcal{K}^{\mathrm{eq}}$, unique up to interdefinability, which is interdefinable with the graph of f.

PROPOSITION 2. *Let $U \in \mathcal{U}$, $f : U \to G$ be a definable function, and let $B := \mathrm{acl}_G(\ulcorner f \urcorner)$. Then $f \in \mathrm{dcl}(B)$.*

PROOF. Consider $\Sigma := \{D \subset U : D, f|_D$ both definable over $B\}$. If $\bigcup \Sigma = U$, then by compactness, f is B-definable, so we may suppose $\bigcup \Sigma \neq U$. Then there is a complete type p over B whose realisations lie in $U \setminus \bigcup \Sigma$. As remarked above, p is the generic type of a unary set V over B. As V is a subtorsor of U, also $V \in \mathcal{U}$. By the following theorem, there is a B-definable function g which agrees with f on part of V.

THEOREM 2. *Let $U \in \mathcal{U}$, f be a definable function to G with domain containing U and $B \subset G$ with $B = \mathrm{acl}_G(B^\ulcorner f^\urcorner)$. Suppose that U is $(\infty\text{-})$definable over B. Then there is a B-definable function g with the same germ on U as f, that is, the set of points where f and g agree contains the set of points of U generic over $B^\ulcorner f^\urcorner$.*

Let $X := \{x: f(x) = g(x)\}$. Then $X \cap V \neq \emptyset$, but X is Bf-definable and coded in G (this needs to be proved, of course), so B-definable. Hence, as p is a complete type over B, $X \supseteq V$. But as g is B-definable, $f|_X$ is B-definable, so $X \in \Sigma$, a contradiction. $\dashv$

The heart of the problem is thus Theorem 2. In the case when U is a closed unary set, the method of proof is to consider the type Q over B of the parameters c used to define the function f. We need to show that, for any $c, c' \in Q$, and for any a in the domain of f generic over Bcc', $f_c(a) = f_{c'}(a)$. To show this, we use quantifier elimination to show that the type of $af(a)$ is determined by sets of polynomials, and these sets of polynomials are R-modules which are therefore coded in G. The modules of polynomials thus play a role akin to that of ideals of polynomials in the case of pure algebraically closed fields. For U an open unary set or the intersection of a chain of unary sets, we consider the germ of f on the generic type of U, approximate U from inside by closed 1-torsors, and piece together the corresponding functions to obtain the function g.

I will finish by describing an example which shows that imaginaries cannot be eliminated in a language which just includes sorts for the balls. Choose $\gamma_1, \gamma_2 \in \Gamma$ with γ_1 generic below 1 and $\gamma_2 < \gamma_1$ generic over γ_1. For $i = 1, 2$ let $U_i = a_i + \gamma_i R \in K/\gamma_i R$ with U_1 generic over $\gamma_1 \gamma_2$ and U_2 generic over $\gamma_1 \gamma_2 U_1$. Let $A_i = \mathrm{red}(U_i) = a_i + \gamma_i R/\gamma_i \mathcal{M}$. Then each A_i is a torsor of $\gamma_i R/\gamma_i \mathcal{M}$, so is isomorphic to the residue field, though the isomorphism depends on the choice of parameter a_i. The set of affine homomorphisms from A_1 to A_2, $\mathrm{Aff}(A_1, A_2)$, provides us with the counterexample we want. By definition, any element h of $\mathrm{Aff}(A_1, A_2)$ is determined by an induced homomorphism $\pi(h)$ in $\mathrm{Hom}(\gamma_1 R/\gamma_1 \mathcal{M}, \gamma_2 R/\gamma_2 \mathcal{M})$ and, for any fixed $a \in A_1$, the image $h(a)$. Since A_1, A_2 are essentially copies of k, $\mathrm{Aff}(A_1, A_2)$ is contained in a set which is internal to k, and hence forms a stable structure, in a sense which can be made precise. As such, $\mathrm{Aff}(A_1, A_2)$ clearly has Morley rank 2, as $\mathrm{Hom}(\gamma_1 R/\gamma_1 \mathcal{M}, \gamma_2 R/\gamma_2 \mathcal{M})$ has rank 1. Now if imaginaries could be eliminated in the ball sorts, a generic element h in $\mathrm{Aff}(A_1, A_2)$ would be coded in the

k-internal structure by an independent pair of elements of strongly minimal sets of the form red(s) for s a ball. However, we are able to show that if c is an element of acl(h) with rank 1, then $c \in$ acl($\pi(h)$), so acl(h) contains a unique rank 1 algebraically closed subset. This contradiction finishes the example.

REFERENCES

[1] S. BUECHLER, *Essential stability theory*, Springer, 1996.

[2] D. HASKELL, E. HRUSHOVSKI, and D. MACPHERSON, *Definable sets in algebraically closed valued fields. Part I: elimination of imaginaries*, preprint, 2002.

[3] W. HODGES, *Model theory*, Cambridge University Press, 1993.

[4] J. HOLLY, *Prototypes for definable subsets of algebraically closed valued fields*, **The Journal of Symbolic Logic**, vol. 62 (1997), pp. 1093–1141.

[5] A. MACINTYRE and P. SCOWCROFT, *On the elimination of imaginaries from certain valued fields*, **Annals of Pure and Applied Logic**, vol. 61 (1993), pp. 241–276.

[6] A. PILLAY, *Groups and fields definable in o-minimal structures*, **Journal of Pure and Applied Algebra**, vol. 53 (1988), pp. 239–255.

[7] A. PRESTEL and P. ROQUETTE, *Formally p-adic fields*, Lecture Notes in Mathematics, vol. 1050, Springer, 1984.

[8] A. ROBINSON, *Complete theories*, North-Holland, 1956.

[9] S. SHELAH, **Classification theory and the number of nonisomorphic models**, North Holland, 1990, (revised edition).

DEPARTMENT OF MATHEMATICS
COLLEGE OF THE HOLY CROSS
WORCESTER, MA, 01610, USA
and
DEPARTMENT OF MATHEMATICS AND STATISTICS
MCMASTER UNIVERSITY
HAMILTON, ON L8S 4K1, CANADA
E-mail: haskell@math.mcmaster.ca

SIMPLE SETS AND Σ_3 IDEALS UNDER m-REDUCIBILITY

KEJIA HO AND FRANK STEPHAN

§1. Introduction. With a typical priority argument, one can show that for any simple set A, there is a simple set B such that $B \not\leq_m A$. Carl Jockusch (personal communication) asked: Is there such a B which is effectively simple? The investigation of this question was the starting point of the current work. We call a class S of c.e. sets *bounded* iff there is an m-incomplete c.e. set A such that every set in S is m-reducible to A. Then a generalized question is: for a natural class of simple sets, is it bounded? Furthermore, we may ask for an arbitrary reducibility r, whether there is a bounded class intersecting every c.e. r-degree; and for which c.e. sets A is the class of all c.e. sets in their Turing degree bounded? The main purpose of this paper is to explore these and related questions. Section 2 deals with nontrivial upper bounds for natural classes of simple sets with respect to m-reducibility. We first give a negative answer to Jockusch's question. Then we show that the class of all maximal sets is bounded. In Section 3 we deal with the question whether some bounded class intersects every c.e. r-degree. We construct a bounded class intersecting every $bwtt$-degree. We show that the semirecursive c.e. sets form a bounded class intersecting every c.e. tt-degree. A parallel result is that there is a wtt-incomplete c.e. set A such that every c.e. Turing degree contains a set $B \leq_m A$. In Section 4, we explore the question of when, for an arbitrary reducibility r, the class of all c.e. sets in a given c.e. r-degree is bounded. We show that for any c.e. set A, the class of all c.e. sets in the Turing degree of A is bounded iff A is low$_2$. One can transfer this result to Q-reducibility, if A is a semirecursive c.e. set. But there is a non-low$_2$ c.e. set A such that the class of all c.e. sets in the Q-degree of A is bounded. Our notation is as in Odifreddi [6] and Soare [7].

DEFINITION 1.1. A set A is *effectively simple* if it is a coinfinite c.e. set and there is a computable function f (called a *bound function* for A) such that $(\forall e)[W_e \subseteq \overline{A} \Rightarrow |W_e| \leq f(e)]$.

DEFINITION 1.2. Let S be a nonempty class of sets. $\mathcal{I}$ is the *m-ideal generated by S* if $\mathcal{I}$ is obtained by closing S downward under "$\leq_m$" and by closing S under joins. That is, $\mathcal{I}$ is the least class of sets under set-inclusion such that:

Logic Colloquium '99
Edited by J. van Eijck, V. van Oostrom, and A. Visser
Lecture Notes in Logic, 17

(1) $A \in \mathcal{S} \Rightarrow A \in \mathcal{I}$, (2) $B \leq_m A$ and $A \in \mathcal{I} \Rightarrow B \in \mathcal{I}$, (3) $A, B \in \mathcal{I} \Rightarrow A \oplus B \in \mathcal{I}$. Every pair (B, C) determines an m-ideal $\mathcal{I}_{B,C} = \{A \mid A \leq_m B$ and $A \leq_m C\}$ provided that it is a nonempty set. A pair (B, C) is called *an exact pair* for $\mathcal{I}$ if $\mathcal{I} = \mathcal{I}_{B,C}$.

§2. **Bounding natural classes of simple sets.** In this section, we study non-trivial upper bounds for natural classes of simple sets with respect to m-reducibility. First we give a negative answer to Jockusch's question of whether for any simple set A, there is an effectively simple set B such that $B \not\leq_m A$.

THEOREM 2.1. *There is a simple set A such that for any effectively simple set B, $B \leq_1 A$.*

PROOF. Let $\{(i_n, j_n, k_n)\}_{n \in N}$ be a computable enumeration without repetition of $C = \{(i, j, k) : (\forall e \leq k)[\varphi_j(e)\downarrow$ and $[|W_{e,k}| > \varphi_j(e) \Rightarrow W_e \cap W_i \neq \emptyset]]\}$. Note that for fixed i, j, the tuples (i, j, k) are enumerated into C for all k iff φ_j is total and W_i is effectively simple or cofinite with bound function φ_j. Here "bound function" is defined for a cofinite set just as it is for an effectively simple set. Let

$$A = \{x : \text{(a) } k_x \in W_{i_x} \text{ or}$$
$$\text{(b) for some } e, x \text{ is the first element } y$$
$$\text{enumerated into } W_e \text{ with } i_y + j_y > e\}.$$

Every effectively simple set is 1-reducible to A. Let W_i be effectively simple with φ_j a bound function for it. For all k, define

$$f(k) = x \text{ for the unique } x \text{ such that } (i_x, j_x, k_x) = (i, j, k).$$

Note that f is well-defined and is a one-one computable function. By (a), $A(f(k)) = 1$ whenever $W_i(k) = 1$. Assume now that $A(f(k)) = 1$ while $W_i(k) = 0$. Then $f(k)$ is enumerated into A by (b) and $f(k)$ is the first element y enumerated into W_e with $i_y + j_y > e$ for some e. Since $i_y = i$ and $j_y = j$, it follows that $W_i(k) = A(f(k))$ for all but at most $i + j$ numbers. Furthermore, the set of all $x \in \overline{A}$ with $i_x + j_x > i + j$ is infinite: consider for example an index $i' > i$ of W_i, and some x with $i_x = i'$, $j_x = j$ and k_x not in the coinfinite set $W_i = W_{i'}$ and also x is not enumerated into A by (b). Note that there are infinitely many such x. Thus one can repair f on the finitely many k with $W_i(k) = 0$ and $A(f(k)) = 1$ by assigning to them distinct numbers x with $i_x + j_x > i + j$ and $A(x) = 0$. This corrected f is then a 1-reduction from the effectively simple set W_i to the set A.

A is simple. Clearly, A is a coinfinite c.e. set by the argument in the previous paragraph. Let g be a computable function such that for any c.e. set W_e and any pair (i, j), $W_{g(i,j,e)} = \{k : (\exists x \in W_e)[(i_x, j_x, k_x) = (i, j, k)]\}$. Assume now that $W_e \cap A = \emptyset$. Note that $W_{g(i,j,e)} \cap W_i = \emptyset$ for otherwise some $x \in W_e$ would be enumerated into A by case (a). Note that $W_{g(i,j,e)}$ is empty for all

i, j with $i + j > e$ since otherwise some element of W_e would be enumerated into A by case (b). For any pair (i, j) with $i + j \leq e$, the set $W_{g(i,j,e)}$ is finite: if W_i is effectively simple or cofinite with bound function φ_j, then $W_{g(i,j,e)}$ has at most $\varphi_j(g(i, j, e))$ elements and it is a finite set; otherwise there are in total only finitely many x with $i_x = i$ and $j_x = j$ and so the set $W_{g(i,j,e)}$ is also finite. Since the map $x \mapsto (i_x, j_x, k_x)$ is one-one, one has that the cardinality of W_e is exactly the sum of all cardinalities $|W_{g(i,j,e)}|$. Only finitely many cardinalities are not 0 and each cardinality is finite. So the cardinality of W_e is the sum of finitely many finite numbers and therefore is finite. Thus each W_e disjoint from A is finite and A is simple. ⊣

The next theorem stands in contrast to Theorem 2.1.

THEOREM 2.2. *The class of all maximal sets is not bounded.*

PROOF. The basic idea of the proof is to show that

$$(\Pi_3, \Sigma_3) \leq_m (\text{maximal}, m\text{-complete})$$

in the style of Soare [7, Chapter XII]. That is, for the Π_3-complete set $A = \{a : W_a \text{ is coinfinite}\}$, we construct an m-reduction f such that for all a, $W_{f(a)}$ is maximal if $a \in A$, and $W_{f(a)}$ is m-complete if $a \notin A$. Assume by way of contradiction that there is an m-incomplete c.e. set C which bounds all maximal sets. One has that $W_{f(a)} \leq_m C$ iff $a \in A$. The condition that $W_{f(a)} \leq_m C$ for a c.e. set C is a Σ_3 condition. Thus $a \in A$ iff $f(a)$ satisfies a Σ_3 condition. This implies that A is also in Σ_3 which contradicts that A is Π_3-complete.

It remains to construct the reduction. This reduction is a modification of the standard construction of maximal sets via markers equipped with e-states as in Soare [7, Section X.3]. Note that for any e, the e-states are ordered lexicographically by first differences. This order is denoted by "$<_{lex}$".

Now one constructs for each a a c.e. set B_a such that B_a is m-complete if $a \in A$ and that B_a is maximal otherwise. The obtained family is uniform in the parameter a in the sense that $\{(a, x) : x \in B_a\}$ is c.e. and thus there is a computable function f with $W_{f(a)} = B_a$. This f is then the desired m-reduction. Recall that K denotes the halting problem.

Construction.

Let Γ_e be a marker and compute, at stage s, the e-state $\beta_{e,x,s}$ of Γ_e under the assumption that the marker sits at position x as defined as follows:

$$\beta_{e,x,s} = W_{0,s}(x) W_{1,s}(x) \ldots W_{g(e,s),s}(x) \text{ where } g(e, s) = |\{0, 1, \ldots, e\} - W_{a,s}|.$$

Stage 0: Placing marker Γ_e on e for all e.

Stage $s + 1$:

- Let $B_{a,s}$ contain all $x \leq s$ such that either no marker or only one marker Γ_e with $e \in K_s$ is sitting on x.
- Compute for all $e, x \leq s$ the e-states $\beta_{e,x,s}$.

- Search for the least e such that there are $x, y, e' \leq s$ with Γ_e sitting on x, $\Gamma_{e'}$ sitting on y, $e < e'$, $e \notin K_s$, $e' \notin K_s$, and $\beta_{e,x,s} <_{lex} \beta_{e,y,s}$.
- If such an e and the corresponding x, y, e' are found then move Γ_e from x to y and move any marker $\Gamma_{e''}$ with $e'' \geq e'$ and $e'' \notin K_s$ to the place which had the marker $\Gamma_{e'''}$ at stage s, where e''' is the least number having the property $e''' > e''$ and $e''' \notin K_s$.
- If no such e is found then all markers stand still at their current places.

End of Construction.

Note that this construction has two main differences from the construction of the maximal set as given by Soare [7, Section X.3]: 1. The e-states depend in their length on the function $g(e, s)$ and might therefore become shorter from time to time. 2. Some but only coinfinitely many markers Γ_e might end up in a position inside the set B_a: these markers are precisely those with $e \in K$. Now fix a.

Every marker Γ_e moves only finitely often. This is shown by induction, assume that for a given e the markers $\Gamma_{e'}$ with $e' < e$ move only finitely often. Now there a stage s such that no marker $\Gamma_{e'}$ with $e' < e$ will move again, that all $e' \leq e$ which are in K are also already in K_s and that all $e' \leq e$ which are in W_a have also already appeared in $W_{a,s}$. So the value $g(e, s)$ equals to $g(e, t)$ for all $t \geq s$. Thus the only change of the e-state of Γ_e can occur due to the fact that Γ_e moves. These moves are not enforced by markers of higher priority since those do no longer move at all. Thus the moves can only happen if Γ_e moves from some place x to a place y with $\beta_{e,x,s} <_{lex} \beta_{e,y,s}$. Since this increases the e-state lexicographically and since it furthermore holds that $\beta_{e,x,s} \leq_{lex} \beta_{e,x,t}$ for all e, x, s, t with $t > s$ there can only be finitely many moves of the marker Γ_e after stage s and thus Γ_e also moves in total only finitely often.

B_a is coinfinite. All markers Γ_e with $e \notin K$ end up in a final position which will never be enumerated into B_a. Since there are never two markers on the same place at the same time, they cover infinitely many places and enforce that the complement of B_a is infinite.

If $k \leq |\overline{W_a}|$ then there is a string $\gamma = c_0 c_1 \ldots c_k$ with $W_0(x) W_1(x) \ldots W_k(x) = \gamma$ for almost all $x \notin B_a$. For every $x \notin B_a$ there is exactly one number $e(x)$ such that the marker $\Gamma_{e(x)}$ will eventually come to sit on x forever; almost all of these satisfy furthermore that there are k non-elements of B_a below $e(x)$ and thus $g(e(x), s) \geq k$ for all s. Among these x there is an x such that $W_0(x) W_1(x) \ldots W_k(x)$ takes the lexicographically minimum value, let γ be this value and fix the corresponding x. Let $U = \{y \notin B_a : e(y) \geq e(x)\}$. Note that $B_a \cup U$ is cofinite since there are only finitely many $y \notin B_a$ such that $e(y) < e(x)$. Furthermore, from the choice of x it follows that $W_0(y) W_1(y) \ldots W_k(y) \geq_{lex} \gamma$ for all $y \in U$. On the other hand the marker $\Gamma_{e(x)}$ never moves to a place $y \in U$, thus there is no stage $s >$

$x + y$ such that $\Gamma_{e(x)}$ and $\Gamma_{e(y)}$ have already arrived at x, y and $\beta_{e(x),x,s} <_{lex}$ $\beta_{e(x),y,s}$. Since, for every s, $W_{0,s}(x) W_{1,s}(x) \ldots W_{k,s}(x) \leq_{lex} \gamma$, there is no s such that $\gamma <_{lex} W_{0,s}(y) W_{1,s}(y) \ldots W_{k,s}(y)$. So it follows that, for all $y \in U$, $W_0(y) W_1(y) \ldots W_k(y) = \gamma$.

If W_a is coinfinite, then B_a is maximal. Take any k such that $B_a \subseteq W_k$. Since $k \leq |\overline{W_a}|$ there is a string $c_0 c_1 \ldots c_k$ and a set $U \subseteq \overline{B_a}$ such that $B_a \cup U$ is cofinite and $W_k(y) = c_k$ on all $y \in U$. It follows that W_k is a finite variant of B_a if $c_k = 0$ and W_k is cofinite if $c_k = 1$. So B_a is maximal.

If W_a is cofinite, then B_a is m-complete. Let $k = |\overline{W_a}|$. There is a set $U \subseteq \overline{B_a}$ and a string γ such that $B_a \cup U$ is cofinite and $W_0(y) W_1(y) \ldots W_k(y) = \gamma$ for all $y \in U$. Let x be that element of U with minimal value $e' = e(x)$. Let E be the set of the markers $e \notin K$ such that either $e = e'$ or their final positions are outside U. The set E is finite. So let s be the first stage such that all markers Γ_e with $e \in E$ have reached their final positions, that all elements of K below $\max(E)$ have been enumerated into K_s and that also $W_{l,s}(y) = W_l(y)$ for all $l \leq k$ and $y \leq \max(\{x\} \cup \overline{(B_a \cup U)})$. Now one can compute for every e the stage $t(e)$ defined as the first stage t such that

- $t \geq s$;
- every marker $\Gamma_{e''}$ with $e'' \leq e + e'$ sits before stage t at some place below t;
- $g(e'', t) \leq k$ for all $e'' \leq e + e'$;
- every $\Gamma_{e''}$ with $e'' \in \{0, 1, \ldots, e + e'\} - E$ has reached a position y such that either $e'' \in K_t$ or $W_{0,s}(y) W_{1,s}(y) \ldots W_{k,s}(y) = \gamma$;

and define that $h(e)$ takes the current value of the marker Γ_e after stage $t(e)$. Now one shows that no marker $\Gamma_{e''}$ with $e'' \leq e + e'$ moves after stage t: Assume that this would be false and that $t' \geq t$ is the first stage where some marker $\Gamma_{e''}$ with $e'' \leq e + e'$ at position z moves to the position y of some marker $\Gamma_{e'''}$; without loss of generality e'' is the least number such that $\Gamma_{e''}$ moves at stage t'. Since the markers belonging to $0, 1, \ldots, e'$ have already found their final position, $e'' > e'$. Furthermore, $e'', e''' \notin K_{t'}$ and $\gamma \leq_{lex} \beta_{e'',z,t'} <_{lex} \beta_{e'',y,t'}$ and $y \leq t'$. But then it holds also that $\gamma = \beta_{e',x,t'} <_{lex} \beta_{e''',x,t'}$ and marker $\Gamma_{e'}$ would move since it has higher priority than marker $\Gamma_{e''}$, in contradiction to the fact that $\Gamma_{e'}$ has already reached its final position. From this contradiction follows that no marker $\Gamma_{e''}$ with $e'' \leq e + e'$ moves after stage $t(e)$, in particular the marker Γ_e does not leave the position $h(e)$. When Γ_e moves to its final position $h(e)$, $h(e)$ is not yet enumerated into B_a. Therefore $h(e)$ will eventually be enumerated into B_a iff e will eventually be enumerated into K. It follows that $K(e) = B_a(h(e))$ and $K \leq_m B_a$ via h. So B_a is m-complete. $\dashv$

§3. **Bounding a set in every c.e. r-degree.** In this section, we investigate whether there are bounded classes intersecting every c.e. r-degree. First let us state a theorem which follows directly from Ambos-Spies, Nies and Shore [1, Theorem 6].

THEOREM 3.1 (Ambos-Spies, Nies and Shore). *Every m-ideal generated by a class of c.e. sets with a Σ_3 index set has an exact pair of c.e. sets.*

PROPOSITION 3.2. *Let S be a class of m-incomplete c.e. sets with a Σ_3 index set. Then S generates an m-ideal which does not contain any creative sets. Moreover, S is bounded.*

PROOF. Lachlan's Universal Set Theorem [4] says that, if $K \leq_m A \times B$ and one of the sets A and B is c.e., then either $K \leq_m A$ or $K \leq_m B$. Since for all nonempty sets A and B, $A \oplus B \leq_m A \times B$, any class of m-incomplete c.e. sets generates an m-ideal which does not contain any creative sets. By Theorem 3.1, the m-ideal generated by the Σ_3 class S has an exact pair of c.e. sets. Since the m-ideal does not contain any creative sets, at least one half of this exact pair, call it A, is not m-complete. Clearly, for any $B \in S$, $B \leq_m A$. ⊣

THEOREM 3.3. *There is an m-incomplete c.e. set A such that every c.e. tt-degree contains a set $B \leq_m A$.*

PROOF. Consider the class of all semirecursive c.e. sets. It is a class of m-incomplete c.e. sets and it has a Σ_3 index set. By Proposition 3.2, there is an m-incomplete c.e. set A such that every set B in this Σ_3 class is m-reducible to A. Jockusch [2, Corollary 3.7(ii)] showed that every c.e. tt-degree contains a semirecursive c.e. set. ⊣

COROLLARY 3.4. *There is an m-incomplete c.e. set A such that every c.e. wtt-degree (Turing degree) contains a set $B \leq_m A$.*

The next theorem establishes the parallel result for *bwtt*-degrees.

THEOREM 3.5. *There is an m-incomplete c.e. set A such that every computably enumerable bwtt-degree contains a set $B \leq_m A$.*

PROOF. Sacks [7, Theorem VII.3.2] showed that K is the disjoint union of two low c.e. sets A_1 and A_2. Let $A = A_1 \oplus A_2$. A is m-incomplete by Lachlan's Universal Set Theorem and the observation that for all nonempty sets A and B, $A \oplus B \leq_m A \times B$. For any c.e. set C, C is m-reducible to K via some computable function f and so one can define $C_k = \{x : f(x) \in A_k\}$ for $k = 1, 2$. So the splitting is carried over to C_1, C_2, that is, C is the disjoint union of C_1 and C_2.

Let $B = C_1 \oplus C_2$. B is m-reducible to A. To see that B is *bwtt*-equivalent to C, note first that $C(x) = \max\{C_1(x), C_2(x)\}$. Conversely, it is sufficient to show how to compute C_1, C_2 from C. Given x, first a query to C is made. If this value is 0, then $C_1(x)$, $C_2(x)$ are both 0. If this value is 1, then C_1 and C_2 are enumerated in parallel until it is known which of the values $C_1(x)$, $C_2(x)$ is 1 and which is 0. ⊣

The above results are false for c.e. positive degrees. By Theorem 2.2, the class of all maximal sets is not bounded. Furthermore, Kummer and Stephan [3, Theorem 5.7] showed that if M is maximal and $M \equiv_p A$ then $M \leq_m A$. So there is no bounded class intersecting every c.e. p-degree. Since any bounded class intersecting every c.e. c-degree or every c.e. d-degree would also intersect every c.e. p-degree, there is no m-incomplete c.e. set which bounds a set in every c.e. c-degree (d-degree). The following theorem gives an affirmative answer to a related question.

THEOREM 3.6. *There is a wtt-incomplete c.e. set A such that every c.e. Turing degree contains a set $B \leq_m A$.*

PROOF. Dekker (see [7, Theorem V.2.5]) showed that there is a computable function f such that for any c.e. set W_e, $W_{f(e)} \equiv_T W_e$ and $W_{f(e)}$ is hypersimple whenever W_e is noncomputable. Now the sequence $\{W_{f(0)}, W_{f(1)}, \dots\}$ generates a *wtt*-ideal; here the term "*wtt*-ideal" is defined analogously to "m-ideal". Since the join of a hypersimple set with other hypersimple or computable sets is again m-equivalent to a hypersimple set, every set in this *wtt*-ideal is *wtt*-reducible to some hypersimple set. Friedberg and Rogers (see Odifreddi [6, Proposition III.8.16]) showed that no hypersimple set is *wtt*-complete. So the sets $W_{f(0)}, W_{f(1)}, \dots$ generate a *wtt*-ideal which does not contain any *wtt*-complete sets. Let $\{W_{f(x),s}\}_{s \in N}$ be a computable enumeration of $W_{f(x)}$ such that if $y \in W_{f(x),s}$, then $x, y < s$. Now define inductively c.e. sets $A = \bigcup_s A_s$ and $C = \bigcup_s C_s$ by letting $A_0 = \emptyset$, $C_0 = \emptyset$, $C_{s+1} = C_s \cup \{(x, y): x, y < s \text{ and } (\exists e \leq x)(\forall z \leq y)[K_s(z) = \varphi_{e,s}^{A_s}(z)]\}$ and $A_{s+1} = A_s \cup \{(x, y): y \in W_{f(x),s}\} \cup C_{s+1}$. Consider the projections $\hat{A}_x = \{y: (x, y) \in A\}$ and $\hat{C}_x = \{y: (x, y) \in C\}$. Note that $\hat{A}_x = W_{f(x)} \cup \hat{C}_x$ and every $\hat{C}_x$ is an initial segment. Namely, either $\hat{C}_x = \{0, 1, \dots, w\}$ for some w or $\hat{C}_x = N$. So for any x, either $\hat{A}_x$ is a finite variant of $W_{f(x)}$ or $\hat{A}_x = N$.

A is *wtt*-complete iff $\hat{C}_x = N$ for some x. If φ_e is a *wtt*-reduction from K to A, then there is, for every y, some stage $s > y$ such that $K_s(z) = \varphi_{e,s}^{A_s}(z)$ for all $z \leq y$. It follows that $\hat{C}_x = N$ for all $x \geq e$. Conversely, if $\hat{C}_x = N$ for some x, then there is, for every y, an $e \leq x$ and a stage $s > y$ such that $K_s(z) = \varphi_{e,s}^{A_s}(z)$ for all $z \leq y$. It is easy to see that there is a fixed $e \leq x$ such that $(\exists^\infty y)(\exists s > y)(\forall z \leq y)[K_s(z) = \varphi_{e,s}^{A_s}(z)]$. It follows that this φ_e is a *wtt*-reduction from K to A.

For any e, $\hat{C}_e$ is finite. If there were infinite sets $\hat{C}_x$, then K would be *wtt*-reducible to A. Furthermore, there would be only finitely many x such that $\hat{C}_x$ is finite and thus $\hat{A}_x$ is a finite variant of $W_{f(x)}$. For all other x, $\hat{C}_x = N$ and thus $\hat{A}_x = N$. So A would be the join of finitely many sets, each a finite variant of $W_{f(x)}$ for some x, together with infinitely many copies of N. So A would be in the *wtt*-ideal generated by the sets $W_{f(0)}, W_{f(1)}, \dots$ and thus would not be *wtt*-complete. This gives the desired contradiction.

A is *wtt*-incomplete and it bounds a set in every c.e. Turing degree. Since for any e, $\hat{C}_e$ is finite, A is *wtt*-incomplete. Moreover, it follows from $\hat{A}_e = W_{f(e)} \cup \hat{C}_e$ that $\hat{A}_e \equiv_T W_{f(e)} \equiv_T W_e$. Note also that for any e, $\hat{A}_e$ is m-reducible to A.   ⊣

§4. Bounding c.e. *r*-degrees.

For the reducibilities $r = 1, m, c, bwtt, btt, d, p, tt$ and wtt, the class of the c.e. sets inside the r-degree of a given r-incomplete c.e. set A has a Σ_3 index set. By Proposition 3.2, any such class is bounded by some m-incomplete c.e. set. So the question is interesting only for the cases of Turing reducibility and Q-reducibility.

THEOREM 4.1. *Let A be c.e. Then the class of all c.e. sets in the Turing degree of A is bounded iff A is low₂.*

PROOF. If A is low₂, then by a result of Yates (see Soare [7, Chapter XII]), the index set of the sets Turing equivalent to A is Σ_3. By Proposition 3.2, the class of the sets Turing equivalent to A is bounded.

Conversely, assume that C is any m-incomplete c.e. set and A is non-low₂. Now it is shown how to construct a c.e. set $B \leq_T A$ which is not m-reducible to C. In particular, $A \oplus B$ is then c.e. and Turing equivalent to A but not m-reducible to C.

Let $g_1(x) = \max\{\varphi_y(z): y, z \leq x$ and $\varphi_y(z)\downarrow\}$. Without loss of generality, one can assume that $\varphi_0(0)$ is defined so that g_1 is a total function. g_1 is computable relative to K. For each φ_e there is a y such that either $\varphi_e(y)$ is undefined or $K(y) \neq C(\varphi_e(y))$ since φ_e cannot m-reduce K to the m-incomplete c.e. set C. Let $g_2(e)$ denote the least such y; g_2 is also computable relative to K. Thus the function g_3 given by

$$g_3(x) = 1 + \max\{g_2(0), g_2(1), \ldots, g_2(g_1(x))\}$$

is also computable relative to K. Let g_4 be the K-computable function

$$g_4(x) = \min\{s > g_3(x): (\forall y \leq g_3(x))[K_s(y) = K(y)]\}.$$

That is, $g_4(x)$ is the least stage after $g_3(x)$ such that all elements of K up to $g_3(x)$ are enumerated into K.

Since A is non-low₂, the set K is not high relative to A. K is not A-high iff K does not compute A-dominant function. Hence for every K-computable g there is an A-computable f not dominated by it. In particular there is a function f_1, computable relative to A, such that $f_1(x) > g_4(x)$ for infinitely many x.

By assumption the set A is c.e. Let $F_2(x)$ be the set of all s such that there is a $y \leq x$ for which the computation of $f_1(y)$ relative to A_s either queries some $z \in A_{s+1} - A_s$ or does not terminate within s stages or outputs some value $t \geq s$. The set $F_2(x)$ of these numbers s is finite. Whether $s \in F_2(x)$ or not is uniformly computable in the two parameters x and s.

Now let $f_2(x) = \max(F_2(x))$. $f_2(x)$ is basically the use funciton of all the computations $f_1(y)$, $y \leq x$. Note that $f_2(x) \geq f_1(x)$ for all x.

The function f_2 is computable relative to A since $f_2(x)$ is the least s with the following property: (1) $s \in F_2(x)$ and $s + 1 \notin F_2(x)$; (2) $A_{s+1}(z) = A(z)$ for all z queried by the computation of any value $f_1(y)$ with $y \leq x$. From $s+1 \notin F_2(x)$ it follows that the computations of $f_1(y)$ for all $y \leq x$ relative to A_{s+1} terminate within $s + 1$ stages with an output of at most s. Furthermore, it follows from the second A-computable condition that these computations use always the correct values and compute indeed $f_1(y)$ and not just some approximation.

Now $\{(x, y): y \leq f_2(x)\} = \{(x, y): (\exists s \geq y)[s \in F_2(x)]\}$ and the latter set is c.e. since the function which computes whether $s \in F_2(x)$ is computable in the two parameters s and x.

Clearly, $f_2(x) > g_4(x) > g_3(x)$ for infinitely many x. Now let

$$B = \{(x, y): y \leq f_2(x) \text{ and } y \in K_{f_2(x)}\}.$$

By the above remarks on f_2, the set B is c.e. and is also computable relative to A. Moreover, if $f_2(x) > g_4(x)$ and $y \leq g_3(x)$, then

$$B(x, y) = K_{f_2(x)}(y) = K_{g_4(x)}(y) = K(y).$$

Assume by way of contradiction that $B \leq_m C$ via some computable function h. Let e be a computable function such that for all x, y, $\varphi_{e(x)}(y) = h(x, y)$. Let $z' = g_2(e(x))$. By the definition of g_2, either $\varphi_{e(x)}(z')$ is undefined or $K(z') \neq C(\varphi_{e(x)}(z'))$. For almost all x, $e(x) \leq g_1(x)$. For these x, it follows that $g_2(e(x)) < g_3(x)$. But for infinitely many of these x, it holds also that $f_2(x) > g_4(x) > g_3(x) > g_2(e(x))$. As a consequence, for some x, there is a $z = g_2(e(x)) < g_3(x)$ with $C(h(x, z)) = C(\varphi_{e(x)}(z)) \neq K(z) = B(x, z)$. So the assumption that h is an m-reduction from B to C is false and $B \nleq_m C$.   ⊣

This result for Turing reducibility can be transferred to Q-reducibility in the case of semirecursive sets. The main reason is that Turing reducibility and Q-reducibility coincide if both sets are c.e. and the set on the right hand side is semirecursive.

COROLLARY 4.2. *Let A be a c.e. semirecursive set. Then the class of all c.e. sets in the Q-degree of A is bounded iff A is low$_2$.*

PROOF. ($\Leftarrow$) A is low$_2$. Let B be any c.e. set. Odifreddi [6, Exercise III.4.3.(f)] states that $B \leq_Q A \Rightarrow B \leq_T A$ if both sets A and B are c.e. Thus every c.e. set in the Q-degree of A is also in the Turing degree of A.

($\Rightarrow$) Now let A be non-low$_2$ but semirecursive and let C be any m-incomplete c.e. set. By Theorem 4.1 there is a c.e. set $B \leq_T A$ which is not m-reducible to C. Following a result of Marchenkov [5] and using the fact that B is c.e. and A is semirecursive, it holds also that $B \leq_Q A$. In particular $A \oplus B$ is c.e., Q-equivalent to A and not m-reducible to C.   ⊣

One might ask whether Theorem 4.1 would hold completely with Turing reducibility replaced by Q-reducibility. The answer is "no" as the next theorem shows. The witness is the Q-degree of a c.e. set A which is neither low$_2$ nor Turing-complete and whose complement is semi-low. Recall that $\overline{A}$ is semi-low iff $\{e: W_e \not\subseteq A\} \leq_T K$; see Soare [7, Exercise IV.4.6].

THEOREM 4.3. *There is a c.e. set A such that the class of c.e. sets in its Q-degree is bounded but A is non-low$_2$.*

PROOF. By Soare [7, Exercise IV.4.11], every c.e. Turing degree contains a c.e. set with semi-low complement. So let A be any c.e. set such that $\overline{A}$ is semi-low and A is neither Turing complete nor low$_2$. Since any Q-complete c.e. set is also Turing complete, the Q-degree of A does not contain K.

Since the complement of A is semi-low, there is a function $f \leq_T K$ such that for all e, $f(e)$ is 1 if $W_e \subseteq A$, and $f(e)$ is 0 otherwise. Let g be a computable function such that for all e' and x,

$$W_{g(e',x)} = \begin{cases} W_{\varphi_{e'}(x)} & \text{if } \varphi_{e'}(x) \text{ is defined;} \\ \emptyset & \text{otherwise.} \end{cases}$$

Using this convention, a set W_e is Q-reducible to A iff $(\exists e')(\forall x)[x \in W_e \Leftrightarrow W_{g(e',x)} \subseteq A]$. Using the definition of f, one obtains the following formula:

$$W_e \leq_Q A \Leftrightarrow (\exists e')(\forall x)[W_e(x) = f(g(e', x))].$$

Since $W_e(x)$ and $f(g(e', x))$ can be evaluated using the oracle K, the whole formula can be translated into a Σ_3 formula. Therefore, the index set $\{e: W_e \leq_Q A\}$ is a nontrivial Σ_3 set. By Proposition 3.2, the class of all c.e. sets Q-reducible to A is bounded. Thus the class of all c.e. sets in the Q-degree of A is bounded.   ⊣

Theorem 4.3 shows that the c.e. sets for which the class of the c.e. sets in their Q-degree is bounded are not characterized by the low$_2$ sets. Since there are c.e. low$_2$ sets which do not have a semi-low complement (see Soare [7, Exercise IV.4.10]), the condition that the complement is semi-low is also not a characterization for those sets for which the c.e. sets in the Q-degree are bounded.

Acknowledgments. We thank Carl G. Jockusch, Jr. and the anonymous referee for many valuable discussions and suggestions on the material of this paper.

REFERENCES

[1] KLAUS AMBOS-SPIES, ANDRÉ NIES, and RICHARD A.SHORE, *The theory of the recursively enumerable weak truth-table degrees is undecidable*, **The Journal of Symbolic Logic**, vol. 57 (1992), pp. 864–874.

[2] CARL G. JOCKUSCH, JR., *Semirecursive sets and positive reducibility*, **Transactions of the American Mathematical Society**, vol. 131 (1968), pp. 420–436.

[3] MARTIN KUMMER and FRANK STEPHAN, *Recursion theoretic properties of frequency computation and bounded queries*, **Information and Computation**, vol. 120 (1995), pp. 59–77.

[4] ALISTAIR H. LACHLAN, *A note on universal sets*, **The Journal of Symbolic Logic**, vol. 31 (1966), pp. 573–574.

[5] S. S. MARCHENKOV, *One class of partial sets*, **Matematiceskie Zametki**, vol. 20 (1976), pp. 473–478 (Russian), English translation available.

[6] PIERGIORGIO ODIFREDDI, *Classical recursion theory*, North-Holland, Amsterdam, 1989.

[7] ROBERT I. SOARE, *Recursively enumerable sets and degrees*, Perspectives in Mathematical Logic, Omega Series, Springer-Verlag, Berlin, 1987.

DEPARTMENT OF MATHEMATICS AND COMPUTER SCIENCE
  WHEATON COLLEGE
    WHEATON, IL 60187, USA
E-mail: K.Joyce.Ho@wheaton.edu

MATHEMATISCHES INSTITUT
  UNIVERSITÄT HEIDELBERG
    69120 HEIDELBERG, GERMANY
E-mail: fstephan@math.uni-heidelberg.de

BOREL IRREDUCIBILITY BETWEEN TWO LARGE FAMILIES
OF BOREL EQUIVALENCE RELATIONS

VLADIMIR KANOVEI[†] AND MICHAEL REEKEN

Abstract. We prove that if $\mathcal{I}$ is a Borel ideal, which includes a dense summable ideal, and E is a Borel equivalence relation that can be obtained from Fin using certain elementary operations like the Fubini product and countable power relation, then $\mathsf{E}_{\mathcal{I}}$ is not Borel reducible to E. The ideals $\mathcal{I}$ in the scope of this theorem include, for instance, all Borel P-ideals except for $\mathcal{I}_3 = 0 \times$ Fin and (trivial variations of) Fin.

§1. The result. Let $\mathfrak{E}$ be the smallest class of Borel equivalence relations (or ERs, for brevity) E on Polish spaces, containing the equality relations on finite and countable sets and closed under the the following transformations:

(1) *countable union* (if it results in a ER) and *countable intersection* of ERs on one and the same space;

(2) *countable disjoint union* $\mathsf{E} = \bigvee_k \mathsf{E}_k$ of ERs E_k on pairwise disjoint spaces $\mathbb{S}_k$, that is, a ER on the union $\bigcup_k \mathbb{S}_k$ defined by: $x\, \mathsf{E}\, y$ iff x, y belong to the same $\mathbb{S}_k$ and $x\, \mathsf{E}_k\, y$;

(3) the *Fubini product* $\mathsf{Fin} \bigotimes_k \mathsf{E}_k$ of ERs E_k on spaces $\mathbb{S}_k$, over the ideal Fin of all finite subsets of $\mathbb{N}$, that is, the ER on the product space $\prod_k \mathbb{S}_k$ defined as follows: $x\, \mathsf{E}\, y$ iff $x(k)\, \mathsf{E}_k\, y(k)$ for all but finite k;

(4) *product* $\mathsf{E} = \prod_k \mathsf{E}_k$ of ERs E_k on spaces $\mathbb{S}_k$, that is, the ER on the product space $\prod_k \mathbb{S}_k$ defined by: $x\, \mathsf{E}\, y$ iff $x(k)\, \mathsf{E}_k\, y(k)$ for all k;

(5) *countable power* ER E^∞ of a ER E on a space $\mathbb{S}$, that is, a ER on $\mathbb{S}^{\mathbb{N}}$ defined as follows: $x\, \mathsf{E}^\infty\, y$ iff $\{[x(k)]_{\mathsf{E}} : k \in \mathbb{N}\} = \{[y(k)]_{\mathsf{E}} : k \in \mathbb{N}\}$, where $[z]_{\mathsf{E}}$ is the E-class of $z \in \mathbb{S}$: thus, it is required that for any k there is l with $x(k)\, \mathsf{E}\, y(l)$ and for any l there is k with $x(k)\, \mathsf{E}\, y(l)$.

2000 *Mathematics Subject Classification.* 03E15.

Key words and phrases. Borel reducibility, P-ideals, turbulence.

[†]Supported by DFG grant 17/108/99, NSF grant DMS 96-19880, and visits to University of Wuppertal and Caltech.

Logic Colloquium '99
Edited by J. van Eijck, V. van Oostrom, and A. Visser
Lecture Notes in Logic, 17

THEOREM 1 (The main result). *If $\mathcal{Z}$ is a nontrivial[1] Borel P-ideal on $\mathbb{N}$ and E is a ER in $\mathfrak{E}$ then $\mathsf{E}_{\mathcal{Z}}$ is not Baire-measurable reducible to E unless $\mathcal{Z}$ is isomorphic[2] to* Fin, *a trivial variation of* Fin, *or $\mathcal{I}_3 = 0 \times$* Fin.

Special notions involved in this theorem (all of them known in this direction of descriptive set theory, see, *e.g.*, [1, 4]) are explained in the next section.

§2. Notation and comments.

Recall that any ideal $\mathcal{I}$ on a set A induces an ER $\mathsf{E}_{\mathcal{I}}$ on 2^A: $x \, \mathsf{E}_{\mathcal{I}} \, y$ iff the set $x \, \Delta \, y = \{i \in A : x(i) \neq y(i)\}$ belongs to $\mathcal{I}$.

- An equivalence relation E on $\mathbb{S}$ is *Borel* or *Baire measurable* (BM, for brevity) *reducible* to a ER E' on $\mathbb{S}'$ if there is a resp. Borel or BM *reduction* E to E', that is, a resp. Borel or BM map $F : \mathbb{S} \to \mathbb{S}'$ such that we have $x \, \mathsf{E} \, y \iff F(x) \, \mathsf{E}' \, F(y)$ for all $x, y \in \mathbb{S}$.
- *P-ideals* are those ideals $\mathcal{Z}$ which satisfy the requirement that for any sequence of sets $x_0, x_1, x_2, \ldots \in \mathcal{Z}$ there is $x \in \mathcal{Z}$ with $x_n \subseteq^* x$ for all n, where $y \subseteq^* x$ means that $y \setminus x$ is finite.

Borel P-ideals (in fact all of them belong to Borel class $\boldsymbol{\Pi}_3^0$) admit different characterizations (see, *e.g.*, the proof of Proposition 2 below) and form an important and widely studied class, which includes, for instance,

(i) the ideal Fin of all finite subsets of $\mathbb{N}$,

(ii) the ideal $\mathcal{I}_3 = 0 \times$ Fin of all sets $x \subseteq \mathbb{N} \times \mathbb{N}$ such that every cross-section $(x)_n = \{k : \langle n, k \rangle \in x\}$ is finite,

(iii) *trivial variations of* Fin, *i.e.*, by Kechris [5], ideals of the form $\mathcal{I} = \{x \subseteq \mathbb{N} : x \cap W \in$ Fin$\}$, where $W \subseteq \mathbb{N}$ is infinite and coinfinite (all of them are isomorphic to each other).

Borel P-ideals also include summable ideals, density ideals, and many more (see Farah [1], Solecki [7, 8]).

The class $\mathfrak{E}$ of ERs contains, for instance, the equality $\mathsf{D}(2^{\mathbb{N}})$ on $2^{\mathbb{N}}$,[3] it also contains the ERs E_{Fin} and $\mathsf{E}_{0 \times \text{Fin}}$ (usually denoted by resp. E_0 and E_3), associated with the ideals Fin and $\mathcal{I}_3 = 0 \times$ Fin,[4] as well as those associated with trivial variations of Fin. Thus the exclusion of Fin, $0 \times$ Fin, and trivial variations of Fin in Theorem 1 is necessary and fully motivated.

Furthermore $\mathfrak{E}$ contains all ERs associated with the *iterated Frechet ideals*, *i.e.*, the smallest family of ERs containing equality relations on finite and countable sets and closed under (3). Class $\mathfrak{E}$ also contains all ERs associated with the *indecomposable ideals* (Farah [1]) $\mathcal{I}_\xi = \{x \subseteq \omega^\xi : \text{otp} \, x < \omega^\xi\}$, $\xi < \omega_1$ (otp x is the order type of a set $x \subseteq$ Ord), yet in this case it takes

---

[1] An ideal $\mathcal{I} \subseteq \mathcal{P}(\mathbb{N})$ is <u>nontrivial</u> if it is not equal to $\mathcal{P}(X)$ for some $X \subseteq \mathbb{N}$.

[2] By <u>isomorphism</u> we mean isomorphism via a bijection between the underlying sets.

[3] To see that $\mathsf{D}(2^{\mathbb{N}})$ belongs to $\mathfrak{E}$ let each E_k be the equality on a 2-element set in (4).

[4] To see that E_{Fin} belongs to $\mathfrak{E}$ let each E_k be the equality on a 2-element set in (3). To see that $\mathsf{E}_{0 \times \text{Fin}}$ belongs to $\mathfrak{E}$ take each E_k to be E_{Fin} in (4).

some effort to find a recursive construction of $\mathcal{I}_\xi$ in terms of the transformations (1)–(4). Class $\mathfrak{E}$ also contains all ERs T_α of Friedman and Stanley [3, 2], obtained from the equality on $\mathbb{N}$ via operations (2) and (5): they can be seen as all (modulo Borel reduction) ERs which, in some broad sense, admit classification by countable structures.

Two earlier related results must be mentioned. Friedman and Stanley announced in [3] (Friedman gives a full proof in [2]) that $\mathsf{E}_{\mathcal{Z}_0}$, the ER associated with the density-0 ideal $\mathcal{Z}_0$, is not BM reducible to any ER of the form T_α (see above). Kechris [5] proved (as a part of the proof of another result on ideals) the particular case $\mathsf{E} = \mathsf{E}_3 = \mathsf{E}_{0 \times \mathrm{Fin}}$ of Theorem 1.

The arguments of Kechris and (implicitly) Friedman were based on ideas of Hjorth's *turbulence theory*.[5] So is our proof, its scheme is to show that a certain stronger form of irreducibility (called: the generic ergodicity) is preserved under the transformations of equivalence relations (1)–(5). Our proof is self-contained and rather elementary, in particular, it makes no use of model theory or facts related to topological group theory, yet it makes use of forcing.

§3. "Special" ideals. Before the main part of the proof of Theorem 1 begins, we are going to simplify the task. This section reduces the problem to ideals which include some kind of summable ideals. Recall that any sequence of reals $r_n \geq 0$ produces an ideal

$$\mathcal{S}_{\{r_n\}} = \left\{ x \subseteq \mathbb{N} : \sum_{n \in x} r_n < +\infty \right\}.$$

Ideals of this kind are called *summable*, and $\mathcal{S}_{\{r_n\}}$ is *nontrivial dense* if $\{r_n\} \to 0$ and $\sum_n r_n = +\infty$ (then $\mathcal{S}_{\{r_n\}} \neq \mathcal{P}(\mathbb{N})$). Say that an ideal $\mathcal{I}$ on $\mathbb{N}$ is *"special"* if there is a nontrivial dense summable ideal $\mathcal{S}_{\{r_n\}}$ with $\mathcal{S}_{\{r_n\}} \subseteq \mathcal{I} \subsetneq \mathcal{P}(\mathbb{N})$.

PROPOSITION 2 (Essentially, Kechris [5]). *Let $\mathcal{Z}$ be a nontrivial Borel P-ideal on $\mathbb{N}$, __not__ isomorphic to one of* Fin, $\mathcal{I}_3 = 0 \times$ Fin, *or a trivial variation of* Fin. *Then there is a set $W \notin \mathcal{Z}$ such that $\mathcal{Z} \restriction W = \{x \cap W : x \in \mathcal{Z}\}$ is isomorphic (via a bijection W onto $\mathbb{N}$) to a "special" Borel ideal.*

PROOF. Recall that a lower semicontinuous (l.s.c.) submeasure on $\mathbb{N}$ is any map $\varphi : \mathcal{P}(\mathbb{N}) \to [0, +\infty]$ satisfying $\varphi(x) \leq \varphi(x \cup y) \leq \varphi(x) + \varphi(y)$ for all $x, y \in \mathcal{P}(\mathbb{N})$, $\varphi(\emptyset) = 0$, and $\varphi(x) = \sup_{n \in \mathbb{N}} \varphi(x \cap [0, n))$ for all $x \in \mathcal{P}(\mathbb{N})$. By Solecki [7, 8], as $\mathcal{Z}$ is a Borel P-ideal, there is an l.s.c. submeasure $\varphi : \mathcal{P}(\mathbb{N}) \to [0, +\infty]$ such that $\mathcal{Z} = \{x \in \mathcal{P}(\mathbb{N}) : \varphi_\infty(x) = 0\}$, where $\varphi_\infty(x) = \lim_{n \to \infty} \varphi(x \cap [n, \infty))$.

---

[5] For instance, Kechris observed, that any nontrivial Borel P-ideal $\mathcal{Z}$, with the same exceptions, induces a *turbulent* Δ-action on $\mathcal{P}(\mathbb{N})$, while $\mathsf{E}_{0 \times \mathrm{Fin}}$ is induced by a continuous action of S_∞, the group of all permutations of $\mathbb{N}$, which is enough, by the first turbulence theorem, for $\mathsf{E}_{\mathcal{Z}}$ to be Borel irreducible to $\mathsf{E}_{0 \times \mathrm{Fin}}$.

Put $U_n = \{k : \varphi(\{k\}) \leq \frac{1}{n}\}$, separately $U_0 = \mathbb{N}$, thus, $U_{n+1} \subseteq U_n$ for all n. We claim that $\inf_{m \in \mathbb{N}} \varphi(U_m) > 0$. Suppose otherwise. Then a set $x \subseteq \mathbb{N}$ belongs to $\mathcal{Z}$ iff $x \setminus U_n$ is finite for any n. If the set $N = \{n : U_n \setminus U_{n+1}$ is infinite$\}$ is empty then easily $\mathcal{Z} = \mathcal{P}(\mathbb{N})$. If N is finite and nonempty then $\mathcal{Z}$ is isomorphic to either Fin (if eventually $U_n = \emptyset$) or a trivial variation of Fin (if U_n is nonempty for all n). If finally N is infinite then easily $\mathcal{Z}$ is isomorphic to $0 \times$ Fin (for instance, if all sets $D_n = U_n \setminus U_{n+1}$ are infinite then $x \in \mathcal{Z}$ iff $x \cap D_n$ is finite for all n). Thus we always have a contradiction to the assumptions of the Proposition.

Thus there is $\varepsilon > 0$ such that $\varphi(U_m) > \varepsilon$ for all m. As φ is l.s.c., we can define an increasing sequence of numbers $n_1 < n_2 < n_3 < \ldots$ and for any l a finite set $w_l \subseteq U_{n_l} \setminus U_{n_{l+1}}$ with $\varphi(w_l) > \varepsilon$. Then $W = \bigcup_l w_l \notin \mathcal{Z}$ and obviously $\{r_k\}_{k \in W} \to 0$ and $\sum_{k \in W} r_k \geq \sum_l \varphi(w_l) = \infty$, where $r_k = \varphi(\{k\})$. Finally, if a set $x \subseteq W$ satisfies $\sum_{k \in x} r_k < +\infty$ then $x \in \mathcal{Z}$: indeed, we have $\varphi_\infty(x) \leq \sum_{k \in x} r_k$ because φ is a l.s.c. submeasure.   ⊣

It follows (indeed, $\mathsf{E}_{\mathcal{Z} \restriction W} \leq_\mathrm{B} \mathsf{E}_{\mathcal{Z}}$) that the next theorem implies Theorem 1:

THEOREM 3. *If $\mathcal{I}$ is a "special" Borel ideal then $\mathsf{E}_{\mathcal{I}}$ is not BM reducible to any* ER E *in* $\mathfrak{E}$.

§4. **Ergodicity and dense summable ideals.** The next preliminary step is to further reduce the task to summable ideals. This involves the following special form of irreducibility.

DEFINITION 4. Let E, F be ERs on Polish spaces, resp., $\mathbb{X}$, $\mathbb{Y}$. A map $\vartheta : \mathbb{X} \to \mathbb{Y}$ is

- *a.e.* E, F-*invariant* if there is a co-meager set $D \subseteq \mathbb{X}$ such that $x \mathrel{\mathsf{E}} x' \implies \vartheta(x) \mathrel{\mathsf{F}} \vartheta(x')$ for all $x, x' \in D$, and E, F-*invariant* if this holds for $D = \mathbb{X}$.
- *a.e.* F-*constant* if there is a co-meager set $D \subseteq \mathbb{S}$ such that $\vartheta(x) \mathrel{\mathsf{F}} \vartheta(x')$ holds for all $x, x' \in D$.

Finally, following Kechris [6, 12.1] and Hjorth [4, 3.6], we say that E is *generically* F-*ergodic* if every BM E, F-invariant map ϑ is a.e. F-constant.   ⊣

REMARK 5. To see that E is generically F-ergodic, it suffices to demonstrate that every <u>Borel</u> and <u>a.e.</u> E, F-invariant map ϑ is a.e. F-constant: indeed, any BM map is continuous, hence, Borel, on a comeager set.   ⊣

LEMMA 6. *Suppose that $\mathcal{I}$ is a nontrivial Borel ideal on $\mathbb{N}$ and E is a Borel* ER. *If $\mathsf{E}_{\mathcal{I}}$ is generically E-ergodic then $\mathsf{E}_{\mathcal{I}}$ is not* BM *reducible to* E.

PROOF. Assume, towards the contrary, that $\vartheta : \mathcal{P}(\mathbb{N}) \to \mathbb{Y}$ (where $\mathbb{Y}$ is the domain of E) is a BM reduction of $\mathsf{E}_{\mathcal{I}}$ to E. There is a co-meager set $D \subseteq \mathcal{P}(\mathbb{N})$ such that $\vartheta \restriction D$ is Borel (even continuous). Let $\vartheta' : \mathcal{P}(\mathbb{N}) \to \mathbb{Y}$ coincide with ϑ on D and be constant on $\mathcal{P}(\mathbb{N}) \setminus D$. Then ϑ' is a Borel a.e. $\mathsf{E}_{\mathcal{I}}$, E-invariant map, therefore, it is a.e. E-constant, so that, by the ergodicity, we have a comeager $E_\mathcal{I}$-equivalence class in $\mathcal{P}(\mathbb{N})$. It follows that $\mathcal{I}$ itself is comeager in $\mathcal{P}(\mathbb{N})$, which easily implies that $\mathcal{I} = \mathcal{P}(\mathbb{N})$, a contradiction to the nontriviality of $\mathcal{I}$.   ⊣

The remainder of the note contains the proof of the following theorem:

THEOREM 7. *For any nontrivial dense summable ideal* $\mathcal{S}_{\{r_n\}}$ *the equivalence relation* $E_{\{r_n\}} = E_{\mathcal{S}_{\{r_n\}}}$ *is generically* E*-ergodic for any* ER E *in* $\mathfrak{E}$.

This implies Theorem 3, hence, Theorem 1. Indeed, first, if $\mathcal{I}$ is a "special" Borel ideal then, by Proposition 2 there is a nontrivial dense summable ideal $\mathcal{S}_{\{r_n\}} \subseteq \mathcal{I}$. The latter is generically E-ergodic for any ER E in $\mathfrak{E}$ by Theorem 7, hence, $E_\mathcal{I}$ itself is generically E-ergodic because now any $E_\mathcal{I}$, E-invariant map is $E_{\{r_n\}}$, E-invariant. It follows that $E_\mathcal{I}$ is BM irreducible to E by Lemma 6, as required.

§5. Preliminaries to the proof.

In the proof of Theorem 7 we shall use the following notation and keep the following agreements.

- For the course of the proof of Theorem 7, we fix a sequence of nonnegative reals $\{r_n\} \to 0$ with $\sum_{n\in\mathbb{N}} r_n = +\infty$.
- $2^\mathbb{N}$ is the Cantor space of all infinite dyadic sequences, with the ordinary product topology.
- $2^{<\omega}$ is the set of all finite dyadic sequences.
- For $u \in 2^{<\omega}$ we define $I_u = \{a \in 2^\mathbb{N} : u \subset a\}$, a basic clopen set in $2^\mathbb{N}$.
- If X is a nonempty open set in a Polish space then the phrase "$P(x)$ holds for a.a. $x \in X$" will mean that $\{x \in X : P(x)\}$ is comeager in X.
- We shall systematically identify sets $X \subseteq \mathbb{N}$ with their characteristic functions, unless it becomes ambiguous. In particular, for $a, b \in 2^\mathbb{N}$ we define $a \, \Delta \, b = \{n : a(n) \neq b(n)\}$ (as a set) and $(a \, \Delta \, b)(n) = |a(n) - b(n)|$ (as an infinite dyadic sequence).
- $E_{\{r_n\}}$ is $E_{\mathcal{S}_{\{r_n\}}}$, thus, for $x, y \in 2^\mathbb{N}$, $x \, E_{\{r_n\}} \, y$ iff $\sum_{x(n)\neq y(n)} r_n < +\infty$.
- In accordance with Remark 5, we shall consider only Borel maps ϑ.

DEFINITION 8. A ER E is $\{r_n\}$-*irreducing* if $E_{\{r_n\}}$ is generically E-ergodic.   ⊣

The method of the proof of Theorem 7 will be to show, by induction on the construction of ERs in $\mathfrak{E}$, that all ERs in $\mathfrak{E}$ are $\{r_n\}$-irreducing.

§6. Base of induction.

To begin with, we prove

LEMMA 9. *If* $\mathbb{S}$ *is a Polish space then* $D_\mathbb{S}$, *the equality on* $\mathbb{S}$, *is* $\{r_n\}$-*irreducing.*

PROOF. First of all, as all (uncountable) Polish spaces are Borel isomorphic, we may assume that $\mathbb{S} = 2^\mathbb{N}$. Furthermore, as any invariant $\vartheta : 2^\mathbb{N} \to 2^\mathbb{N}$ splits into a family of invariant "coordinate" maps $\vartheta_k : 2^\mathbb{N} \to \{0, 1\}$, we can suppose that $\mathbb{S} = \{0, 1\}$, a two-element discrete set. Assume, towards the contrary, that an a.e. $E_{\{r_n\}}$, $D_\mathbb{S}$-invariant Borel map $\vartheta : 2^\mathbb{N} \to \{0, 1\}$ is <u>not</u> a.e. constant.

By definition there is a dense G_δ set $D \subseteq \mathcal{P}(\mathbb{N})$ with $a\; \mathsf{E}_{\{r_n\}}\; b \Longrightarrow \vartheta(a) = \vartheta(b)$ for all $a, b \in D$.

According to the contrary assumption, there exist sequences $u, v \in 2^{<\omega}$ such that $\vartheta(a) = 0$ and $\vartheta(b) = 1$ for a.a. (in the sense of category) $a \in I_u$ and a.a. $b \in I_v$. We can assume that $\vartheta(a) = 0$ and $\vartheta(b) = 1$ actually for all $a \in D \cap I_u$ and $b \in D \cap I_v$, and that $u, v \in 2^m$ for one and the same m. (Clearly $u \neq v$.)

Let $D = \bigcap_n D_n$, where each D_n is open dense. Put $r = \sum_{i < m, u(i) \neq v(i)} r_i$. Using the assumption $\{r_n\} \to 0$, we can easily define an increasing sequence $m = m_0 < m_1 < m_2 < \ldots$ of natural numbers, and $u = u_0 \subset u_1 \subset u_2 \subset \ldots$ and $v = v_0 \subset v_1 \subset v_2 \subset \ldots$ of tuples $u_n, v_n \in 2^{m_i}$ with $\sum_{i < m_n, u_n(i) \neq v_n(i)} r_i < r+1$ and $I_{u_n} \cup I_{v_n} \subseteq D_n$ for all n. Then $a = \bigcup_n u_n \in D \cap I_u$, $b = \bigcup_n v_n \in D \cap I_v$, and $a\; \mathsf{E}_{\{r_n\}}\; b$, but $\vartheta(a) = x \neq y = \vartheta(b)$, which is a contradiction. $\dashv$

It remains to demonstrate that the property of $\{r_n\}$-irreducibility is preserved under the transformations of ERs which produce the class $\mathfrak{E}$.

§7. Inductive step of the Fubini product.
In this section, we show that the Fubini product preserves $\{r_n\}$-irreducibility.

LEMMA 10. *Suppose that E_k, $k \in \mathbb{N}$, are Borel $\{r_n\}$-irreducing ERs on Polish spaces $\mathbb{S}_k$. Then $\mathsf{E} = \mathrm{Fin} \bigotimes_{k \in \mathbb{N}} \mathsf{E}_k$ is $\{r_n\}$-irreducing as well.*

PROOF. Let $\mathbb{S} = \prod_k \mathbb{S}_k$, so that E is a ER on $\mathbb{S}$. Let $\vartheta \colon 2^{\mathbb{N}} \to \mathbb{S}$ be a Borel (according to Remark 5) function. It splits in the sequence of Borel functions

$$\vartheta_k(x) = \vartheta(x)(k) \colon 2^{\mathbb{N}} \to \mathbb{S}_k.$$

Suppose that ϑ is $\mathsf{E}_{\{r_n\}}$, E-invariant on a dense G_δ set $D \subseteq 2^{\mathbb{N}}$, so that

$$a\; \mathsf{E}_{\{r_n\}}\; b \Longrightarrow \exists k_0 \forall k \geq k_0 (\vartheta_k(a)\; \mathsf{E}_k\; \vartheta_k(b))$$

for all $a, b \in D$. Our plan is to show that almost all ϑ_k are a.e. $\mathsf{E}_{\{r_n\}}, \mathsf{E}_k$-invariant.

In that we'll make use of two topologies. The first one is the ordinary product topology on $2^{\mathbb{N}}$. The other one is the topology on the set $\mathbb{S}_{\{r_n\}} \subseteq 2^{\mathbb{N}}$, generated by the metric $d_{\{r_n\}}(a, b) = \varphi_{\{r_n\}}(a \vartriangle b)$ on $\mathbb{S}_{\{r_n\}}$, where

$$\varphi_{\{r_n\}}(X) = \sum_{n \in X} r_n \text{ for } X \subseteq \mathbb{N}, \text{ so that } \mathbb{S}_{\{r_n\}} = \{X \colon \varphi_{\{r_n\}}(X) < +\infty\};$$

It is easy to verify (even in a much more general case, see [7]) that $d_{\{r_n\}} \restriction \mathbb{S}_{\{r_n\}}$ is a Polish (i.e., complete separable) metric on $\mathbb{S}_{\{r_n\}}$. The $d_{\{r_n\}}$-topology is stronger than the product topology of $2^{\mathbb{N}}$ on $\mathbb{S}_{\{r_n\}}$, yet it yields the same Borel subsets of $\mathbb{S}_{\{r_n\}}$ as the product topology. Sets of the form

$$\mathcal{U}_\varepsilon(t) = \{z \in \mathbb{S}_{\{r_n\}} \colon d_{\{r_n\}}(z, t^*) < \varepsilon\}, \quad t \in 2^{<\omega},$$

where $t^* \in 2^{\mathbb{N}}$ denotes the extension of $t \in 2^{<\omega}$ (a finite sequence) by infinitely many zeros, and ε is a positive rational, provide a base of the $d_{\{r_n\}}$-topology on $\mathcal{S}_{\{r_n\}}$, and the countable set $\{t^* : t \in 2^{<\omega}\}$ is $d_{\{r_n\}}$-dense in $\mathcal{S}_{\{r_n\}}$.

Below, let $\mathbb{P} = 2^{<\omega}$ be the ordinary Cohen forcing for $2^{\mathbb{N}}$, and let $\mathbb{P}_{\{r_n\}}$ be the Cohen forcing for $\langle \mathcal{S}_{\{r_n\}}; d_{\{r_n\}} \rangle$, which consists of all sets of the form $\mathcal{U}_\varepsilon(t)$, where $t \in 2^{<\omega}$ and $\varepsilon > 0$ is rational. (Smaller sets are stronger conditions.) Let us fix a countable transitive model $\mathfrak{M}$ of a big enough fragment of **ZFC**,[6] which contains all relevant objects or their codes, in particular, the sequence $\{r_n\}$ and a code of the Borel map ϑ.

CLAIM 11. *Suppose that* $\langle a, z \rangle \in 2^{\mathbb{N}} \times \mathcal{S}_{\{r_n\}}$ *is* $\mathbb{P} \times \mathbb{P}_{\{r_n\}}$*-generic over* $\mathfrak{M}$. *Then* $b = a \, \Delta \, z$ *is* $\mathbb{P}$*-generic over* $\mathfrak{M}$.

PROOF OF THE CLAIM. Actually, b is $\mathbb{P}$-generic even over $\mathfrak{M}[z]$: indeed, a is such by the product forcing lemma, and, for any fixed z, the map $a \longmapsto a \, \Delta \, z$ is a homeomorphism.   ⊣

Fix $u \in 2^{<\omega}$. Then by the invariance of ϑ and Claim 11 there is another sequence $v \in 2^{<\omega}$ with $u \subset v$, a number k_0, and a non-empty $d_{\{r_n\}}$-nbhd $\mathcal{U}_\varepsilon(t)$ in $\mathcal{S}_{\{r_n\}}$ (a condition in $\mathbb{P}_{\{r_n\}}$), where $\varepsilon > 0$ and $t \in 2^{<\omega}$, such that $\vartheta_k(a) \, \mathsf{E}_k$ $\vartheta_k(a \, \Delta \, z)$ holds for any $\mathbb{P} \times \mathbb{P}_{\{r_n\}}$-generic over $\mathfrak{M}$ pair $\langle a, z \rangle$ of $a \in I_v$ and $z \in \mathcal{U}_\varepsilon(t)$ and any $k \geq k_0$. We can assume that the length $\mathrm{lh}\, v$ is big enough for $i \geq \mathrm{lh}\, v \implies r_i < \varepsilon$.

CLAIM 12. *If* $a, b \in I_v$ *are* $\mathbb{P}$*-generic over* $\mathfrak{M}$ *and* $a \, \mathsf{E}_{\{r_n\}} \, b$ *then* $\vartheta_k(a) \, \mathsf{E}_k \, \vartheta_k(b)$ *holds for all* $k \geq k_0$.

PROOF OF THE CLAIM. First consider the case when $\varphi_{\{r_n\}}(a \, \Delta \, b) < \varepsilon$. Take any $z \in Z = \mathcal{U}_\varepsilon(t)$ with $\varphi_{\{r_n\}}(z \, \Delta \, t) < \varepsilon - \varphi_{\{r_n\}}(a \, \Delta \, b)$, $\mathbb{P}_{\{r_n\}}$-generic over $\mathfrak{M}[a, b]$.[7] (This is possible as $r_n \to 0$.) Then z is $\mathbb{P}_{\{r_n\}}$-generic over $\mathfrak{M}[a]$, hence, $\langle a, z \rangle$ is $\mathbb{P} \times \mathbb{P}_{\{r_n\}}$-generic over $\mathfrak{M}$ by the product forcing lemma, thus, $\vartheta_k(a) \, \mathsf{E}_k \, \vartheta_k(a \, \Delta \, z)$. Moreover, $z' = z \, \Delta \, (a \, \Delta \, b)$ still belongs to Z and is $\mathbb{P}_{\{r_n\}}$-generic over $\mathfrak{M}[a, b]$, so that $\vartheta_k(b) \, \mathsf{E}_k \, \vartheta_k(b \, \Delta \, z')$ by the same argument. Yet we have $a \, \Delta \, z = b \, \Delta \, z'$.

Now consider the general case. By definition $X = a \, \Delta \, b$ satisfies $\sum_{n \in X} r_n = \varphi_{\{r_n\}}(X) < +\infty$, moreover, $\min X \geq \mathrm{lh}\, v$, hence, by the choice of v, all r_j with $j \in X$ satisfy $r_j < \varepsilon$. In this case X has the form $X = \{j_1, \dots, j_n\} \cup X'$, where $\varphi_{\{r_n\}}(X') < \varepsilon$, and $r_{j_m} < \varepsilon$ for all m. Define $a_m = a \, \Delta \, \{j_1, \dots, j_m\}$ for $m = 1, \dots, n$. Then $a = a_0, a_1, a_2, \dots, a_m, a_{m+1} = b$ is a chain of $\mathbb{P}$-generic,

---

[6] For instance, the first one million of **ZFC** axioms plus the Replacement for Σ_{100} formulas.

[7] By $\mathfrak{M}[a, b]$, we understand a countable transitive model of the same fragment of **ZFC** as mentioned in Footnote 6, which contains a, b, and all sets in $\mathfrak{M}$. This model may contain more ordinals than $\mathfrak{M}$ because the pair $\langle a, b \rangle$ of two generic elements is not necessarily generic. On the contrary, the models $\mathfrak{M}[a]$ and $\mathfrak{M}[b]$ are ordinary generic extensions of $\mathfrak{M}$, containing the same ordinals as $\mathfrak{M}$.

over $\mathfrak{M}$, elements of I_v with $d_{\{r_n\}}(a_m, a_{m+1}) < \varepsilon$ for all m, so that we can apply the particular case considered above. ⊣

To summarize, we have shown that for any $u \in 2^{<\omega}$ there exist k_0 and $v \in 2^{<\omega}$ with $u \subset v$ such that $\vartheta_k(a)$ is $\mathsf{E}_{\{r_n\}}, \mathsf{E}_k$-invariant for all $k \geq k_0$ and all sufficiently $\mathbb{P}$-generic $a \in I_v$, so that ϑ_k is a.e. $\mathsf{E}_{\{r_n\}}, \mathsf{E}_k$-invariant on I_v. Therefore, as all E_k are $\{r_n\}$-irreducing, the map ϑ_k is a.e. E_k-constant on I_v, for each $k \geq k_0$. (Indeed, we can trivially extend $\vartheta_k \restriction I_v$ on the whole domain $2^{\mathbb{N}}$ so that the invariance is preserved.) Thus ϑ is a.e. E-constant on I_v as well.

In other words, for each $u \in 2^{<\omega}$ there is $v \in 2^{<\omega}$ with $u \subset v$ such that ϑ is a.e. E-constant on I_v. To complete the proof of Lemma 10, it remains to demonstrate that these E-constants cannot be different. Thus assume that $s, t \in 2^{<\omega}$, $\mathrm{lh}\, s = \mathrm{lh}\, t = m$, $x, y \in \mathbb{S}$, and $\vartheta(a)\, \mathsf{E}\, x$ for a.a. $a \in I_s$ while $\vartheta(b)\, \mathsf{E}\, y$ for a.a. $b \in I_t$. We have to show that $x\, \mathsf{E}\, y$. Indeed, the same construction as in the proof of Lemma 9 yields $a \in D \cap I_s$ and $b \in D \cap I_t$ with $a\, \mathsf{E}_{\{r_n\}}\, b$ such that $\vartheta(a)\, \mathsf{E}\, x$ and $\vartheta(b)\, \mathsf{E}\, y$. Then $\vartheta(a)\, \mathsf{E}\, \vartheta(b)$ by the invariance of ϑ, hence, $x\, \mathsf{E}\, y$, as required. ⊣

§8. Other inductive steps.
In this section, we show that all other operations over ERs, defined in §1, also preserve $\{r_n\}$-irreducibility.

LEMMA 13. *Suppose that* $\mathsf{E}_1, \mathsf{E}_2, \mathsf{E}_3, \ldots$ *are Borel* $\{r_n\}$-*irreducing ERs on a Polish space* $\mathbb{S}$, *and* $\mathsf{E} = \bigcup_k \mathsf{E}_k$ *is a ER. Then* E *also is* $\{r_n\}$-*irreducing.*

PROOF. Let $\vartheta \colon 2^{\mathbb{N}} \to \mathbb{S}$ be a Borel $\mathsf{E}_{\{r_n\}}, \mathsf{E}$-invariant map. For each $u \in 2^{<\omega}$, by the invariance of ϑ, there exist: $v \in 2^{<\omega}$ with $u \subset v$, a number k, and a non-empty $d_{\{r_n\}}$-nbhd $Z = \mathcal{U}_\varepsilon(t)$ in $\mathbb{S}_{\{r_n\}}$ (a condition in $\mathbb{P}_{\{r_n\}}$), where $\varepsilon > 0$ and $t \in 2^{<\omega}$, such that $\vartheta(a)\, \mathsf{E}_k\, \vartheta(a \bigtriangleup z)$ for any $\mathbb{P} \times \mathbb{P}_{\{r_n\}}$-generic, over $\mathfrak{M}$, pair $\langle a, z \rangle$ of $a \in I_v$ and $z \in Z$. We can assume that $\mathrm{lh}\, v$ is big enough for $i \geq \mathrm{lh}\, v \Longrightarrow r_i < \varepsilon$. Then, similarly to Claim 12, it is true that $\vartheta(a)\, \mathsf{E}_k\, \vartheta(b)$ for any pair of $\mathbb{P}$-generic, over $\mathfrak{M}$, elements $a, b \in I_v$. It follows, as in the proof of Lemma 10, that ϑ is a.e. E_k-constant on I_v, hence, a.e. E-constant on I_v as well. It remains to show that these E-constants are equal to each other, which is demonstrated as in the end of the proof of Lemma 10. ⊣

COROLLARY 14. *Let* $\mathsf{E}_1, \mathsf{E}_2, \mathsf{E}_3, \ldots$ *be Borel* $\{r_n\}$-*irreducing ERs on disjoint Polish spaces* $\mathbb{S}_1, \mathbb{S}_2, \mathbb{S}_3, \ldots$ *. Then* $\mathsf{E} = \bigvee_k \mathsf{E}_k$ *also is* $\{r_n\}$-*irreducing.*

PROOF. Apply Lemma 13 for the relations E'_k defined on $\mathbb{S} = \bigcup_k \mathbb{S}_k$ as follows: $x\, \mathsf{E}'_k\, y$ iff either $x = y$ or $x, y \in \mathbb{S}_k$ and $x\, \mathsf{E}_k\, y$. ⊣

LEMMA 15. *Let* $\mathsf{E}_1, \mathsf{E}_2, \mathsf{E}_3, \ldots$ *be Borel* $\{r_n\}$-*irreducing ERs on a Polish space* $\mathbb{S}$. *Then* $\mathsf{E} = \bigcap_k \mathsf{E}_k$ *also is* $\{r_n\}$-*irreducing.*

PROOF. Any $\mathsf{E}_{\{r_n\}}, \mathsf{E}$-invariant map is $\mathsf{E}_{\{r_n\}}, \mathsf{E}_k$-invariant for all k. ⊣

COROLLARY 16. *Let* $\mathsf{E}_1, \mathsf{E}_2, \mathsf{E}_3, \ldots$ *be Borel* $\{r_n\}$-*irreducing ERs on Polish spaces* $\mathbb{S}_1, \mathbb{S}_2, \mathbb{S}_3, \ldots$ *. Then* $\mathsf{E} = \prod_k \mathsf{E}_k$ *also is* $\{r_n\}$-*irreducing.*

PROOF. Apply Lemma 15 for the relations $x \, \mathsf{E}'_k \, y$ iff $x(k) \, \mathsf{E}_k \, y(k)$ on $\prod_k \mathbb{S}_k$.

$\dashv$

LEMMA 17. *If a Borel ER* E *on a Polish space* $\mathbb{S}$ *is* $\{r_n\}$-*irreducing then* E^∞ *is also* $\{r_n\}$-*irreducing.*

PROOF. Suppose that a Borel map $\vartheta \colon 2^{\mathbb{N}} \to \mathbb{S}^{\mathbb{N}}$ is a.e. $\mathsf{E}_{\{r_n\}}$, E^∞-invariant, that is, $a \, \mathsf{E}_{\{r_n\}} \, b \implies \vartheta(a) \, \mathsf{E}^\infty \, \vartheta(b)$ for all $a, b \in D$, where $D \subseteq 2^{\mathbb{N}}$ is a dense G_δ set. Let $\vartheta_k(a) = \vartheta(a)_k$. The invariance of ϑ can be reformulated as follows:

$$a \, \mathsf{E}_{\{r_n\}} \, b \implies \forall k \exists l (\vartheta_k(a) \, \mathsf{E} \, \vartheta_l(b)) \wedge \forall l \exists k (\vartheta_k(a) \, \mathsf{E} \, \vartheta_l(b))$$

for all $a, b \in D$. As in the proof of Lemma 10, for any k and any $u \in 2^{<\omega}$ there are: a number l, a sequence $v \in 2^{<\omega}$ with $u \subset v$, and a $d_{\{r_n\}}$-nbhd $Z = \mathcal{U}_\varepsilon(t)$ in $\mathbb{S}_{\{r_n\}}$ such that $\vartheta_k(a) = \vartheta_l(a \, \Delta \, z)$ for any $\mathbb{P} \times \mathbb{P}_{\{r_n\}}$-generic, over $\mathfrak{M}$, pair $\langle a, z \rangle$ of $a \in I_v$ and $z \in Z$. The same argument (Claim 12) shows that $\vartheta_k(a) \, \mathsf{F} \, \vartheta_l(b)$ whenever $a, b \in I_v$ are $\mathbb{P}$-generic over $\mathfrak{M}$.

Thus for any $u \in 2^{<\omega}$ there is $v \in 2^{<\omega}$ with $u \subset v$ such that $\vartheta_k(a)$ is $\mathsf{E}_{\{r_n\}}$, E-invariant for all generic, over $\mathfrak{M}$, elements $a \in I_v$, in other words, ϑ_k is a.e. $\mathsf{E}_{\{r_n\}}$, E-invariant on I_v. It follows, as above, that ϑ_k is a.e. E-constant on I_v. It follows that there is a dense G_δ set $D' \subseteq D$ and a countable set $Y = \{y_j \colon j \in \mathbb{N}\} \subseteq \mathbb{S}$ such that, for all k and $a \in D'$, we have $\vartheta_k(a) \, \mathsf{E} \, y_j$ for some j. Let $\xi_k(a)$ be the least such an index j, thus, $\xi_k(a)$ is defined for all $a \in D'$ and all k, and each $\xi_k \colon D' \to \mathbb{N}$ is a Borel map. Now, by the invariance of ϑ,

$$a \, \mathsf{E}_{\{r_n\}} \, b \implies \{\xi_k(a) \colon k \in \mathbb{N}\} = \{\xi_k(b) \colon k \in \mathbb{N}\}$$

for all $a, b \in D'$. Lemma 9 then implies that there is a set $\Xi \subseteq \mathbb{N}$ such that $\{\xi_k(a) \colon k \in \mathbb{N}\} = \Xi$ for a.a. $a \in D'$. We conclude that ϑ, the given function, is a.e. E^∞-constant on D', as required. $\dashv$

COROLLARY 18. *All ERs in* $\mathfrak{E}$ *are* $\{r_n\}$-*irreducing.* $\dashv$

PROOF. Apply Lemma 9 and the results of this Section. $\dashv$

This ends the proof of Theorems 7, 3, and 1. $\dashv$(*Main Theorem*)

§9. A corollary and a question. Recall that an ER E is *countable* iff all E-equivalence classes are countable. E is *essentially countable* iff E is Borel reducible to a countable Borel ER.

COROLLARY 19. *If* $\mathcal{Z}$ *is a nontrivial Borel P-ideal and* $\mathsf{E}_{\mathcal{Z}}$ *is an essentially countable ER then* $\mathcal{Z}$ *is* Fin *or a trivial variation of* Fin.

This is true even for Borel ideals which are not P-ideals: in such a general form the result appears in [7, Corollary 4.2]. To derive this generalization from Corollary 19, we can use the following two results:

(1) If a Borel ideal $\mathcal{Z}$ is not a P-ideal then $E_{\mathrm{Fin} \times 0}$ is Borel reducible to $E_{\mathcal{Z}}$ (Solecki [7, 8]. The ideal $\mathrm{Fin} \times 0 = \mathcal{I}_1$, consists of all sets $x \subseteq \mathbb{N} \times \mathbb{N}$ such that $x \subseteq \{0, 1, \ldots, m\} \times \mathbb{N}$ for some m.)

(2) $E_{\mathrm{Fin} \times 0}$ is not essentially countable.

PROOF OF COROLLARY 19. We first prove that

(3) Any countable Borel ER E on a Polish space $\mathbb{S}$ is Borel reducible to $D(\mathbb{S})^{\infty}$.

This is enough to prove the corollary. Indeed, as $D(\mathbb{S})^{\infty}$ clearly belongs to $\mathfrak{E}$, the ideal $\mathcal{Z}$ is either $0 \times \mathrm{Fin}$ or Fin or a trivial modification of Fin by Theorem 1. Yet the first option is impossible as it is known that

(4) $E_{0 \times \mathrm{Fin}}$ is not essentially countable.

To prove (3) note that, by a classical theorem of descriptive set theory, $E = \bigcup_n E_n$, where each $E_n \colon \mathbb{S} \to \mathbb{S}$ is a Borel map (identified with its graph). For any $x \in \mathbb{S}$ let $\vartheta(x) \in \mathbb{S}^{\mathbb{N}}$ be defined by $\vartheta(x)_n = E_n(x)$. Then $\{\vartheta(x)_n \colon n \in \mathbb{N}\} = [x]_E$, so that ϑ is a Borel reduction E to $D(\mathbb{S})^{\infty}$, as required.   ⊣

QUESTION 20. Which ideals except for P-ideals satisfy Theorem 1? There is an interesting Borel ideal whose relations in terms of Borel reducibility are not yet clear. The *Weyl ideal* $\mathcal{Z}_W$ consists of those sets $x \subseteq \mathbb{N}$ which satisfy

$$\lim_{n \to +\infty} \sup_k \frac{\#(x \cap [k, k + n))}{n} = 0.$$

Despite a semblance of the density-0 ideal $\mathcal{Z}_0$, $\mathcal{Z}_W$ has quite different properties, in particular, it is not a P-ideal. Most likely, $E_{\mathcal{Z}_W}$ is not Borel reducible to any ER in $\mathfrak{E}$, but how to prove this claim?

Acknowledgements. The authors are thankful to I. Farah, G. Hjorth, and A. S. Kechris for useful discussions.

REFERENCES

[1] ILIJAS FARAH, *Analytic quotients: theory of liftings for quotients over analytic ideals on the integers*, **Memoirs of the American Mathematical Society**, vol. 148 (2000), no. 702, pp. xvi+177.

[2] HARVEY FRIEDMAN, *Borel and Baire reducibility*, **Fundamenta Mathematicae**, vol. 164 (2000), no. 1, pp. 61–69.

[3] HARVEY FRIEDMAN and LEE STANLEY, *A Borel reducibility theory for classes of countable structures*, **The Journal of Symbolic Logic**, vol. 54 (1989), no. 3, pp. 894 – 914.

[4] GREG HJORTH, *Classification and orbit equivalence relations*, Mathematical surveys and monographs, vol. 75, American Mathematical Society, 2000.

[5] ALEXANDER S. KECHRIS, *Rigidity properties of Borel ideals on the integers*, **Topology and Applications**, vol. 85 (1998), no. 1-3, pp. 195–205.

[6] ——, *Actions of Polish groups and classification problems*, **Analysis and logic**, London Mathematical Society Lecture Note Series, Cambridge University Press, 2001, to appear.

[7] SŁAWOMIR SOLECKI, *Analytic ideals*, **The Bulletin of Symbolic Logic**, vol. 2 (1996), no. 3, pp. 339–348.

[8] ——, *Analytic ideals and their applications*, **Annals of Pure and Applied Logic**, vol. 99 (1999), no. 1-3, pp. 51–72.

MOSCOW CENTER FOR CONTINUOUS MATHEMATICAL EDUCATION
  BOL. VLASEVSKI 11, MOSCOW 121002, RUSSIA
E-mail: kanovei@math.uni-wuppertal.de

DEPARTMENT OF MATHEMATICS
  UNIVERSITY OF WUPPERTAL
    WUPPERTAL 42097, GERMANY
E-mail: reeken@math.uni-wuppertal.de

LINEAR LOGIC AS A FRAMEWORK FOR SPECIFYING
SEQUENT CALCULUS

DALE MILLER AND ELAINE PIMENTEL

Abstract. In recent years, intuitionistic logic and type systems have been used in numerous computational systems as frameworks for the specification of natural deduction proof systems. As we shall illustrate here, linear logic can be used similarly to specify the more general setting of sequent calculus proof systems. Linear logic's meta theory can be used also to analyze properties of a specified object-level proof system. We shall present several example encodings of sequent calculus proof systems using the Forum presentation of linear logic. Since the object-level encodings result in logic programs (in the sense of Forum), various aspects of object-level proof systems can be automated.

§1. **Introduction.** Logics and type systems have been exploited in recent years as frameworks for the specification of deduction in a number of logics. Such *meta logics* or *logical frameworks* have generally been based on intuitionistic logic in which quantification at (non-predicate) higher-order types is available. Identifying a framework that allows the specification of a wide range of logics has proved to be most practical since a single implementation of such a framework can then be used to provide various degrees of automation of object-logics. For example, Isabelle [26] and λProlog [25] are implementations of an intuitionistic logic subset of Church's Simple Theory of Types, while Elf [27] is an implementation of a dependently typed λ-calculus [16]. These computer systems have been used as meta languages to automate various aspects of various logics.

Features of a meta-logic are often directly inherited by any object-logic. This inheritance can be, at times, a great asset. For example, if the meta-logic is rich enough to include λ-bindings in its syntax and to provide α and β conversion as part of its equality of syntax (as is the case for the systems mentioned above), the object-logics immediately inherit such simple and declarative treatments of binding constructs and substitutions. On the other hand, features of the meta-logic can limit the kinds of object-logics that can be directly and naturally encoded. For example, the structural rules of an intuitionistic meta-logic (weakening and contraction) are also inherited

Logic Colloquium '99
Edited by J. van Eijck, V. van Oostrom, and A. Visser
Lecture Notes in Logic, 17

making it difficult to have natural encodings of any logic for which these structural rules are not intended. Also, intuitionistic logic does not have an involutive negation, making it difficult to address directly dualities in object-logics.

In this paper, we make use of linear logic as a meta-logic and find that we can specify a variety of proof systems for object-level systems. By making use of classical linear logic, we are able to capture not only natural deduction proof systems but also many sequent calculus proof systems. We will present our scheme for encoding proof systems in linear logic and show several examples of making such specifications. Since the encodings of such logical systems are natural and direct the rich meta-theory of linear logic can be used to drawing conclusions about the object-level proof systems, and we illustrate such reasoning as well.

This paper is organized as follows: Section 2 gives an introduction to linear logic and Forum. Section 3 shows the representation of sequents and inference rules, while Forum encodings of the well-known proof systems for linear, classical and intuitionistic logics are presented in Section 4. Using the meta-theory, it is possible to prove the collapsing of some modal prefixes for the specified classical and intuitionistic systems. In Section 5 other sequent calculus for these logics are encoded where modal prefixes collapse less dramatically. In order to show how to represent systems that make use of polarities, Section 6 presents an encoding of the so called Logic of Unity (LU) proof system. Section 7 provides an overview of how one might proof search for both Forum and encoded object-level proof systems. We conclude and discuss some future research directions in Section 8.

Our main purpose in this paper is to illustrate via examples how linear logic can be used to both specify and reason about object-level sequent proof systems. We shall do this largely by presenting a series of examples. More extensive discussion of the material in this paper can be found in the PhD dissertation of the second author [28].

§2. **Overview of Linear Logic and Forum.** Linear Logic [13] uses the following logical connectives: the exponentials ! and ?; $\otimes$, $\bindnasrepma$, $\perp$, and 1 for the multiplicative conjunction, disjunction, false, and true; &, $\oplus$, 0, $\top$ for the additive version of these connectives; $\multimap$ for linear implication, and $\forall$ and $\exists$ for universal and existential quantification. We shall assume that the reader is familiar with the sequent calculus presentation of linear logic and with its basic properties.

2.1. The Forum presentation of linear logic. The connectives of linear logic can be classified as *synchronous* and *asynchronous* [2] depending on whether or not the right introduction rule for that connective needs to "synchronize" with its surrounding context. The de Morgan dual of a connective in one of these classes yields a connective in the other class. Given this division of connectives, Miller proposed in [23] the *Forum* presentation of linear logic in which formulas are build using only the asynchronous connectives, namely, ?, $\mathcal{B}$, $\perp$, &, $\top$, $-\!\circ$, and $\forall$, along with the intuitionistic version of implication ($B \Rightarrow C$ denotes $!B -\!\circ C$). The synchronous connectives are implicitly available in Forum since a two sided sequent is used: connectives appearing on the left of the sequent arrow behave synchronously. Proof search in the Forum presentation of linear logic resembles the search involved in logic programming [24, 22]: introducing asynchronous connectives corresponds to goal-directed search and introducing synchronous connectives corresponds to backchaining over logic program clauses.

Forum is a presentation of all of linear logic since it contains a complete set of connectives. The connectives missing from Forum are directly definable using the following logical equivalences:

$$B^\perp \equiv B -\!\circ \perp \qquad 0 \equiv \top -\!\circ \perp \qquad 1 \equiv \perp -\!\circ \perp \qquad \exists x.B \equiv (\forall x.B^\perp)^\perp$$
$$!B \equiv (B \Rightarrow \perp) -\!\circ \perp \qquad B \oplus C \equiv (B^\perp \,\&\, C^\perp)^\perp \qquad B \otimes C \equiv (B^\perp \,\mathcal{B}\, C^\perp)^\perp$$

The collection of connectives in Forum is not minimal. For example, ? and $\mathcal{B}$, can be defined in terms of the remaining connectives:

$$?B \equiv (B -\!\circ \perp) \Rightarrow \perp \quad \text{and} \quad B \,\mathcal{B}\, C \equiv (B -\!\circ \perp) -\!\circ C.$$

Here, the equivalence $B \equiv C$ means that the universal closure of the expression $(B -\!\circ C) \,\&\, (C -\!\circ B)$ is provable in linear logic.

To help make the connection between proof search in Forum and logic programming, it is useful to introduce the notions of goal and clause into Forum. A formula is a *Forum clause* if it is of the form

$$\forall \bar{y}(G_1 \hookrightarrow \cdots \hookrightarrow G_m \hookrightarrow G_0), \quad (m \geq 0)$$

where $G_0, \ldots, G_m$ are arbitrary Forum formulas and occurrences of $\hookrightarrow$ are either occurrences of $-\!\circ$ or $\Rightarrow$. A formula of Forum is a *flat goal* if it does not contain occurrences of $-\!\circ$ and $\Rightarrow$ and all occurrences of ? have atomic scope. A Forum clause is a *flat clause* if $G_0, \ldots, G_m$ are flat goals. It is possible to also add the restriction that the formula G_0, the *head* of the clause, is of the form $B_1 \,\mathcal{B}\, \ldots \,\mathcal{B}\, B_n$, where $n \geq 0$ and each B_i is an atom. If $n = 0$ then we write the head as simply $\perp$ and say that the head is *empty*. A flat clause is essentially a clause of the LinLog system [2] except that heads of flat clauses may be empty. It will be the case that all formulas used to specify sequent calculus inferences rules in this paper will be flat Forum clauses.

As in Church's Simple Theory of Types [6], both terms and formulas are built using a simply typed λ-calculus. We assume the usual rules of α, β, and η-conversion and we identify terms and formulas up to α-conversion. A term is λ-normal if it contains no β and no η redexes. All terms are λ-convertible to a term in λ-normal form, and such a term is unique up to α-conversion. The substitution notation $B[t/x]$ denotes the λ-normal form of the β-redex

$(\lambda x.B)t$. Following [6], we shall also assume that formulas of Forum have type o.

2.2. Proof system for Forum. The proof system for Forum, $\mathcal{F}$, is given in Figure 1. Sequents in $\mathcal{F}$ have the form

$$\Sigma: \Psi; \Delta \longrightarrow \Gamma; \Upsilon \quad \text{and} \quad \Sigma: \Psi; \Delta \xrightarrow{B} \Gamma; \Upsilon,$$

where Σ is a signature, Δ and Γ are multisets of formulas, Ψ and Υ are sets of formulas, and B is a formula. All formulas in sequents are composed of the

$$\frac{}{\Sigma: \Psi; \Delta \longrightarrow \top, \Gamma; \Upsilon} \; \top R$$

$$\frac{\Sigma: \Psi; \Delta \longrightarrow B, \Gamma; \Upsilon \quad \Sigma: \Psi; \Delta \longrightarrow C, \Gamma; \Upsilon}{\Sigma: \Psi; \Delta \longrightarrow B \& C, \Gamma; \Upsilon} \; \& R$$

$$\frac{\Sigma: \Psi; \Delta \longrightarrow \Gamma; \Upsilon}{\Sigma: \Psi; \Delta \longrightarrow \bot, \Gamma; \Upsilon} \; \bot R \qquad \frac{\Sigma: \Psi; \Delta \longrightarrow B, C, \Gamma; \Upsilon}{\Sigma: \Psi; \Delta \longrightarrow B \,\mathscr{X}\, C, \Gamma; \Upsilon} \; \mathscr{X} R$$

$$\frac{\Sigma: \Psi; B, \Delta \longrightarrow C, \Gamma; \Upsilon}{\Sigma: \Psi; \Delta \longrightarrow B \multimap C, \Gamma; \Upsilon} \; \multimap R \qquad \frac{\Sigma: B, \Psi; \Delta \longrightarrow C, \Gamma; \Upsilon}{\Sigma: \Psi; \Delta \longrightarrow B \Rightarrow C, \Gamma; \Upsilon} \; \Rightarrow R$$

$$\frac{y: \tau, \Sigma: \Psi; \Delta \longrightarrow B[y/x], \Gamma; \Upsilon}{\Sigma: \Psi; \Delta \longrightarrow \forall_\tau x.B, \Gamma; \Upsilon} \; \forall R \qquad \frac{\Sigma: \Psi; \Delta \longrightarrow \Gamma; B, \Upsilon}{\Sigma: \Psi; \Delta \longrightarrow ? B, \Gamma; \Upsilon} \; ? R$$

$$\frac{\Sigma: B, \Psi; \Delta \xrightarrow{B} \mathcal{A}; \Upsilon}{\Sigma: B, \Psi; \Delta \longrightarrow \mathcal{A}; \Upsilon} \; decide! \qquad \frac{\Sigma: \Psi; \Delta \longrightarrow \mathcal{A}, B; B, \Upsilon}{\Sigma: \Psi; \Delta \longrightarrow \mathcal{A}; B, \Upsilon} \; decide?$$

$$\frac{\Sigma: \Psi; \Delta \xrightarrow{B} \mathcal{A}; \Upsilon}{\Sigma: \Psi; B, \Delta \longrightarrow \mathcal{A}; \Upsilon} \; decide$$

$$\frac{}{\Sigma: \Psi; \cdot \xrightarrow{A} A; \Upsilon} \; initial \qquad \frac{}{\Sigma: \Psi; \cdot \xrightarrow{A} \cdot; A, \Upsilon} \; initial?$$

$$\frac{}{\Sigma: \Psi; \cdot \xrightarrow{\bot} \cdot; \Upsilon} \; \bot L \qquad \frac{\Sigma: \Psi; \Delta \xrightarrow{B_i} \mathcal{A}; \Upsilon}{\Sigma: \Psi; \Delta \xrightarrow{B_1 \& B_2} \mathcal{A}; \Upsilon} \; \& L_i \qquad \frac{\Sigma: \Psi; B \longrightarrow \cdot; \Upsilon}{\Sigma: \Psi; \cdot \xrightarrow{? B} \cdot; \Upsilon} \; ? L$$

$$\frac{\Sigma: \Psi; \Delta_1 \xrightarrow{B} \mathcal{A}_1; \Upsilon \quad \Sigma: \Psi; \Delta_2 \xrightarrow{C} \mathcal{A}_2; \Upsilon}{\Sigma: \Psi; \Delta_1, \Delta_2 \xrightarrow{B \mathscr{X} C} \mathcal{A}_1, \mathcal{A}_2; \Upsilon} \; \mathscr{X} L \qquad \frac{\Sigma: \Psi; \Delta \xrightarrow{B[t/x]} \mathcal{A}; \Upsilon}{\Sigma: \Psi; \Delta \xrightarrow{\forall_\tau x.B} \mathcal{A}; \Upsilon} \; \forall L$$

$$\frac{\Sigma: \Psi; \Delta_1 \longrightarrow \mathcal{A}_1, B; \Upsilon \quad \Sigma: \Psi; \Delta_2 \xrightarrow{C} \mathcal{A}_2; \Upsilon}{\Sigma: \Psi; \Delta_1, \Delta_2 \xrightarrow{B \multimap C} \mathcal{A}_1, \mathcal{A}_2; \Upsilon} \; \multimap L$$

$$\frac{\Sigma: \Psi; \cdot \longrightarrow B; \Upsilon \quad \Sigma: \Psi; \Delta \xrightarrow{C} \mathcal{A}; \Upsilon}{\Sigma: \Psi; \Delta \xrightarrow{B \Rightarrow C} \mathcal{A}; \Upsilon} \; \Rightarrow L$$

FIGURE 1. The $\mathcal{F}$ proof system. The rule $\forall R$ has the proviso that y is not declared in the signature Σ, and the rule $\forall L$ has the proviso that t is a Σ-term of type τ. In $\& L_i$, $i = 1$ or $i = 2$.

asynchronous connectives listed above (together with $\Rightarrow$) and contain at most the non-logical symbols present in Σ (such formulas are called Σ-*formulas*).

The intended meanings of these two sequents in linear logic are

$$! \Psi, \Delta \longrightarrow \Gamma, ? \Upsilon \quad \text{and} \quad ! \Psi, \Delta, B \longrightarrow \Gamma, ? \Upsilon,$$

respectively. In the proof system of Figure 1, the only right rules are those for sequents of the form $\Sigma : \Psi ; \Delta \longrightarrow \Gamma ; \Upsilon$. The syntactic variable $\mathcal{A}$ in Figure 1 denotes a multiset of atomic formulas. Left rules are applied only to the formula B that labels the sequent arrow in $\Sigma : \Psi ; \Delta \xrightarrow{B} \mathcal{A} ; \Upsilon$.

We use the turnstile symbol as the mathematical-level judgment that a sequent is provable: that is, $\Delta \vdash \Gamma$ means that the two-sided sequent $\Delta \longrightarrow \Gamma$ has a linear logic proof. The following correctness theorem for $\mathcal{F}$ is given in [23] and is based on the focusing result of Andreoli in [2].

THEOREM 2.1. *Let Σ be a signature, Δ and Γ be multisets of Σ-formulas, and Ψ and Υ be sets of Σ-formulas. The sequent $\Sigma : \Psi ; \Delta \longrightarrow \Gamma ; \Upsilon$ has a proof in $\mathcal{F}$ if and only if $! \Psi, \Delta \vdash \Gamma, ? \Upsilon$.*

We shall use the term *backchaining* to refer to an application of either the *decide* or the *decide*! inference rule followed by a series of applications of left-introduction rules (reading a proof bottom-up). This notion of backchaining generalizes the usual notion found in the logic programming literature.

When presenting examples of Forum code we often use $\circ\!-$ and $\Rightarrow$ to be the converses of $\multimap$ and $\Leftarrow$ since they provide a more natural operational reading of clauses (similar to the use of $:-$ in Prolog). We will assume that when parsing expressions, $\bindnasrepma$ and & bind tighter than $\circ\!-$ and $\Leftarrow$.

Multiset rewriting can be captured naturally in proof search. Consider, for example, the clause

$$a \bindnasrepma b \circ\!- c \bindnasrepma d \bindnasrepma e.$$

and the sequent $\Sigma : \Psi ; \Delta \longrightarrow a, b, \Gamma ; \Upsilon$, where the clause displayed above is a member of Ψ. A proof for this sequent can then end with the following inference rules.

$$
\cfrac{
  \cfrac{
    \cfrac{\Sigma : \Psi ; \Delta \longrightarrow c, d, e, \Gamma ; \Upsilon}
          {\Sigma : \Psi ; \Delta \longrightarrow c, d \bindnasrepma e, \Gamma ; \Upsilon}}
        {\Sigma : \Psi ; \Delta \longrightarrow c \bindnasrepma d \bindnasrepma e, \Gamma ; \Upsilon}
  \qquad
  \cfrac{\cfrac{\Sigma : \Psi ; \cdot \xrightarrow{a} a ; \Upsilon \quad \Sigma : \Psi ; \cdot \xrightarrow{b} b ; \Upsilon}{\Sigma : \Psi ; \cdot \xrightarrow{a \bindnasrepma b} a, b ; \Upsilon}}{}
}
{
  \cfrac{\Sigma : \Psi ; \Delta \xrightarrow{c \bindnasrepma d \bindnasrepma e - \circ a \bindnasrepma b} a, b, \Gamma ; \Upsilon}
        {\Sigma : \Psi ; \Delta \longrightarrow a, b, \Gamma ; \Upsilon}
}
$$

We can interpret this proof fragment as a reduction of the multiset a, b, Γ to the multiset c, d, e, Γ by backchaining on the clause displayed above.

Of course, a clause may have multiple, top-level implications. In this case, the surrounding context must be manipulated properly to prove the sub-goals that arise in backchaining. Consider a clause of the form

$$G_1 \multimap G_2 \Rightarrow G_3 \multimap G_4 \Rightarrow B_1 \bindnasrepma B_2$$

labeling the sequent arrow in the sequent $\Sigma: \Psi; \Delta \longrightarrow B_1, B_2, \mathcal{A}; \Upsilon$. An attempt to prove this sequent would then lead to attempt to prove the four sequents

$$\Sigma: \Psi; \Delta_1 \longrightarrow G_1, \mathcal{A}_1; \Upsilon \qquad \Sigma: \Psi; \cdot \longrightarrow G_2; \Upsilon$$
$$\Sigma: \Psi; \Delta_2 \longrightarrow G_3, \mathcal{A}_2; \Upsilon \qquad \Sigma: \Psi; \cdot \longrightarrow G_4; \Upsilon$$

where Δ is the multiset union of Δ_1 and Δ_2, and $\mathcal{A}$ is the multiset union of $\mathcal{A}_1$ and $\mathcal{A}_2$. In other words, those subgoals immediately to the left of an $\Rightarrow$ are attempted with empty bounded contexts: the bounded contexts, here Δ and $\mathcal{A}$, are divided up to be used to prove those goals immediately to the left of $-\circ$.

2.3. Applications of Forum. Forum specifications have been presented for the operational semantics of programming languages containing side effects, concurrency features, references, exceptions, continuations, and objects [3, 5, 8, 23]. Chirimar [5] used Forum to present the semantics of a RISC processor and Chakravarty [4] used it to specify the logical and operational semantics of a parallel programming language. A specification of a sequent calculus for intuitionistic logic was given by Miller in [23]: that example was improved by Ricci [30], where a proof system for classical logic was also given. The examples in [23, 30] are significantly generalized in this paper.

§3. Representing sequents and inference rules. Since we now wish to represent one logic and proof system within another, we need to distinguish between the meta-logic, namely, linear logic as presented by Forum, and the various object-logics for which we wish to specify sequent proof systems. Formulas of the object-level will be identified with meta-level terms of type *bool*. Object-level logical connectives will be introduced as needed and as constructors of this type.

A two-sided sequent $\Delta \longrightarrow \Gamma$ is generally restricted so that Δ and Γ are either lists, multisets, or sets of formulas. Sets are used if all three structural rules (exchange, weakening, contraction) are implicit; multisets are used if exchange is implicit; and lists are used if no structural rule is implicit. Since our goal here is to encode object-level sequents into meta-level sequents as directly as possible, and since contexts in Forum are either multisets or sets, we will not be able to represent sequents that make use of lists. It is unlikely, for example, that non-commutative object-logics can be encoded into our linear logic meta theory along the lines we describe below.

3.1. Three schemes for encoding sequents. Consider the well known, two-sided sequent proof systems for classical, intuitionistic, and linear logic. A convenient distinction between these logics can be described, in part, by where the structural rules of thinning and contraction can be applied. In classical logic, these structural rules are allowed on both sides of the sequent arrow; in intuitionistic logic, no structural rules are allowed on the right of the sequent arrow; and in linear logic, they are not allowed on either sides of the arrow.

Thus a classical sequent is a pairing of two sets; a linear logic sequent is a pairing of two multisets; and an intuitionistic sequent is the pairing of a set (for the left-hand side) and a multiset (for the right-hand side). This discussion suggests the following representation of sequents in these three systems. Let *bool* be the type of object-level propositional formulas and let $\lfloor \cdot \rfloor$ and $\lceil \cdot \rceil$ be two meta-level predicates, both of type $bool \to o$.

We will identify three schemes for encoding sequents. The *linear scheme* encodes the (object-level) sequent $B_1, \ldots, B_n \longrightarrow C_1, \ldots, C_m$ $(n, m \geq 0)$ by the meta-level formula $\lfloor B_1 \rfloor \,\mathbin{\mathscr{R}}\ldots \mathbin{\mathscr{R}} \lfloor B_n \rfloor \mathbin{\mathscr{R}} \lceil C_1 \rceil \mathbin{\mathscr{R}}\ldots \mathbin{\mathscr{R}} \lceil C_m \rceil$ or by the Forum sequent

$$\Sigma: \cdot;\cdot \longrightarrow \lfloor B_1 \rfloor, \ldots, \lfloor B_n \rfloor, \lceil C_1 \rceil, \ldots, \lceil C_m \rceil; \cdot.$$

The *intuitionistic scheme* encodes $B_1, \ldots, B_n \longrightarrow C_1, \ldots, C_m$, where $n, m \geq 0$, with the meta-level formula $?\lfloor B_1 \rfloor \mathbin{\mathscr{R}}\ldots \mathbin{\mathscr{R}} ?\lfloor B_n \rfloor \mathbin{\mathscr{R}} \lceil C_1 \rceil \mathbin{\mathscr{R}}\ldots \mathbin{\mathscr{R}} \lceil C_m \rceil$ or by the Forum sequent

$$\Sigma: \cdot;\cdot \longrightarrow \lceil C_1 \rceil, \ldots, \lceil C_m \rceil; \lfloor B_1 \rfloor, \ldots, \lfloor B_n \rfloor.$$

Often intuitionistic sequents are additionally restricted to having one formula on the right. Finally, the *classical scheme* encodes the sequent $B_1, \ldots, B_n \longrightarrow C_1, \ldots, C_m$ $(n, m \geq 0)$ as the meta-level formula

$$?\lfloor B_1 \rfloor \mathbin{\mathscr{R}}\ldots \mathbin{\mathscr{R}} ?\lfloor B_n \rfloor \mathbin{\mathscr{R}} ?\lceil C_1 \rceil \mathbin{\mathscr{R}}\ldots \mathbin{\mathscr{R}} ?\lceil C_m \rceil$$

or by the Forum sequent

$$\Sigma: \cdot;\cdot \longrightarrow \cdot; \lfloor B_1 \rfloor, \ldots, \lfloor B_n \rfloor, \lceil C_1 \rceil, \ldots, \lceil C_m \rceil.$$

The $\lfloor \cdot \rfloor$ and $\lceil \cdot \rceil$ predicates are used to identify which object-level formulas appear on which side of the sequent arrow, and the ? modal is used to mark the formulas to which weakening and contraction can be applied.

3.2. Encoding additive and multiplicative inference rules. We first illustrate how to encode object-level inference rules using the linear scheme.

Consider the specification of the logical inference rules for object-level conjunction, represented here as the infix constant $\wedge$ of type $bool \to bool \to bool$. Consider first the additive inference rules for this connective.

$$\frac{\Delta, A \longrightarrow \Gamma}{\Delta, A \wedge B \longrightarrow \Gamma} \wedge L_1 \qquad \frac{\Delta, B \longrightarrow \Gamma}{\Delta, A \wedge B \longrightarrow \Gamma} \wedge L_2 \qquad \frac{\Delta \longrightarrow \Gamma, A \quad \Delta \longrightarrow \Gamma, B}{\Delta \longrightarrow \Gamma, A \wedge B} \wedge R$$

These three inference rules can be specified in Forum using the clauses

$$(\wedge L_1) \quad \lfloor A \wedge B \rfloor \circ\!\!-\ \lfloor A \rfloor. \qquad (\wedge R) \quad \lceil A \wedge B \rceil \circ\!\!-\ \lceil A \rceil \,\&\, \lceil B \rceil.$$
$$(\wedge L_2) \quad \lfloor A \wedge B \rfloor \circ\!\!-\ \lfloor B \rfloor.$$

Let Ψ be a set of formulas that contains the three clauses. The Forum sequent

$$\Sigma: \Psi;\cdot \longrightarrow \lfloor B_1 \rfloor, \ldots, \lfloor B_n \rfloor, \lceil C_1 \rceil, \ldots, \lceil C_m \rceil; \cdot$$

can be the conclusion of a *decide*! rule that selected $(\wedge R)$ rule only if one of the $\lceil \cdot \rceil$-atoms, say $\lceil C_1 \rceil$, is of the form $A \wedge B$ and the sequent

$$\Sigma: \Psi; \cdot \longrightarrow \lfloor B_1 \rfloor, \dots, \lfloor B_n \rfloor, \lceil A \rceil \mathbin{\&} \lceil B \rceil, \lceil C_2 \rceil, \dots, \lceil C_m \rceil; \cdot$$

is provable. This formula is provable if and only if the two sequents

$$\Sigma: \Psi; \cdot \longrightarrow \lfloor B_1 \rfloor, \dots, \lfloor B_n \rfloor, \lceil A \rceil, \lceil C_2 \rceil, \dots, \lceil C_m \rceil; \cdot$$

and

$$\Sigma: \Psi; \cdot \longrightarrow \lfloor B_1 \rfloor, \dots, \lfloor B_n \rfloor, \lceil B \rceil, \lceil C_2 \rceil, \dots, \lceil C_m \rceil; \cdot$$

are provable in Forum. Thus, backchaining on the $(\wedge R)$ clause above can be used to reduce the problem of finding an object-level proof of

$$B_1, \dots, B_n \longrightarrow A \wedge B, C_2, \dots, C_m$$

to the problem of finding object-level proofs for

$$B_1, \dots, B_n \longrightarrow A, C_2, \dots, C_m \quad \text{and} \quad B_1, \dots, B_n \longrightarrow B, C_2, \dots, C_m.$$

Thus, we have successfully captured this right introduction rule for conjunction using *decide*! with the clause corresponding to $(\wedge R)$. A similar and simpler argument shows how left introduction for $\wedge$ is also correctly encoded using the two clauses for $(\wedge L)$. Notice that the two clauses for left introduction could be written equivalently in linear logic as the one formula

$$\lfloor A \wedge B \rfloor \circ\!\!-\ \lfloor A \rfloor \oplus \lfloor B \rfloor.$$

(Although $\oplus$ is not a connective of Forum, we shall use it in this fashion in order to write two Forum clauses as one formula.) Thus, these additive rules make use of two (dual) meta-level additive connectives: $\&$ and $\oplus$.

Now consider encoding the multiplicative version of conjunction introduction.

$$\frac{\Delta, A, B \longrightarrow \Gamma}{\Delta, A \wedge B \longrightarrow \Gamma} \wedge L \qquad \frac{\Delta_1 \longrightarrow \Gamma_1, A \quad \Delta_2 \longrightarrow \Gamma_2, B}{\Delta_1, \Delta_2 \longrightarrow \Gamma_1, \Gamma_2, A \wedge B} \wedge R$$

It is an easy matter to check that the following two clauses encode these two inference rules.

$$(\wedge L) \quad \lfloor A \wedge B \rfloor \circ\!\!-\ \lfloor A \rfloor \mathbin{\rotatebox[origin=c]{180}{\&}} \lfloor B \rfloor. \qquad (\wedge R) \quad \lceil A \wedge B \rceil \circ\!\!-\ \lceil A \rceil \circ\!\!-\ \lceil B \rceil.$$

Notice that the clause for right introduction could be written equivalently in linear logic as

$$\lceil A \wedge B \rceil \circ\!\!-\ \lceil A \rceil \otimes \lceil B \rceil.$$

Thus, these multiplicative rules make use of two (dual) meta-level multiplicative connectives: $\otimes$ and $\mathbin{\rotatebox[origin=c]{180}{\&}}$.

Consider now using the classical scheme for representing sequents and consider writing the additive version of the $(\wedge R)$ rule as

$$\lceil A \wedge B \rceil \circ\!\!-\ ?\lceil A \rceil \mathbin{\&} ?\lceil B \rceil.$$

In that case, backchaining on this clause would reduce proof search of the sequent

$$\Sigma: \Psi; \cdot \longrightarrow \cdot; \lfloor B_1 \rfloor, \ldots, \lfloor B_n \rfloor, \lceil C_1 \rceil, \ldots, \lceil C_m \rceil$$

(where for some i, C_i is $A \wedge B$) to finding proofs for the two sequents

$$\Sigma: \Psi; \cdot \longrightarrow ?\lceil A \rceil; \lfloor B_1 \rfloor, \ldots, \lfloor B_n \rfloor, \lceil C_1 \rceil, \ldots, \lceil C_m \rceil$$

and

$$\Sigma: \Psi; \cdot \longrightarrow ?\lceil B \rceil; \lfloor B_1 \rfloor, \ldots, \lfloor B_n \rfloor, \lceil C_1 \rceil, \ldots, \lceil C_m \rceil$$

which in turn are provable if and only if the sequents

$$\Sigma: \Psi; \cdot \longrightarrow \cdot; \lfloor B_1 \rfloor, \ldots, \lfloor B_n \rfloor, \lceil A \rceil, \lceil C_1 \rceil, \ldots, \lceil C_m \rceil$$

and

$$\Sigma: \Psi; \cdot \longrightarrow \cdot; \lfloor B_1 \rfloor, \ldots, \lfloor B_n \rfloor, \lceil B \rceil, \lceil C_1 \rceil, \ldots, \lceil C_m \rceil$$

are provable.

If we had used, instead, the multiplicative encoding of conjunctive introduction,

$$\lceil A \wedge B \rceil \;\circ\!\!-\; ?\lceil A \rceil \otimes ?\lceil B \rceil.$$

a slightly different meta-level proof would have made the same reduction.

It seems natural to consider using a question mark in the head of a clause describing a right or left-introduction rule for classical logic (or just the left-introduction rule for intuitionistic logic). For example, the $(\wedge R)$ rule could have been encoded as

$$?\lceil A \wedge B \rceil \;\circ\!\!-\; ?\lceil A \rceil \;\&\; ?\lceil B \rceil.$$

This encoding style was used in [23, 30], for example. We shall prefer, instead, to encode inference rules without occurrences of question marks in the head of clauses since the structure of meta-level proofs often does not correspond to the structure of object-level proofs. For example, although the sequent $A \wedge B \wedge C \longrightarrow B$ is provable in classical logic, there is no equivalent object-level proof for the proof displayed below. Here, signatures are not displayed in sequents, $\Gamma = \lceil B \rceil, \lfloor A \wedge B \wedge C \rfloor$, and Ψ is a set of formulas that contains the clause displayed above and *Initial* (see Section 3.4).

$$
\cfrac{
\cfrac{
\cfrac{
\cfrac{\Psi; \cdot \xrightarrow{\lfloor A \wedge B \rfloor} \cdot; \lfloor A \wedge B \rfloor, \Gamma}
{\cfrac{\Psi; \lfloor A \wedge B \rfloor \longrightarrow \cdot; \lfloor A \wedge B \rfloor, \Gamma}
{\Psi; \lfloor A \wedge B \rfloor \longrightarrow ?\lfloor A \wedge B \rfloor; \Gamma}}
\quad
\cfrac{
\cfrac{\Psi; \cdot \xrightarrow{\lfloor A \wedge B \wedge C \rfloor} \cdot; \Gamma}
{\Psi; \lfloor A \wedge B \wedge C \rfloor \longrightarrow \cdot; \Gamma}}
{\Psi; \cdot \xrightarrow{?\lfloor A \wedge B \wedge C \rfloor} \cdot; \Gamma}
}
{\cfrac{\Psi; \lfloor A \wedge B \rfloor \xrightarrow{?\lfloor A \wedge B \wedge C \rfloor \circ\!-?\lfloor A \wedge B \rfloor} \cdot; \Gamma}
{\cfrac{\Psi; \lfloor A \wedge B \rfloor \longrightarrow \cdot; \Gamma}
{\Psi; \cdot \xrightarrow{?\lfloor A \wedge B \rfloor} \cdot; \Gamma}}}
\quad
\cfrac{
\cfrac{
\cfrac{\Psi; \cdot \xrightarrow{\lceil B \rceil} \cdot; \lfloor B \rfloor, \Gamma}
{\cfrac{\Psi; \lceil B \rceil \longrightarrow \cdot; \lfloor B \rfloor, \Gamma}
{\Psi; \cdot \xrightarrow{?\lceil B \rceil} \cdot; \lfloor B \rfloor, \Gamma}}
\quad
\cfrac{\Psi; \cdot \xrightarrow{\lfloor B \rfloor} \cdot; \lfloor B \rfloor, \Gamma}
{\cfrac{\Psi; \lfloor B \rfloor \longrightarrow \cdot; \lfloor B \rfloor, \Gamma}
{\Psi; \cdot \xrightarrow{?\lfloor B \rfloor} \cdot; \lfloor B \rfloor, \Gamma}}
}
{\cfrac{\Psi; \cdot \xrightarrow{?\lceil B \rceil \otimes ?\lfloor B \rfloor} \cdot; \lfloor B \rfloor, \Gamma}
{\cfrac{\Psi; \cdot \longrightarrow \cdot; \lfloor B \rfloor, \Gamma}
{\Psi; \cdot \longrightarrow ?\lfloor B \rfloor; \Gamma}}}
}
{\cfrac{\Psi; \cdot \xrightarrow{?\lfloor A \wedge B \rfloor \circ\!-?\lfloor B \rfloor} \cdot; \Gamma}
{\Psi; \cdot \longrightarrow \cdot; \Gamma}} \; decide!
$$

The fact that "focus" is lost when a question mark is encountered on a formula labeling a sequent arrow means that it is much harder to control the structure of meta-level proofs and to relate them to object-level proofs.

3.3. Encoding quantifier introduction rules. Using the quantification of higher-order types that is available in Forum, it is a simple matter to encode the inference rules for object-level quantifiers. For example, if we use the linear scheme for representing sequents, then the left and right introduction rules for object-level universal quantifier can be written as

$$(\forall L) \quad \lfloor \forall B \rfloor \; \circ\!\!- \; \lfloor Bx \rfloor. \qquad\qquad (\forall R) \quad \lceil \forall B \rceil \; \circ\!\!- \; \forall x \lceil Bx \rceil.$$

Here, the symbol $\forall$ is used for both meta-level and object-level quantification: at the object-level $\forall$ has the type $(i \to bool) \to bool$. Thus the variable B above has the type $i \to bool$. Consider the Forum sequent $\Sigma : \Psi; \cdot \longrightarrow \lceil \forall B \rceil, \Theta; \cdot$ where Ψ contains the above two clauses. Using *decide*! with the clause for $(\forall R)$ would cause the search for a proof of the above sequent to be reduced to the search for a proof of the sequent $\Sigma, y : i : \Psi; \cdot \longrightarrow \lceil By \rceil, \Theta; \cdot$ where y is not present in the signature Σ. Here, the meta-level eigen-variable y also serves the role of an object-level eigen-variable. Dually, consider the Forum sequent $\Sigma : \Psi; \cdot \longrightarrow \lfloor \forall B \rfloor, \Theta; \cdot$. Using the *decide*! with the clause for $(\forall L)$ would cause proof search to reduce this sequent to the sequent $\Sigma : \Psi; \cdot \longrightarrow \lfloor Bt \rfloor, \Theta; \cdot$ where t is a Σ-term of type i. If we restrict appropriately the use of the type i by constants in Σ, then Σ-terms of type i can be identified with object-level terms.

Notice that the clause for $(\forall L)$ is logically equivalent to the formula

$$\lfloor \forall B \rfloor \; \circ\!\!- \; \exists x \lfloor Bx \rfloor.$$

Thus, these quantifier rules make use of two (dual) meta-level quantifiers.

3.4. The cut and initial rules. Up to this point, all the Forum clauses used to specify an inference figure have been such that the head of the clause has been an atom. Clauses specifying the cut and initial rules will have heads of rather different structure.

Consider specifying the initial rule (the one asserting that the sequent $B \longrightarrow B$ is provable) using the linear scheme for encoding sequents. The clause

$$(Initial) \qquad \lfloor B \rfloor \; \mathbin{⅋} \; \lceil B \rceil.$$

will properly encode this rule. Notice that this clause has a head with two atoms (and an empty body). Similarly, the cut rule

$$\frac{\Delta_1 \longrightarrow \Gamma_1, B \quad \Delta_2, B \longrightarrow \Gamma_2}{\Delta_1, \Delta_2 \longrightarrow \Gamma_1, \Gamma_2} \; Cut$$

can be specified simply as the clause

$$(Cut) \qquad \bot\!\!-\!\!\circ \; \lfloor B \rfloor \; \circ\!\!- \; \lceil B \rceil.$$

Dually to the initial rule, this clause has an empty head and two bodies.

3.5. Advantages of such encodings. The encoding of an object-level proof system as Forum clauses has certain advantages over encoding them as inference figures. For example, the Forum specifications do not deal with context explicitly and instead they focus on the formulas that are directly involved in the inference rule. The distinction between making the inference rule additive or multiplicative is achieved in inference rule figures by explicitly presenting contexts and either splitting or copying them. The Forum clause representation achieves the same distinction using meta-level additive or multiplicative connectives. Object-level quantifiers can be handled directly using the meta-level quantification. Similarly, the structural rules of contraction and thinning can be captured together using the ? modal. Finally, since the encoding of proof systems is natural and direct, we hope to be able to use the rich meta-theory of linear logic to help in drawing conclusions about object-level proof systems. An example of this kind of meta-level reason will be illustrated in Section 4.4 where a sequent calculus presentation of intuitionistic logic is transformed into a natural deduction presentation by rather simple linear logic equivalences.

Since the encodings of object-level encodings result in logic programs (in the sense of Forum) and since there is significant knowledge and tools available to provide automatic and interactive tools to compute with those logic programs, encodings such as those described here can be important for the automation of various proof systems (see Section 7).

There are, of course, some disadvantages to using linear logic as a meta-theory, the principle one being that it will not be possible to capture all proof systems, such as those for non-commutativity. As we shall see, however, significant and interesting proof systems can be encoded into linear logic and for these systems, broad avenues of meta-level reasoning and automation should be available.

§4. Linear, classical, and intuitionistic logics.
In this section, we present Forum encodings of well-known proof systems for linear, classical, and intuitionistic logics. Object-level linear logic will be encoded reusing the same symbols that appear at the meta-level, namely, $!$, $?$, $\otimes$, $\wp$, $\perp$, 1, $\&$, $\oplus$, 0, $\top$, $\multimap$, $\forall_l$, and $\exists_l$. Classical logic is encoded using $\wedge$, $\vee$, $\Rightarrow$, f_c, t_c, $\forall_c$, and $\exists_c$ for conjunction, disjunction, implication, false, true, and universal and existential quantification, respectively, while intuitionistic logic is encoded with $\cap$, $\cup$, $\supset$, f_i, t_i, $\forall_i$, and $\exists_i$ for conjunction, disjunction, implication, false, true, and universal and existential quantification, respectively.

We use the type i to denote object-level individuals and *bool* to denote object-level formulas (our object-logics will all be first-order). All binary connectives have type $bool \to bool \to bool$ and will be written as infix. Object-level constants representing quantification are all of the second order type

$(i \rightarrow bool) \rightarrow bool$: we abbreviated expressions such as $\forall_i (\lambda x.B)$ as simply $\forall_i x B$.

The three signatures Σ_l, Σ_c, and Σ_j will denote the signatures for the object-logics for linear, classical, and intuitionistic, respectively. We assume that each of these signatures also contains the two predicates $\lfloor \cdot \rfloor$ and $\lceil \cdot \rceil$.

4.1. Three proof systems. Let LL, LK, and LJ denote the be the set of clauses displayed in Figures 2, 3, and 4, respectively.

PROPOSITION 4.1. *The following three correctness statements hold.*

1. *The sequent* $B_1, \ldots, B_n \longrightarrow C_1, \ldots, C_m$ $(m, n \geq 0)$ *has a linear logic proof* [13] *iff* $\Sigma_l : LL; \cdot \longrightarrow \lfloor B_1 \rfloor, \ldots, \lfloor B_n \rfloor, \lceil C_1 \rceil, \ldots, \lceil C_m \rceil; \cdot$ *has a $\mathcal{F}$-proof.*

2. *The sequent* $B_1, \ldots, B_n \longrightarrow C_1, \ldots, C_m$ $(m, n \geq 0)$ *has an LK-proof* [12] *iff* $\Sigma_c : LK; \cdot \longrightarrow \cdot; \lfloor B_1 \rfloor, \ldots, \lfloor B_n \rfloor, \lceil C_1 \rceil, \ldots, \lceil C_m \rceil$ *has a $\mathcal{F}$-proof.*

3. *The sequent* $B_1, \ldots, B_n \longrightarrow B_0$ *has an LJ-proof* [12] *if and only if the sequent* $\Sigma_j : LJ; \cdot \longrightarrow \lceil B_0 \rceil; \lfloor B_1 \rfloor, \ldots, \lfloor B_n \rfloor$ *has a $\mathcal{F}$-proof and the sequent* $B_1, \ldots, B_n \longrightarrow$ *has an LJ-proof iff* $\Sigma_j : LJ; \cdot \longrightarrow \cdot; \lfloor B_1 \rfloor, \ldots, \lfloor B_n \rfloor$ *has a $\mathcal{F}$-proof* $(n \geq 0)$.

Proofs are by structural induction of over proof structures. In all cases, proofs in Forum match closely proofs in the corresponding object-logic.

4.2. Modular presentations of classical and intuitionistic logics. The essential difference between the theories LJ and LK is the different set of occurrences of the ? modal. Consider the theories LK_0 and LJ_0 given in Figures 5 and 6. These result from removing the cut and initial rules as well as deleting from the introduction rules of the corresponding LK and LJ theories the ? modal. Define the two new theories

$$LJ' = LJ_0 \cup \{Cut, Initial, Pos_2\} \text{ and } LK' = LK_0 \cup \{Cut, Initial, Pos_2, Neg_2\},$$

where the additional formulas are defined in Figure 7. While LJ' is a strengthening of LJ, they can both prove the same object-level, intuitionistic sequents. Similarly for LK' and LK.

PROPOSITION 4.2. *The following two correctness statements hold.*

1. *Let* $B_0, \ldots, B_n$ *(for $n \geq 0$) be object-level, intuitionistic formulas. Then*

$$\Sigma_j : LJ; \longrightarrow \lceil B_0 \rceil; \lfloor B_1 \rfloor, \ldots, \lfloor B_n \rfloor$$

has a $\mathcal{F}$-proof if and only if $\Sigma_j : LJ'; \longrightarrow \lfloor B_1 \rfloor, \ldots, \lfloor B_n \rfloor, \lceil B_0 \rceil;$ *has a $\mathcal{F}$-proof.*

2. *Let* $B_1, \ldots, B_n, C_1, \ldots, C_m$ *(for $n, m \geq 0$) be object-level, classical formulas. Then* $\Sigma_c : LK; \longrightarrow; \lfloor B_1 \rfloor, \ldots, \lfloor B_n \rfloor, \lceil C_1 \rceil, \ldots, \lceil C_m \rceil$ *has a $\mathcal{F}$-proof if and only if* $\Sigma_c : LK'; \longrightarrow \lfloor B_1 \rfloor, \ldots, \lfloor B_n \rfloor, \lceil C_1 \rceil, \ldots, \lceil C_m \rceil;$ *has a $\mathcal{F}$-proof.*

PROOF. We prove the first of these cases since the second is similar.
A consequence of LJ' is the equivalence $\lfloor B \rfloor \equiv ? \lfloor B \rfloor$. Thus we can rewrite

$(\multimap L)$	$\lfloor A \multimap B\rfloor \mathbin{\circ\!-} \lceil A\rceil \mathbin{\circ\!-} \lfloor B\rfloor.$	$(\multimap R)$	$\lceil A \multimap B\rceil \mathbin{\circ\!-} \lfloor A\rfloor \,⅋\, \lceil B\rceil.$
$(\otimes L)$	$\lfloor A \otimes B\rfloor \mathbin{\circ\!-} \lfloor A\rfloor \,⅋\, \lfloor B\rfloor.$	$(\otimes R)$	$\lceil A \otimes B\rceil \mathbin{\circ\!-} \lceil A\rceil \mathbin{\circ\!-} \lceil B\rceil.$
$(\&L_1)$	$\lfloor A \,\&\, B\rfloor \mathbin{\circ\!-} \lfloor A\rfloor.$	$(\&R)$	$\lceil A \,\&\, B\rceil \mathbin{\circ\!-} \lceil A\rceil \,\&\, \lceil B\rceil.$
$(\&L_2)$	$\lfloor A \,\&\, B\rfloor \mathbin{\circ\!-} \lfloor B\rfloor.$	$(\oplus R_1)$	$\lceil A \oplus B\rceil \mathbin{\circ\!-} \lceil A\rceil.$
$(\oplus L)$	$\lfloor A \oplus B\rfloor \mathbin{\circ\!-} \lfloor A\rfloor \,\&\, \lfloor B\rfloor.$	$(\oplus R_2)$	$\lceil A \oplus B\rceil \mathbin{\circ\!-} \lceil B\rceil.$
$(⅋L)$	$\lfloor A \,⅋\, B\rfloor \mathbin{\circ\!-} \lfloor A\rfloor \mathbin{\circ\!-} \lfloor B\rfloor.$	$(⅋R)$	$\lceil A \,⅋\, B\rceil \mathbin{\circ\!-} \lceil A\rceil \,⅋\, \lceil B\rceil.$
$(!L)$	$\lfloor !B\rfloor \mathbin{\circ\!-} ?\lfloor B\rfloor.$	$(!R)$	$\lceil !B\rceil \Leftarrow \lceil B\rceil.$
$(?L)$	$\lfloor ?B\rfloor \Leftarrow \lfloor B\rfloor.$	$(?R)$	$\lceil ?B\rceil \mathbin{\circ\!-} ?\lceil B\rceil.$
$(\forall_l L)$	$\lfloor \forall_l B\rfloor \mathbin{\circ\!-} \lfloor Bx\rfloor.$	$(\forall_l R)$	$\lceil \forall_l B\rceil \mathbin{\circ\!-} \forall x \lceil Bx\rceil.$
$(\exists_l L)$	$\lfloor \exists_l B\rfloor \mathbin{\circ\!-} \forall x \lfloor Bx\rfloor.$	$(\exists_l R)$	$\lceil \exists_l B\rceil \mathbin{\circ\!-} \lceil Bx\rceil.$
$(1L)$	$\lfloor 1\rfloor \mathbin{\circ\!-} \bot.$	$(1R)$	$\lceil 1\rceil \Leftarrow \top.$
$(\bot L)$	$\lfloor \bot\rfloor \Leftarrow \top.$	$(\bot R)$	$\lceil \bot\rceil \mathbin{\circ\!-} \bot.$
$(0L)$	$\lfloor 0\rfloor \mathbin{\circ\!-} \top.$	$(\top R)$	$\lceil \top\rceil \mathbin{\circ\!-} \top.$
(Cut)	$\bot \mathbin{\circ\!-} \lfloor B\rfloor \mathbin{\circ\!-} \lceil B\rceil.$	$(Initial)$	$\lfloor B\rfloor \,⅋\, \lceil B\rceil.$

FIGURE 2. Forum specification of the LL sequent calculus.

$(\Rightarrow L)$	$\lfloor A \Rightarrow B\rfloor \mathbin{\circ\!-} ?\lceil A\rceil \mathbin{\circ\!-} ?\lfloor B\rfloor.$	$(\Rightarrow R)$	$\lceil A \Rightarrow B\rceil \mathbin{\circ\!-} ?\lfloor A\rfloor \,⅋\, ?\lceil B\rceil.$
$(\wedge L_1)$	$\lfloor A \wedge B\rfloor \mathbin{\circ\!-} ?\lfloor A\rfloor.$	$(\wedge R)$	$\lceil A \wedge B\rceil \mathbin{\circ\!-} ?\lceil A\rceil \,\&\, ?\lceil B\rceil.$
$(\wedge L_2)$	$\lfloor A \wedge B\rfloor \mathbin{\circ\!-} ?\lfloor B\rfloor.$	$(\vee R_1)$	$\lceil A \vee B\rceil \mathbin{\circ\!-} ?\lceil A\rceil.$
$(\vee L)$	$\lfloor A \vee B\rfloor \mathbin{\circ\!-} ?\lfloor A\rfloor \,\&\, ?\lfloor B\rfloor.$	$(\vee R_2)$	$\lceil A \vee B\rceil \mathbin{\circ\!-} ?\lceil B\rceil.$
$(\forall_c L)$	$\lfloor \forall_c B\rfloor \mathbin{\circ\!-} ?\lfloor Bx\rfloor.$	$(\forall_c R)$	$\lceil \forall_c B\rceil \mathbin{\circ\!-} \forall x\, ?\lceil Bx\rceil.$
$(\exists_c L)$	$\lfloor \exists_c B\rfloor \mathbin{\circ\!-} \forall x\, ?\lfloor Bx\rfloor.$	$(\exists_c R)$	$\lceil \exists_c B\rceil \mathbin{\circ\!-} ?\lceil Bx\rceil.$
$(f_c L)$	$\lfloor f_c\rfloor \mathbin{\circ\!-} \top.$	$(t_c R)$	$\lceil t_c\rceil \mathbin{\circ\!-} \top.$
(Cut)	$\bot \mathbin{\circ\!-} ?\lfloor B\rfloor \mathbin{\circ\!-} ?\lceil B\rceil.$	$(Initial)$	$\lfloor B\rfloor \,⅋\, \lceil B\rceil.$

FIGURE 3. Forum specification of the LK sequent calculus.

$(\supset L)$	$\lfloor A \supset B\rfloor \mathbin{\circ\!-} \lceil A\rceil \mathbin{\circ\!-} ?\lfloor B\rfloor.$	$(\supset R)$	$\lceil A \supset B\rceil \mathbin{\circ\!-} ?\lfloor A\rfloor \,⅋\, \lceil B\rceil.$
$(\cap L_1)$	$\lfloor A \cap B\rfloor \mathbin{\circ\!-} ?\lfloor A\rfloor.$	$(\cap R)$	$\lceil A \cap B\rceil \mathbin{\circ\!-} \lceil A\rceil \,\&\, \lceil B\rceil.$
$(\cap L_2)$	$\lfloor A \cap B\rfloor \mathbin{\circ\!-} ?\lfloor B\rfloor.$	$(\cup R_1)$	$\lceil A \cup B\rceil \mathbin{\circ\!-} \lceil A\rceil.$
$(\cup L)$	$\lfloor A \cup B\rfloor \mathbin{\circ\!-} ?\lfloor A\rfloor \,\&\, ?\lfloor B\rfloor.$	$(\cup R_2)$	$\lceil A \cup B\rceil \mathbin{\circ\!-} \lceil B\rceil.$
$(\forall_i L)$	$\lfloor \forall_i B\rfloor \mathbin{\circ\!-} ?\lfloor Bx\rfloor.$	$(\forall_i R)$	$\lceil \forall_i B\rceil \mathbin{\circ\!-} \forall x \lceil Bx\rceil.$
$(\exists_i L)$	$\lfloor \exists_i B\rfloor \mathbin{\circ\!-} \forall x\, ?\lfloor Bx\rfloor.$	$(\exists_i R)$	$\lceil \exists_i B\rceil \mathbin{\circ\!-} \lceil Bx\rceil.$
$(f_i L)$	$\lfloor f_i\rfloor \mathbin{\circ\!-} \top.$	$(t_i R)$	$\lceil t_i\rceil \mathbin{\circ\!-} \top.$
(Cut)	$\bot \mathbin{\circ\!-} ?\lfloor B\rfloor \mathbin{\circ\!-} \lceil B\rceil.$	$(Initial)$	$\lfloor B\rfloor \,⅋\, \lceil B\rceil.$

FIGURE 4. Specification of the LJ sequent calculus.

the clauses of LJ' into those of LJ by inserting the $?$ modal. Thus, assuming $\Sigma_j : LJ; \longrightarrow \lceil B_0\rceil; \lfloor B_1\rfloor, \ldots, \lfloor B_n\rfloor$ has a $\mathcal{F}$-proof, we can use cut-elimination to conclude $\Sigma_j: LJ'; \longrightarrow \lceil B_0 \rceil; \lfloor B_1 \rfloor, \ldots, \lfloor B_n \rfloor$ has a $\mathcal{F}$-proof. Using the Pos_2 clauses of LJ' n-times, we can conclude that $\Sigma_j: LJ'; \longrightarrow \lfloor B_1 \rfloor, \ldots, \lfloor B_n \rfloor, \lceil B_0 \rceil$; has a $\mathcal{F}$-proof.

To prove the converse, we prove the following lemma by induction on the height of proofs in Forum: Let $\mathcal{L}_1$ and $\mathcal{L}_2$ be multisets of left-atoms and let R be a right-atom. Then if $\Sigma_j: LJ'; \longrightarrow R, \mathcal{L}_1; \mathcal{L}_2$ has a $\mathcal{F}$-proof, then $\Sigma_j: LJ; \longrightarrow R; \mathcal{L}_1, \mathcal{L}_2$ has a Forum proof. The proof proceeds by examining each case for how this sequent could be proved.   □

An immediate corollary of this Proposition and the correctness of LJ and LK is the correctness of LJ' and LK': namely, $\Sigma_j: LJ'; \longrightarrow \lfloor B_1 \rfloor, \ldots, \lfloor B_n \rfloor$, $\lceil B_0 \rceil$; has a $\mathcal{F}$-proof if and only if the sequent $B_1, \ldots, B_n \longrightarrow B_0$ has an LJ-proof and $\Sigma_c: LK'; \longrightarrow \lfloor B_1 \rfloor, \ldots, \lfloor B_n \rfloor, \lceil C_1 \rceil, \ldots, \lceil C_n \rceil$; has an $\mathcal{F}$-proof if and only if the sequent $B_1, \ldots, B_n \longrightarrow C_1, \ldots, C_n$ has an LK-proof.

$(\Rightarrow L)$	$\lfloor A \Rightarrow B \rfloor \circ\!\!- \lceil A \rceil \circ\!\!- \lfloor B \rfloor.$	$(\Rightarrow R)$	$\lceil A \Rightarrow B \rceil \circ\!\!- \lfloor A \rfloor \,\mathfrak{N}\, \lceil B \rceil.$
$(\wedge L_1)$	$\lfloor A \wedge B \rfloor \circ\!\!- \lfloor A \rfloor.$	$(\wedge R)$	$\lceil A \wedge B \rceil \circ\!\!- \lceil A \rceil \,\&\, \lceil B \rceil.$
$(\wedge L_2)$	$\lfloor A \wedge B \rfloor \circ\!\!- \lfloor B \rfloor.$	$(\vee R_1)$	$\lceil A \vee B \rceil \circ\!\!- \lceil A \rceil.$
$(\vee L)$	$\lfloor A \vee B \rfloor \circ\!\!- \lfloor A \rfloor \,\&\, \lfloor B \rfloor.$	$(\vee R_2)$	$\lceil A \vee B \rceil \circ\!\!- \lceil B \rceil.$
$(\forall_c L)$	$\lfloor \forall_c B \rfloor \circ\!\!- \lfloor Bx \rfloor.$	$(\forall_c R)$	$\lceil \forall_c B \rceil \circ\!\!- \forall x \lceil Bx \rceil.$
$(\exists_c L)$	$\lfloor \exists_c B \rfloor \circ\!\!- \forall x \lfloor Bx \rfloor.$	$(\exists_c R)$	$\lceil \exists_c B \rceil \circ\!\!- \lceil Bx \rceil.$
$(f_c L)$	$\lfloor f_c \rfloor \circ\!\!- \top.$	$(t_c R)$	$\lceil t_c \rceil \circ\!\!- \top.$

FIGURE 5. LK_0: The introduction rules of LK with the ? dropped.

$(\supset L)$	$\lfloor A \supset B \rfloor \circ\!\!- \lceil A \rceil \circ\!\!- \lfloor B \rfloor.$	$(\supset R)$	$\lceil A \supset B \rceil \circ\!\!- \lfloor A \rfloor \,\mathfrak{N}\, \lceil B \rceil.$
$(\cap L_1)$	$\lfloor A \cap B \rfloor \circ\!\!- \lfloor A \rfloor.$	$(\cap R)$	$\lceil A \cap B \rceil \circ\!\!- \lceil A \rceil \,\&\, \lceil B \rceil.$
$(\cap L_2)$	$\lfloor A \cap B \rfloor \circ\!\!- \lfloor B \rfloor.$	$(\cup R_1)$	$\lceil A \cup B \rceil \circ\!\!- \lceil A \rceil.$
$(\cup L)$	$\lfloor A \cup B \rfloor \circ\!\!- \lfloor A \rfloor \,\&\, \lfloor B \rfloor.$	$(\cup R_2)$	$\lceil A \cup B \rceil \circ\!\!- \lceil B \rceil.$
$(\forall_i L)$	$\lfloor \forall_i B \rfloor \circ\!\!- \lfloor Bx \rfloor.$	$(\forall_i R)$	$\lceil \forall_i B \rceil \circ\!\!- \forall x \lceil Bx \rceil.$
$(\exists_i L)$	$\lfloor \exists_i B \rfloor \circ\!\!- \forall x \lfloor Bx \rfloor.$	$(\exists_i R)$	$\lceil \exists_i B \rceil \circ\!\!- \lceil Bx \rceil.$
$(f_i L)$	$\lfloor f_i \rfloor \circ\!\!- \top.$	$(t_i R)$	$\lceil t_i \rceil \circ\!\!- \top.$

FIGURE 6. LJ_0: The introduction rules of LJ with the ? dropped.

(Pos_1)	$\lceil B \rceil \multimap !\lceil B \rceil.$	(Neg_1)	$\lfloor B \rfloor \multimap !\lfloor B \rfloor.$
(Pos_2)	$\lfloor B \rfloor \circ\!\!- ?\lfloor B \rfloor.$	(Neg_2)	$\lceil B \rceil \circ\!\!- ?\lceil B \rceil.$
(Cut)	$\perp \circ\!\!- \lfloor B \rfloor \circ\!\!- \lceil B \rceil.$	$(Initial)$	$\lfloor B \rfloor \,\mathfrak{N}\, \lceil B \rceil.$

FIGURE 7. Some named formulas.

Notice that the inference rules for LJ_0 and LK_0 are identical except for a systematic renaming of logical constants. Thus one way to modularly describe the distinction between intuitionistic and classical logics is that the former logic assumes Pos_2 while the latter logic assumes both Pos_2 and Neg_2. This description amounts to saying that contraction is allowed on the right and left in classical proofs but only on the left in intuitionistic proofs.

4.3. Collapsing of modal prefixes. Note that the following equivalences are provable from the various encodings of proof systems:

1. The Cut and Initial rules of LL prove the equivalence $\lceil B \rceil \equiv \lfloor B \rfloor^\perp$.
2. The Cut and Initial rules of LK prove the following equivalences:
   $\lceil B \rceil \equiv \lfloor B \rfloor^\perp, ?\lfloor B \rfloor \equiv \lfloor B \rfloor, ?\lceil B \rceil \equiv \lceil B \rceil, ?\lceil B \rceil \equiv (?\lfloor B \rfloor)^\perp, ?\lceil B \rceil \equiv !\lceil B \rceil,$
   $?\lfloor B \rfloor \equiv !\lfloor B \rfloor, !\lfloor B \rfloor \equiv \lfloor B \rfloor$, and $?\lceil B \rceil \equiv \lceil B \rceil$. As an example proof of such an equivalence, the *Cut* rule is equivalent to $(?\lfloor B \rfloor)^\perp \circ\!\!- ?\lceil B \rceil$. On the other hand,

$$\frac{\dfrac{\overline{\lfloor B \rfloor \longrightarrow \lfloor B \rfloor}}{\lfloor B \rfloor \longrightarrow ?\lfloor B \rfloor} \quad \overline{\bot \longrightarrow \bot}}{\dfrac{\lfloor B \rfloor, ?\lfloor B \rfloor \!-\!\!\circ \bot \longrightarrow \bot \quad \dfrac{\overline{\lceil B \rceil \longrightarrow \lceil B \rceil}}{\lceil B \rceil \longrightarrow ?\lceil B \rceil}}{\lfloor B \rfloor \,⅋\, \lceil B \rceil, ?\lfloor B \rfloor \!-\!\!\circ \bot \longrightarrow ?\lceil B \rceil}}$$

That is, the *Initial* rule implies $(?\lfloor B \rfloor)^\perp \!-\!\!\circ ?\lceil B \rceil$. Hence $(?\lfloor B \rfloor)^\perp \equiv ?\lceil B \rceil$ follows from them both.

3. The Cut and Initial rules of LJ prove the equivalences $\lceil B \rceil \equiv (?\lfloor B \rfloor)^\perp$, $\lceil B \rceil \equiv !\lceil B \rceil, \lfloor B \rfloor \equiv ?\lfloor B \rfloor$, and $\lceil B \rceil \equiv \lfloor B \rfloor^\perp$.

Thus, the cut and initial rules show the (not surprising fact) that $\lfloor \cdot \rfloor$ and $\lceil \cdot \rceil$ are duals of each other. In the cases of LJ and LK, however, that duality also forces additional equivalences that cause the collapse of some of modals. As it is well known, linear logic has 7 distinct modalities, namely: the empty modality, !, ?, ?!, !?, !?!, and ?!?. Given the LK theory, however, all those modals collapse into just two when applied to a $\lfloor \cdot \rfloor$-atom or a $\lceil \cdot \rceil$-atom and in LJ, these modals collapse to four when applied to either the $\lceil \cdot \rceil$-atoms or the $\lfloor \cdot \rfloor$-atoms.

Such a collapse is certainly undesirable when specifications rely on proof search: we would like to have a lot of distinctions available to help us understanding how formulas are to be used within object-level proofs. It would be far more interesting to have proof systems for intuitionistic and classical logics, for example, in which these modals would not generally collapse. Recent advances in understanding sequent calculus for these logics provide just such proof systems. We illustrate some of them in Section 5.

4.4. Natural deduction. To illustrate an application of using meta-level reasoning to draw conclusions about an object-logic, we show how a specification for natural deduction in intuitionistic logic can be derived from a sequent calculus specification of intuitionistic logic. For simplicity, we consider a minimal logic fragment of intuitionistic logic involving only $\supset$, $\cap$, and $\forall_i$: let LM be the subset of LJ from Figure 4 containing Cut, $Initial$, and introduction rules for those three connectives. (The disjoint sums are addressed in [23].)

Given the equivalences arising from the cut and initial rules in LJ listed in Section 4.3, the specification for $(\supset L)$ is equivalent to the following formulas.

$$? \lfloor B \rfloor \multimap \lceil A \rceil \multimap \lfloor A \supset B \rfloor \equiv \lfloor B \rfloor \multimap \lceil A \rceil \multimap \lfloor A \supset B \rfloor$$
$$\equiv \lceil B \rceil^{\perp} \multimap \lceil A \rceil \multimap \lceil A \supset B \rceil^{\perp}$$
$$\equiv \lceil A \supset B \rceil \multimap \lceil A \rceil \multimap \lceil B \rceil$$

The later can be recognized as a specification of the $\supset$ elimination rule. Similarly, the specification for $(\supset R)$ is equivalent to the following formulas.

$$? \lfloor A \rfloor \,\invamp\, \lceil B \rceil \multimap \lceil A \supset B \rceil \equiv \lceil A \rceil^{\perp} \,\invamp\, \lceil B \rceil \multimap \lceil A \supset B \rceil$$
$$\equiv (! \lceil A \rceil)^{\perp} \,\invamp\, \lceil B \rceil \multimap \lceil A \supset B \rceil$$
$$\equiv (\lceil A \rceil \Rightarrow \lceil B \rceil) \multimap \lceil A \supset B \rceil$$

Continuing in such a manner, we can systematically replace all occurrences of $\lfloor \cdot \rfloor$ with occurrences of $\lceil \cdot \rceil$, as listed in Figure 8. The clauses in this figure, named NM, can easily be seen as specifying the introduction and elimination rules for this particular fragment of minimal logic. The usual specification of natural deduction rules for minimal logic [11, 16] has intuitionistic implications replacing the top-level linear implications in Figure 8, but as observed in [17], the choice of which implication to use for these top-level occurrences does not change the set of atomic formulas that are provable.

$$
\begin{array}{llll}
(\supset I) & \lceil A \supset B \rceil \;\circ\!\!- \; \lceil A \rceil \Rightarrow \lceil B \rceil. & (\supset E) & \lceil B \rceil \;\circ\!\!- \; \lceil A \rceil \;\circ\!\!- \; \lceil A \supset B \rceil. \\
(\forall_i I) & \lceil \forall_i B \rceil \;\circ\!\!- \; \forall x \lceil B x \rceil. & (\forall_i E) & \lceil B x \rceil \;\circ\!\!- \; \lceil \forall_i B \rceil. \\
(\cap I) & \lceil A \cap B \rceil \;\circ\!\!- \; \lceil A \rceil \,\&\, \lceil B \rceil. & (\cap E_1) & \lceil A \rceil \;\circ\!\!- \; \lceil A \cap B \rceil. \\
& & (\cap E_2) & \lceil B \rceil \;\circ\!\!- \; \lceil A \cap B \rceil.
\end{array}
$$

FIGURE 8. Specification of the NM natural deduction calculus.

As a result of this rather natural connection between clauses in LM and NM, the following Propositions have rather direct proofs (see [23] for details).

PROPOSITION 4.3. $! LM \equiv ![(\&NM) \,\&\, Initial \,\&\, Cut]$.

PROPOSITION 4.4. *If B is an object-level formula, then $NM \vdash \lceil B \rceil$ if and only if $LM \vdash \lceil B \rceil$.*

As a consequence of the last Proposition and the correctness of representation of LM and NM, we can conclude that a formula B has a sequent calculus proof if and only if it has a natural deduction proof.

§5. More refined uses of modals. For the sake of presenting examples in this section, we shall consider the fragments of intuitionistic and classical logics that involve just implication and universal quantification. Gentzen's LJ system for these two connectives is reproduced in Figure 9.

$$(\supset L) \quad \lfloor A \supset B \rfloor \circ\!- \lceil A \rceil \circ\!- ?\lfloor B \rfloor. \qquad (\supset R) \quad \lceil A \supset B \rceil \circ\!- ?\lfloor A \rfloor \,\mathfrak{N}\, \lceil B \rceil.$$
$$(\forall_i L) \qquad \lfloor \forall_i B \rfloor \circ\!- ?\lfloor Bx \rfloor. \qquad\quad (\forall_i R) \qquad \lceil \forall_i B \rceil \circ\!- \forall x \lceil Bx \rceil.$$
$$(Cut) \qquad\quad \bot \circ\!- \lceil A \rceil \circ\!- ?\lfloor A \rfloor. \qquad (Initial) \quad \lfloor A \rfloor \,\mathfrak{N}\, \lceil A \rceil.$$

FIGURE 9. The $\{\supset, \forall_i\}$-fragment of *LJ*

It is well known that proof search in the intuitionistic logic of these connectives can be focused, in the sense that left-introduction rules are only applied to a distinguished formula (such focusing is a justification for backchaining in logic programming). Danos et al. [7] present the focused formulation of intuitionistic logic called *ILU* and displayed in Figure 10. Here, sequents have the form $\Pi; \Gamma \longrightarrow A$ where Γ and Π denote multisets, and Π containing at most one formula. The *ILU* proof system can be encoded in Forum by representing such sequents as $\Sigma: \cdot; \cdot \longrightarrow \lfloor \Pi \rfloor, \lceil A \rceil; \lfloor \Gamma \rfloor$ and its inference rules as in Figure 11. (If Γ is a multiset or set of object-level formulas, we write $\lfloor \Gamma \rfloor$ and $\lceil \Gamma \rceil$ to be the corresponding multiset or set of meta-level formulas resulting from applying the corresponding predicate to all formulas in Γ.)

Proofs in ILU are focused in a sense that the left rules $(\supset L)$ and $(\forall_i L)$ can only be applied to formulas in the left linear context Π (in Forum, this is the $\lfloor \cdot \rfloor$-formula without the ?-modal prefix). This restriction, which is enforced using modals in the Forum encoding, constrains proof search significantly.

The two cut rules for ILU, *head-cut* and *mid-cut*, are encoded as two formulas in Figure 11 in such as way that the first implies the second: that is,

$$(\lfloor A \rfloor \multimap \lceil A \rceil \multimap \bot) \Rightarrow (?\lfloor A \rfloor \multimap \lceil A \rceil \Rightarrow \bot).$$

is provable in linear logic. As a result, we shall refer to the head-cut as *the* cut rule. Observe that from the *Cut* and *Initial* rules of ILU, we can prove the equivalence $\lceil B \rceil \equiv \lfloor B \rfloor^{\perp}$ but we cannot prove any equivalences between linear logic modals. Note also that *ILU* is equivalent to the neutral fragment of intuitionistic implicational logic of *LU* (see Section 6), although it was formulated in order to obtain a sequent calculus for an *inductive decoration strategy* (see [7] for the definition) of intuitionistic logic into linear logic.

Two sequent calculi, *LKQ* and *LKT*, which provide a focused kind of proof system for classical logic are also presented in [7]. Sequents of the calculus *LKQ* (Figure 12), written as $\Gamma \longrightarrow \Delta; \Pi$ are encoded as Forum sequents $\Sigma: \cdot; \cdot \longrightarrow \lceil \Pi \rceil; \lfloor \Gamma \rfloor, \lceil \Delta \rceil$ where Π represents a multiset containing at most one

$$\frac{}{A;\cdot \longrightarrow A}\ initial$$

$$\frac{\Pi;\Gamma_1 \longrightarrow A \quad A;\Gamma_2 \longrightarrow B}{\Pi;\Gamma_1,\Gamma_2 \longrightarrow B}\ head\text{-}cut \qquad \frac{;\Gamma_1 \longrightarrow A \quad \Pi;A,\Gamma_2 \longrightarrow B}{\Pi;\Gamma_1,\Gamma_2 \longrightarrow B}\ mid\text{-}cut$$

$$\frac{\Pi;\Gamma \longrightarrow A}{\Pi;\Gamma,B \longrightarrow A}\ WL \qquad \frac{\Pi;\Gamma,B,B \longrightarrow A}{\Pi;\Gamma,B \longrightarrow A}\ CL \qquad \frac{B;\Gamma \longrightarrow A}{\cdot;B,\Gamma \longrightarrow A}\ D$$

$$\frac{\cdot;\Gamma \longrightarrow A \quad B;\Gamma' \longrightarrow C}{A \supset B;\Gamma,\Gamma' \longrightarrow C}\ \supset L \qquad \frac{\Pi;\Gamma,A \longrightarrow B}{\Pi;\Gamma \longrightarrow A \supset B}\ \supset R$$

$$\frac{A[x/t];\Gamma \longrightarrow B}{\forall_l x A;\Gamma \longrightarrow B}\ \forall_i L \qquad \frac{\Pi;\Gamma \longrightarrow A[x/y]}{\Pi;\Gamma \longrightarrow \forall_l x A}\ \forall_l R$$

FIGURE 10. The sequent calculus *ILU*

$$
\begin{array}{llll}
(\supset L) & \lfloor A \supset B \rfloor \Leftarrow \lceil A \rceil \multimap \lfloor B \rfloor. & (\supset R) & \lceil A \supset B \rceil \multimap ?\lfloor A \rfloor \,\bindnasrepma\, \lceil B \rceil. \\
(\forall_i L) & \lfloor \forall_i B \rfloor \Leftarrow \lfloor Bx \rfloor. & (\forall_i R) & \lceil \forall_i B \rceil \multimap \forall x \lceil Bx \rceil. \\
(Head\text{-}cut) & \bot \multimap \lceil A \rceil \multimap \lfloor A \rfloor. & (Initial) & \lfloor A \rfloor \,\bindnasrepma\, \lceil A \rceil. \\
(Mid\text{-}cut) & \bot \Leftarrow \lceil A \rceil \multimap ?\lfloor A \rfloor. & &
\end{array}
$$

FIGURE 11. Specification of the calculus *ILU*

formula. Note that the rules are the same as the ones for the positive classical implicational fragment of *LU* (see Section 6) i.e., rules defined for *positive* formulas. However, *LKQ* cannot be identified with any proper fragment of *LU* since positive polarity is not preserved by the connectives $\Rightarrow$ and $\forall_c$. Sequents of the *LKT* proof system (Figure 13), written as $\Gamma \longrightarrow \Delta;\Pi$, are encoded as $\Sigma: \cdot;\cdot \longrightarrow \lfloor \Pi \rfloor; \lfloor \Gamma \rfloor, \lceil \Delta \rceil$, where again Π is a multiset containing at most one formula. Observe that *LKT* is a classical equivalent of *ILU*; that is, the intuitionistic calculus is obtained from *LKT* by the usual restriction of having exactly one formula on the right side of the sequent. *LKT* is equivalent to the negative fragment of classical implicational logic of *LU*.

In both *LKQ* and *LKT* systems there is a collapse of modal prefixes:

1. The *Cut* and *Initial* rules of *LKQ* prove the equivalence $\lceil B \rceil \equiv !\lceil B \rceil$ and the modalities collapse to four when applied to $\lceil \cdot \rceil$-atoms. Thus, in *LKQ* the formula Pos_1 holds.

2. The *Cut* and *Initial* rules of *LKT* prove the equivalence $\lfloor B \rfloor \equiv !\lfloor B \rfloor$ and the modalities collapse to four when applied to $\lfloor \cdot \rfloor$-atoms. Thus, in *LKT* the formula Neg_1 holds.

We present one final example encoding of a proof system, by picking a system that deviates from the previous one in a few details. An optimized version of (the implicational fragment of) *LJ* is presented in Lincoln et al. [19] (see also [9]). Their system, called *IIL** (Figure 14) does not contain contraction

$(\Rightarrow L)$ $\lfloor A \Rightarrow B \rfloor \Leftarrow \lceil A \rceil \Leftarrow ? \lfloor B \rfloor$. $(\Rightarrow R)$ $\lceil A \Rightarrow B \rceil \Leftarrow ? \lfloor A \rfloor \,\mathbf{\mathfrak{N}} ? \lceil B \rceil$.
$(\forall_c L)$ $\lfloor \forall_c B \rfloor \Leftarrow ? \lfloor Bx \rfloor$. $(\forall_c R)$ $\lceil \forall_c B \rceil \Leftarrow \forall x \, ? \lceil Bx \rceil$.
(Cut) $\perp \circ\!\!-\, \lceil A \rceil \circ\!\!-\, ? \lfloor A \rfloor$. $(Initial)$ $\lfloor A \rfloor \,\mathbf{\mathfrak{N}}\, \lceil A \rceil$.
   $\perp \circ\!\!-\, ? \lceil A \rceil \Leftarrow ? \lfloor A \rfloor$.

FIGURE 12. The calculus LKQ

$(\Rightarrow L)$ $\lfloor A \Rightarrow B \rfloor \Leftarrow ? \lceil A \rceil \circ\!\!-\, \lfloor B \rfloor$. $(\Rightarrow R)$ $\lceil A \Rightarrow B \rceil \circ\!\!-\, ? \lfloor A \rfloor \,\mathbf{\mathfrak{N}} ? \lceil B \rceil$.
$(\forall_c L)$ $\lfloor \forall_c B \rfloor \Leftarrow \lfloor Bx \rfloor$. $(\forall_c R)$ $\lceil \forall_c B \rceil \circ\!\!-\, \forall x \, ? \lceil Bx \rceil$.
(Cut) $\perp \circ\!\!-\, ? \lceil A \rceil \circ\!\!-\, \lfloor A \rfloor$. $(Initial)$ $\lfloor A \rfloor \,\mathbf{\mathfrak{N}}\, \lceil A \rceil$.
   $\perp \Leftarrow ? \lceil A \rceil \circ\!\!-\, ? \lfloor A \rfloor$.

FIGURE 13. The calculus LKT

or cut rules, and weakening is only allowed at the leafs of a proof; that is, when the *Initial* rule is applied (to atomic formulas). A key property of IIL^* is that the *principal* formula is not duplicated in the premises of any of the rules. This suggests the encoding $\Sigma: \cdot; \cdot \longrightarrow \lceil D \rceil, \lfloor \Gamma \rfloor; \cdot$ for the IIL^* sequent $\Gamma \longrightarrow D$. It also requires encoding the *Initial* and $\supset L$ rules differently than we have seen so far: the *Initial* rule uses the additive true, $\top$, to allow weakening, and the $\supset L$ rules uses the additive conjunction, $\&$, to copy the left context and uses a two headed clause so that the right context is not copied but is placed in the correct sequent of the premise.

$(Initial)$ $\lfloor A \rfloor \,\mathbf{\mathfrak{N}}\, \lceil A \rceil \circ\!\!-\, \top \circ\!\!-\, atomic(A)$.
$(\supset R)$ $\lceil B \supset C \rceil \circ\!\!-\, \lceil B \rceil \,\mathbf{\mathfrak{N}}\, \lfloor C \rfloor$.
$(\supset 1L)$ $\lfloor A \supset B \rfloor \,\mathbf{\mathfrak{N}}\, \lceil D \rceil \circ\!\!-\, \lceil A \rceil \& (\lfloor B \rfloor \,\mathbf{\mathfrak{N}}\, \lceil D \rceil) \circ\!\!-\, atomic(A)$.
$(\supset 2L)$ $\lfloor (A \supset B) \supset C \rfloor \,\mathbf{\mathfrak{N}}\, \lceil D \rceil \circ\!\!-\, (\lfloor B \supset C \rfloor \,\mathbf{\mathfrak{N}}\, \lceil A \supset B \rceil) \& (\lfloor C \rfloor \,\mathbf{\mathfrak{N}}\, \lceil D \rceil)$.

FIGURE 14. The calculus IIL^* where $atomic(\cdot)$ is a predicate of type $bool \rightarrow o$ defined to hold for all atomic formulas.

§6. **Using polarities in proof systems.** In [14], Girard introduced the sequent system LU (logic of unity) in which classical, intuitionistic, and linear logics appear as fragments. In this logic, all three of these logics keep their own characteristics but they can also communicate via formulas containing connectives mixing these logics. The key to allowing these logics to share one proof system lies in using *polarities*. In terms of the encoding we have presented here, this corresponds to restricting the use of Pos_2 and Neg_2 rules to *positive* and *negative* formulas respectively and to split the rules for classical, intuitionistic, and linear connectives into cases, depending on the polarities of the subformulas involved.

We proceed to encode LU into Forum as follows. The LU sequent $\Gamma; \Gamma' \longrightarrow \Delta'; \Delta$ is encoded as the Forum sequent

$$\Sigma: \cdot; \cdot \longrightarrow \lfloor \Gamma \rfloor, \lceil \Delta \rceil; \lfloor \Gamma' \rfloor, \lceil \Delta' \rceil.$$

(Notice the different convention used between LU sequents and Forum sequents with regard to which zones in a sequent allow structural rules.) To encode the polarity of object-level, LU formulas, we introduce three meta-level predicates, $pos(\cdot)$, $neg(\cdot)$, and $neu(\cdot)$, all of type $bool \rightarrow o$. We can now encode the Identity and Structure rules on [14, page 206]. The Cut and Initial rules are encoded just as in linear logic. The Cut_2 and Cut_3 rules are, respectively,

$$\bot \circ\!\!- \; ?\lceil B \rceil \; \Leftarrow \lfloor B \rfloor \qquad \text{and} \qquad \bot \Leftarrow \lceil B \rceil \; \circ\!\!- \; ?\lfloor B \rfloor.$$

Both of these formulas, however, are consequences of the Cut rule and are not needed in our encoding of LU. Similarly, the first structural rules are all simple consequences of using exponentials in encoding sequents. Finally, the fact that structural rules are allowed for positive and negative formulas is given as

$$(Neg) \quad \lceil N \rceil \; \circ\!\!- \; ?\lceil N \rceil \; \Leftarrow neg(N).$$
$$(Pos) \quad \lfloor P \rfloor \; \circ\!\!- \; ?\lfloor P \rfloor \; \Leftarrow pos(P).$$

Notice that if we use Cut and Initial to eliminate, say, $\lfloor \cdot \rfloor$ for $\lceil \cdot \rceil$ (as we did in Section 4.4), then the only non-trivial inference rules among those coding Identity and Structure rules in this presentation of LU are the (Neg) and (Pos) rules.

The calculus for linear connectives in *LU* is equivalent to the usual one (see Fig. 2) and the rules do not depend on the polarities. Figure 15 specifies some polarities for classical and intuitionistic connectives (polarities for linear logic connectives can be given similarly). Many of the LU inference rules for classical and intuitionistic connectives are specified in Figure 16. The full encoding of the *LU* proof system is not given here, but most of it is contained in the union of the clauses in Figures 2, 15, and 16. Observe that the use of Forum to encode the *LU* proof system provides a reduced set of rules (compare e.g. 8 disjunction rules versus 24 that appear in [14]). Compaction is due partially to the fact that if B is positive then $\lceil B \rceil \equiv !\lceil B \rceil$ and $\lfloor B \rfloor \equiv ?\lfloor B \rfloor$ and if B is negative then $\lceil B \rceil \equiv ?\lceil B \rceil$ and $\lfloor B \rfloor \equiv !\lfloor B \rfloor$ (thus the expression $A \Leftarrow \lfloor B \rfloor$ might also be equivalent to $A \circ\!\!- \lfloor B \rfloor$). Further compaction occurs by departing from Forum syntax slightly and allowing occurrences of $\oplus$ and $!$ in the body of clauses. Such occurrences can be easily removed to form

$$notpos(B) \Leftarrow neg(B) \oplus neu(B).$$
$$notneg(B) \Leftarrow pos(B) \oplus neu(B).$$
$$notneu(B) \Leftarrow pos(B) \oplus neg(B).$$
$$pos(A \wedge B) \Leftarrow pos(A).$$
$$pos(A \wedge B) \Leftarrow pos(B).$$
$$neg(A \wedge B) \Leftarrow neg(A) \Leftarrow neg(B).$$
$$neg(A \vee B) \Leftarrow neg(A).$$
$$neg(A \vee B) \Leftarrow neg(B).$$
$$pos(A \vee B) \Leftarrow notneg(A) \Leftarrow notneg(B).$$
$$neg(A \Rightarrow B) \Leftarrow pos(A).$$
$$neg(A \Rightarrow B) \Leftarrow neg(B).$$
$$pos(A \Rightarrow B) \Leftarrow notpos(A) \Leftarrow notneg(B).$$
$$neg(A \supset B) \Leftarrow neg(B).$$
$$neg(\forall_c B).$$
$$pos(\exists_c B).$$
$$neu(A \wedge B) \Leftarrow neu(A) \Leftarrow neu(B).$$
$$neu(A \wedge B) \Leftarrow neu(A) \Leftarrow neg(B).$$
$$neu(A \wedge B) \Leftarrow neg(A) \Leftarrow neu(B).$$
$$neu(A \Rightarrow B) \Leftarrow notneg(B).$$

FIGURE 15. Positive and negative polarities for intuitionistic and classical logic. See Table 2 in [14].

Forum clauses: for example, the formula $(a \oplus !\, b) \multimap c$ is logically equivalent to $(a \multimap c) \,\&\, (b \Rightarrow c)$.

§7. Automation of proof systems.

Since the specifications of proof systems are given as clauses in Forum and since Forum can be seen as an *abstract logic programming language* in the sense of [24], it is natural to ask if it is possible to turn these specifications into implementations.

One might attempt to do this using one of the available implementations of Forum [18, 20, 31]. It is, however, a simple matter to turn the specification of Forum given in Figure 1 into a naive interpreter using a logic programming language such as λProlog [25]. We will not present the details of such an implementation except to describe three aspects of it. First, it can be structured such that one inference rule in Figure 1 is translated to one λProlog clause: the resulting implementation is thus rather compact and declarative. Second, the quantification and substitution aspects of the object-logics can be captured directly using λProlog's higher-order features: using a first-order logic programming language such as Prolog would have complicated the implementation significantly. Third, a counter can be used in the clauses of this interpreter to count the number of times a *decide*! or a *decide* rule is used along

Identity and structure

$\lfloor B \rfloor \,\mathbin{⅋}\, \lceil B \rceil$.

$\qquad \perp \multimap \lfloor B \rfloor \multimap \lceil B \rceil$.

$\qquad \lceil N \rceil \multimap ?\lceil N \rceil \; \Leftarrow neg(N)$.

$\qquad \lfloor P \rfloor \multimap ?\lfloor P \rfloor \; \Leftarrow pos(P)$.

Conjunction

$\lceil u \wedge v \rceil \Leftarrow \lceil u \rceil \Leftarrow \lceil v \rceil \qquad\qquad \Leftarrow pos(u) \oplus pos(v)$.

$\lceil u \wedge v \rceil \multimap \lceil u \rceil \,\&\, \lceil v \rceil \qquad\qquad \Leftarrow notpos(u) \Leftarrow notpos(v)$.

$\lfloor u \wedge v \rfloor \multimap ?\lfloor u \rfloor \,\mathbin{⅋}\, ?\lfloor v \rfloor \qquad\quad\, \Leftarrow pos(u) \oplus pos(v)$.

$\lfloor u \wedge v \rfloor \multimap \lfloor u \rfloor \oplus \lfloor v \rfloor \qquad\qquad \Leftarrow notpos(u) \Leftarrow notpos(v)$.

Intuitionistic implication

$\lceil u \supset v \rceil \multimap ?\lfloor u \rfloor \,\mathbin{⅋}\, \lceil v \rceil$.

$\lfloor u \supset v \rfloor \Leftarrow \lceil u \rceil \multimap \lfloor v \rfloor$.

Quantifiers

$\lceil \forall_c u \rceil \multimap \forall x \, ?\lceil ux \rceil$.

$\lfloor \forall_c u \rfloor \Leftarrow \lfloor ux \rfloor$.

$\lceil \exists_c u \rceil \Leftarrow \lceil ux \rceil$.

$\lfloor \exists_c u \rfloor \multimap \forall x \, ?\lfloor ux \rfloor$.

Disjunction

$\lceil u \vee v \rceil \multimap \,!\lceil u \rceil \oplus \,!\lceil v \rceil \qquad\qquad\quad\; \Leftarrow notneg(u) \Leftarrow notneg(v)$.

$\lceil u \vee v \rceil \multimap ?\lceil u \rceil \,\mathbin{⅋}\, ?\lceil v \rceil \qquad\qquad\; \Leftarrow (pos(u) \,\&\, neg(v)) \oplus (neg(u) \,\&\, notneu(v))$.

$\lceil u \vee v \rceil \multimap \lceil u \rceil \,\mathbin{⅋}\, ?!\lceil v \rceil \qquad\qquad\; \Leftarrow neg(u) \Leftarrow neu(v)$.

$\lceil u \vee v \rceil \multimap ?!\lceil u \rceil \,\mathbin{⅋}\, \lceil v \rceil \qquad\qquad\; \Leftarrow neu(u) \Leftarrow neg(v)$.

$\lfloor u \vee v \rfloor \multimap ?\lfloor u \rfloor \,\&\, ?\lfloor v \rfloor \qquad\qquad\; \Leftarrow notneg(u) \Leftarrow notneg(v)$.

$\lfloor u \vee v \rfloor \Leftarrow \lfloor u \rfloor \Leftarrow \lfloor v \rfloor \qquad\qquad\quad \Leftarrow (pos(u) \,\&\, neg(v)) \oplus (neg(u) \,\&\, notneu(v))$.

$\lfloor u \vee v \rfloor \multimap \lfloor u \rfloor \Leftarrow ?\lfloor v \rfloor \qquad\qquad\; \Leftarrow neg(u) \Leftarrow neu(v)$.

$\lfloor u \vee v \rfloor \Leftarrow ?\lfloor u \rfloor \multimap \lfloor v \rfloor \qquad\qquad\; \Leftarrow neu(u) \Leftarrow neg(v)$.

Classical implication

$\lceil u \Rightarrow v \rceil \multimap ?\lfloor u \rfloor \,\mathbin{⅋}\, ?\lceil v \rceil \qquad\qquad \Leftarrow (neg(u) \,\&\, neg(v)) \oplus (pos(u) \,\&\, notneu(v))$.

$\lceil u \Rightarrow v \rceil \multimap \lceil v \rceil \oplus \lfloor u \rfloor \qquad\qquad\; \Leftarrow neg(u) \Leftarrow pos(v)$.

$\lfloor u \Rightarrow v \rfloor \multimap \lceil u \rceil \,\&\, \lfloor v \rfloor \qquad\qquad\; \Leftarrow neg(u) \Leftarrow pos(v)$.

$\lfloor u \Rightarrow v \rfloor \Leftarrow \lceil u \rceil \Leftarrow \lfloor v \rfloor \qquad\qquad \Leftarrow (neg(u) \,\&\, neg(v)) \oplus (pos(u) \,\&\, notneu(v))$.

FIGURE 16. LU rules

a particular Forum proof branch. This counter can be used to limit the size of object-level proofs that are searched and in this way, the search for object-level proofs can be controlled in a simple fashion. In general, object-level proofs can be arbitrarily large, so setting a counter such as this is certainly not a complete proof strategy. It is the case, however, that if there is a proof of height h in the object-level, then the interpreter will find a proof if the counter is set to this value. For a number of proofs that we claim below, the value of this counter is often rather small.

To use this prover to prove object-level formulas, one would initialize the prover with the encoding of an object-level proof system and the encoding of the object-level formula. For example, attempting to prove the Forum sequent $LK \Rightarrow \lceil B \rceil$ for some object-level classical logic formula B would correspond to attempting to prove B using the rules of the classical sequent calculus LK. In particular, the single formula intended as LK is the &-conjunction of the universal closure of the clauses listed in Figure 3.

This prover can also be used determine if one collection of inference rules linearly entail other inference rules and equivalences. In particular, all the following can be proved automatically by setting the counter mentioned above to the value 3.

1. The clauses in Figure 3 encoding LK entails the clauses in Figure 4 encoding LJ, at least when these set of clauses are rewritten to use the same set of object-level constants.
2. The clauses in Figure 9 encoding a fragment of LJ entails the clauses in Figure 11 encoding the focused version of LJ called ILU.
3. The forward direction of Proposition 4.3 is easily proved: $\vdash LM \Rightarrow NM$.
4. All the equivalences mentioned in Section 4.3 that arise from the different cut and initial rules used in linear, intuitionistic, and classical logics have simple proofs.

Of course, if such entailments hold, they have immediate consequences for the object-logics that they encode. For example, from the first point above, we know that any (object-level) formula provable in ILU is also provable in LJ.

§8. **Conclusion and future work.** In this paper, we showed one way that linear logic can be used to specify some sequent calculus proof systems. We presented several examples of such an encoding and argued that such meta-logical encodings can have numerous advantages over the more standard inference figure approach. Since the encodings of the object-level proof systems are natural and direct, the rich meta-theory of linear logic can be used to draw conclusions about object-level proof systems. Because the object-level encodings result in logic programs (in the sense of Forum), the proof systems mentioned in this paper can be easily implemented and some of their properties can be automatically checked.

There is clearly much more to do now that the feasibility of using linear logic in this specification task is clear.

We have not discussed how proof objects can be specified in this setting: adding λ-calculus representations of calculi with natural deduction proofs can probably be done as it is done using an intuitionistic logic meta-theory [10] but such "single-conclusion" proofs would not work in the general sequent calculus setting.

There have been various proposals for non-commutativity variants of classical linear logic [1, 15, 29]: it would be interesting to see if these can be used to capture non-commutative object-level logics in a manner done here.

One reason to use a well understood meta-logic for specification is that it should offer ways to automate many things about inference rules. For example, it seems quite likely that the question whether or not one proof system's encoding linearly entails another proof system's encoding should be decidable, at least in many cases. It is also likely that at least important parts of the proof of cut-elimination for the encoded logic might similarly be automated.

Finally, most interesting proofs that relate provability in two proof systems generally require induction. It seems natural to consider adding to linear logic forms of induction along the lines found in [21, 28].

Acknowledgments. Miller has been funded in part by NSF grants CCR-9803971, INT-9815645, and INT-9815731. Pimentel has been funded by CAPES grant BEX0523/99-2. Pimentel was a visitor at Penn State from September 1999 to January 2001.

REFERENCES

[1] V. MICHELE ABRUSCI and PAUL RUET, *Non-commutative logic I: The multiplicative fragment*, **Annals of Pure and Applied Logic**, vol. 101 (1999), no. 1, pp. 29–64.

[2] JEAN-MARC ANDREOLI, *Logic programming with focusing proofs in linear logic*, **Journal of Logic and Computation**, vol. 2 (1992), no. 3, pp. 297–347.

[3] MICHELE BUGLIESI, GIORGIO DELZANNO, LUIGI LIQUORI, and MAURIZIO MARTELLI, *A linear logic calculus of objects*, **Proceedings of the Joint International Conference and Symposium on Logic Programming** (M. Maher, editor), MIT Press, September 1996.

[4] MANUEL M. T. CHAKRAVARTY, *On the massively parallel execution of declarative programs*, **Ph.D. thesis**, Technische Universität Berlin, Fachbereich Informatik, February 1997.

[5] JAWAHAR CHIRIMAR, *Proof theoretic approach to specification languages*, **Ph.D. thesis**, University of Pennsylvania, February 1995.

[6] ALONZO CHURCH, *A formulation of the simple theory of types*, **The Journal of Symbolic Logic**, vol. 5 (1940), pp. 56–68.

[7] VICENT DANOS, JEAN-BAPTISTE JOINET, and HAROLD SCHELLINX, *LKQ and LKT: sequent calculi for second order logic based upon dual linear decompositions of classical implication*, **Workshop on linear logic** (Girard, Lafont, and Regnier, editors), London Mathematical Society Lecture Notes 222, Cambridge University Press, 1995, pp. 211–224.

[8] GIORGIO DELZANNO and MAURIZIO MARTELLI, *Objects in Forum*, **Proceedings of the International Logic Programming Symposium**, 1995.

[9] ROY DYCKHOFF, *Contraction-free sequent calculi for intuitionistic logic*, **The Joural of Symbolic Logic**, vol. 57 (1992), no. 3, pp. 795–807.

[10] AMY FELTY, *A logic program for transforming sequent proofs to natural deduction proofs*, **Extensions of Logic Programming: International Workshop, Tübingen** (Peter Schroeder-Heister, editor), LNAI, vol. 475, Springer-Verlag, 1991, pp. 157–178.

[11] AMY FELTY and DALE MILLER, *Specifying theorem provers in a higher-order logic programming language*, **Ninth International Conference on Automated Deduction** (Argonne, IL), Springer-Verlag, May 1988, pp. 61–80.

[12] GERHARD GENTZEN, *Investigations into logical deductions*, **The Collected Papers of Gerhard Gentzen** (M. E. Szabo, editor), North-Holland Publishing Co., Amsterdam, 1969, pp. 68–131.

[13] JEAN-YVES GIRARD, *Linear logic*, **Theoretical Computer Science**, vol. 50 (1987), pp. 1–102.

[14] ———, *On the unity of logic*, **Annals of Pure and Applied Logic**, vol. 59 (1993), pp. 201–217.

[15] ALESSIO GUGLIELMI and LUTZ STRAßBURGER, *Non-commutativity and MELL in the calculus of structures*, **CSL 2001** (L. Fribourg, editor), LNCS, vol. 2142, 2001, pp. 54–68.

[16] ROBERT HARPER, FURIO HONSELL, and GORDON PLOTKIN, *A framework for defining logics*, **Journal of the ACM**, vol. 40 (1993), no. 1, pp. 143–184.

[17] JOSHUA HODAS and DALE MILLER, *Logic programming in a fragment of intuitionistic linear logic*, **Information and Computation**, vol. 110 (1994), no. 2, pp. 327–365.

[18] JOSHUA HODAS, KEVIN WATKINS, NAOYUKI TAMURA, and KYOUNG-SUN KANG, *Efficient implementation of a linear logic programming language*, **Proceedings of the 1998 Joint International Conference and Symposium on Logic Programming** (Joxan Jaffar, editor), 1998, pp. 145 – 159.

[19] PATRICK LINCOLN, ANDRE SCEDROV, and NATARAJAN SHANKAR, *Linearizing intuitionistic implication*, **Annals of Pure and Applied Logic**, (1993), pp. 151–177.

[20] P. LÓPEZ and E. PIMENTEL, *The UMA Forum linear logic programming language*, implementation, January 1998.

[21] RAYMOND MCDOWELL and DALE MILLER, *Cut-elimination for a logic with definitions and induction*, **Theoretical Computer Science**, vol. 232 (2000), pp. 91–119.

[22] DALE MILLER, *The π-calculus as a theory in linear logic: Preliminary results*, **Proceedings of the 1992 Workshop on Extensions to Logic Programming** (E. Lamma and P. Mello, editors), LNCS, no. 660, Springer-Verlag, 1993, pp. 242–265.

[23] ———, *Forum: A multiple-conclusion specification language*, **Theoretical Computer Science**, vol. 165 (1996), no. 1, pp. 201–232.

[24] DALE MILLER, GOPALAN NADATHUR, FRANK PFENNING, and ANDRE SCEDROV, *Uniform proofs as a foundation for logic programming*, **Annals of Pure and Applied Logic**, vol. 51 (1991), pp. 125–157.

[25] GOPALAN NADATHUR and DALE MILLER, *An Overview of λProlog*, **Fifth International Logic Programming Conference** (Seattle), MIT Press, August 1988, pp. 810–827.

[26] LAWRENCE C. PAULSON, *The foundation of a generic theorem prover*, **Journal of Automated Reasoning**, vol. 5 (1989), pp. 363–397.

[27] FRANK PFENNING, *Elf: A language for logic definition and verified metaprogramming*, **Fourth Annual Symposium on Logic in Computer Science** (Monterey, CA), June 1989, pp. 313–321.

[28] ELAINE GOUVÊA PIMENTEL, *Lógica linear e a especificação de sistemas computacionais*, **Ph.D. thesis**, Universidade Federal de Minas Gerais, Belo Horizonte, M.G., Brasil, December 2001, (written in English).

[29] CHRISTIAN RETORÉ, *Pomset logic: a non-commutative extension of classical linear logic*, **Proceedings of TLCA**, vol. 1210, 1997, pp. 300–318.

[30] GIORGIA RICCI, *On the expressive powers of a logic programming presentation of linear logic (FORUM)*, **Ph.D. thesis**, Department of Mathematics, Siena University, December 1998.

[31] CHRISTIAN URBAN, *Forum and its implementations*, **Master's thesis**, University of St. Andrews, December 1997.

COMPUTER SCIENCE AND ENGINEERING, 220 POND LAB,
    PENNSYLVANIA STATE UNIVERSITY
        UNIVERSITY PARK, PA 16802-6106, USA
    E-mail: dale@cse.psu.edu

DEPARTAMENTO DE CIÊNCIA DA COMPUTAÇÃO,
    UNIVERSIDADE FEDERAL DE MINAS GERAIS
        BELO HORIZONTE, M.G. BRASIL
    E-mail: pimentel@dcc.ufmg.br

KRIPKE MODELS OF CERTAIN SUBTHEORIES
OF HEYTING ARITHMETIC

TOMASZ POŁACIK

§1. Introduction. It is remarkable that, in comparison to the variety of results concerning models of classical arithmetic, there is still very little known about Kripke models of intuitionistic arithmetic. On the other hand, the usefulness of Kripke models in proving certain properties of Heyting Arithmetic, HA, has been already shown with a strong evidence in the article [5] of C. Smoryński. Of course, in model-theoretic investigations a variety of examples is of a great importance. However, the construction of Kripke models introduced in Smoryński's article is the only one known so far which provide us with nontrivial examples of Kripke models of HA and, to our knowledge, any other examples are not known. Using Smoryński's method, we obtain models of HA which are trees of finite depth whose all but the terminal worlds are copies of the standard model of Peano Arithmetic, PA. The use of the copies of the standard model in this construction seems to be crucial (however, see [5] for a discussion) and, in fact, as the example [8] of A. Visser and D. Zambella shows, the standard model cannot be replaced by an arbitrary model of PA.

Although Smoryński's construction can be obviously generalized to the case of models of subtheories of HA (we only have to modify the assumptions on the terminal worlds), in general, constructing Kripke models of subtheories of Heyting Arithmetic constitutes a separate problem. Recently, some attempts to find other methods of constructing Kripke models in the case of some weak subtheories of HA were made by K. Wehmeier, cf. [9, 10]. This paper is our modest contribution towards filling the existing gap in this research. Our main idea presented here relies on considering Kripke models whose appropriate worlds are related to each other stronger than in arbitrary models. So, we introduce partially-elementary extension models and show that such models are very useful for constructing models of subtheories of HA. We also remark on cofinal-extension Kripke models which, if PA-normal, are always models of a classical theory containing PA.

Logic Colloquium '99
Edited by J. van Eijck, V. van Oostrom, and A. Visser
Lecture Notes in Logic, 17
136

§2. Preliminaries. We begin with fixing the notation and recalling basic notions and facts which we use in this paper. We consider intuitionistic first order arithmetic in the language L containing the constant symbol 0 (for the number zero), the unary function symbol S (for the successor function) and a function symbol for each primitive recursive function. Heyting Arithmetic, HA, is the system whose specific axioms are equations defining all the primitive recursive functions and the induction schema: $A0 \wedge \forall x(Ax \rightarrow A(Sx)) \rightarrow \forall x Ax$. Obviously, PA can be obtained by adding to HA the Principle of Excluded Middle or another appropriate schema.

As our basic arithmetical theory we take Primitive Recursive Arithmetic, PRA. Roughly speaking, PRA can be viewed as the quantifier-free fragment of HA and one of its main feature is that it proves the Principle of Excluded Middle for quantifier-free formulae; see e.g., [6] for details. All theories which will be considered in this paper are assumed to contain PRA; moreover, all models are assumed to be models of PRA. In the sequel we will consider subtheories of HA which result in allowing only the formulae of a class Γ to appear in induction axioms. These theories will be denoted by $i\Gamma$; by $I\Gamma$ we will denote their classical counterparts, i.e., the theories $i\Gamma + \{\forall \vec{x}(A \vee \neg A): \text{ for all } \vec{x} \text{ and } A\}$. We will use the well-known (classical) hierarchy of prenex arithmetic formulae; recall that Δ_0 denotes the class of all bounded formulae and for any $n > 0$ the classes Σ_n and Π_n consist of all formulae which result in prefixing to a bounded formula n alternating blocks of quantifiers beginning with a block of existential and universal quantifiers respectively.

A *Kripke model* for a first order intuitionistic arithmetical theory in the language L can be viewed as a triple $\mathcal{K} = \langle K, \preceq, \{\mathfrak{M}_\alpha : \alpha \in K\} \rangle$ where $\langle K, \preceq \rangle$ is a partially ordered set of *nodes*, called the *frame* of $\mathcal{K}$, and $\{\mathfrak{M}_\alpha : \alpha \in K\}$ is a family of classical models for the language L, called the *worlds* of the model $\mathcal{K}$. We assume additionally that if $\alpha \preceq \beta$ then $\mathfrak{M}_\alpha$ is a submodel of $\mathfrak{M}_\beta$. It is known that in models of theories containing PRA, for all α, β with $\alpha \preceq \beta$, the world $\mathfrak{M}_\alpha$ is in fact a Δ_0-elementary submodel of $\mathfrak{M}_\beta$ (i.e., $\mathfrak{M}_\alpha$ is a submodel of $\mathfrak{M}_\beta$ and for any bounded formula A with parameters from $\mathfrak{M}_\alpha$, we have $\mathfrak{M}_\alpha \models A$ if and only if $\mathfrak{M}_\beta \models A$). In the usual way we define the forcing relation $\Vdash$ on the model $\mathcal{K}$; see e.g., [6].

§3. T-normal models. Let T be a first-order theory. Following [2], we say that a Kripke model $\mathcal{K}$ is T-*normal* if all the worlds of $\mathcal{K}$ are (classical) models of T. When one starts thinking of constructing Kripke models of an intuitionistic theory, say T, probably the first idea is to construct the model using classical structures which are models of classical counterparts of the theory T, i.e., using T-normal models. This is a natural idea but, in case of arithmetic, works only for weak theories.

It is not difficult to see that every PRA-normal Kripke model is a model of PRA and conversely, every model of PRA is PRA-normal. So, constructing models of PRA is very easy. We can also easily construct Kripke models of $i\Sigma_1$ since every $I\Sigma_1$-normal model is a model of $i\Sigma_1$, cf. [2]. However, in contrast to the previous case, we do not know whether every model of $i\Sigma_1$ is $I\Sigma_1$-normal. Note that by the recent result of A. Visser (cf. [7]), the positive answer to this problem would imply existence of a semi-positive axiomatization of $i\Sigma_1$. Following the idea of using T-normal models, we pass to the next case: models of $i\Pi_1$. In this case, by constructing an appropriate counterexample, S. Buss showed in [2] that a $I\Pi_1$-normal (even: PA-normal) Kripke model need not be a model of $i\Pi_1$. From this fact it follows that, in general, the idea of considering T-normal models in construction Kripke models of arithmetical theories cannot be applied to theories containing $i\Pi_1$. However, since the model constructed by S. Buss is infinite, one could still expect that every finite PA-normal model would be a model of HA. Unfortunately, it is not the case. An example of a finite PA-normal Kripke model which is not a model of $i\Sigma_2$ was given by A. Visser and D. Zambella in [8]. Moreover, the only two worlds of this model are chosen in this way that the top world is an end-extension of the root.

As we have seen, if we want to construct Kripke models of arithmetical theories, it is not enough to collect a family of classical models of some sufficiently strong classical theory and order them with the submodel relation — evidently, we have to take into account other properties. We have also seen that finiteness of the frame and end-extension relation between the worlds of a model cannot guarantee us that a given, say PA-normal, model would be a model of a sufficiently strong subtheory of Heyting Arithmetic.

§4. **Cofinal extensions.** Thinking about other model-theoretic relations between the worlds of a Kripke model of an arithmetical theory, one naturally considers cofinal extensions. Recall that we say that $\mathfrak{N}$ is a *cofinal extension* of $\mathfrak{M}$, if $\mathfrak{M}$ is a submodel of $\mathfrak{N}$ and for every $a \in N$ there is $b \in M$ such that $\mathfrak{N} \models a \leq b$. We will call a Kripke model $\mathcal{K}$ a *cofinal-extension model* if for every nodes α and β of $\mathcal{K}$, whenever $\alpha \preceq \beta$ then $\mathfrak{M}_\beta$ is a cofinal extension of $\mathfrak{M}_\alpha$.

We show that PA-normal cofinal extension models are not suitable for constructing nontrivial models of arithmetical theories. First, let us note that if the worlds of a given Kripke model are not merely the appropriate extensions but elementary-extensions then the theory of the model in question is classical. More precisely, if $\mathcal{K}$ is a model in which $\mathfrak{M}_\alpha \prec \mathfrak{M}_\beta$ whenever $\alpha \preceq \beta$ then $\mathcal{K} \Vdash T$ if and only if $\mathcal{K}$ is T-normal; moreover, if $\mathcal{K}$ is T-normal then $\mathcal{K} \Vdash T + \{\forall \vec{x}(A \vee \neg A): \text{for all } \vec{x} \text{ and } A\}$.

Let us cite the following result concerning models of Peano Arithmetic:

GAIFMAN'S THEOREM. *Let* $\mathfrak{M} \models$ PA *and let* $\mathfrak{N} \models$ PRA *be a cofinal extension of* $\mathfrak{M}$. *Then* $\mathfrak{M} \prec_{\Delta_0} \mathfrak{N}$ *if and only if* $\mathfrak{N} \models$ PA *if and only if* $\mathfrak{M} \prec \mathfrak{N}$.

Now, since in every Kripke model of a theory containing PRA we have $\mathfrak{M}_\alpha \prec_{\Delta_0} \mathfrak{M}_\beta$ for $\alpha \preceq \beta$, by Gaifman's Theorem, we get

THEOREM 1. *Let* $\mathcal{K} \Vdash$ PRA *be a cofinal-extension model. Then, if* $\mathfrak{M}_\alpha \models$ PA *then, for every* β *with* $\beta \succeq \alpha$, $\mathfrak{M}_\beta \models$ PA; *moreover,* $\mathfrak{M}_\alpha \prec \mathfrak{M}_\beta$. *In particular, for every cofinal-extension,* PA-*normal Kripke model* $\mathcal{K}$ *we have* $\mathcal{K} \Vdash$ PA.

Thus, PA-normal cofinal Kripke models are trivial ones in the sense that their theories are always classical. As a result, looking for a method of constructing non-trivial models of arithmetical theories we should exclude both elementary and cofinal extensions.

§5. Partially-elementary extensions.

Recall that in every Kripke model of an arithmetical theory containing PRA, the relation between its worlds is stronger that in the general case, namely, $\mathfrak{M}_\beta$ is a Δ_0-elementary extension of $\mathfrak{M}_\alpha$ for $\alpha \preceq \beta$. One can ask what happens when in a given Kripke model of PRA the relation in question is even stronger. This leads to considering partially-elementary extension models for which we prove the main results of this paper.

DEFINITION. We say that a Kripke model $\mathcal{K}$ is an Σ_n-*elementary extension model* if for every nodes α and β with $\alpha \preceq \beta$ the world $\mathfrak{M}_\beta$ is an Σ_n-elementary extension model of the world $\mathfrak{M}_\alpha$ (which means that $\mathfrak{M}_\alpha$ is a submodel of $\mathfrak{M}_\beta$ and for every Σ_n formula A with parameters from $\mathfrak{M}_\alpha$ we have $\mathfrak{M}_\alpha \models A$ if and only if $\mathfrak{M}_\beta \models A$).

It is known that for every formula A in the prenex normal form, if A is forced at a node α then A is true in $\mathfrak{M}_\alpha$ It is easy to see that we can prove an analogue of this fact for all $\wedge$, $\vee$-combinations of prenex formulae. However, in general, we cannot prove that if A is in the prenex normal form, then whenever $\mathfrak{M}_\alpha \models A$ then $\alpha \Vdash A$. Confining ourselves to the class of partially-elementary extension models, we are looking for a possibly most general class of formulae for which the this property holds. Moreover, we would welcome a class which is intuitionistically more natural than the class of prenex formulae. In order to satisfy theses conditions we define the classes Σ_n^B and Π_n^B which properly contain the usual classes Σ_n and Π_n respectively:

DEFINITION. We say that the formula A is in Σ_{n+1}^B if and only if A is a brouwerian combination of sentences in Σ_{n+1} with the proviso that every subformula of the formula A of the form $B \to C$ has $B \in \Sigma_n$ and $C \in \Sigma_{n+1}$. The class Π_n^B is defined analogously.

Of course, modulo provable equivalence, the class Π_n^B is contained in Σ_{n+1}^B. Similarly for Σ_n^B and Π_{n+1}^B. Observe also that over classical predicate logic, the classes Σ_n^B and Σ_n coincide; the same is true for Π_n^B and Π_n. It should be noted that the classes Σ_n^B and Π_n^B, although more suitable for intuitionistic logic than the classes of prenex formulae, do not exhaust (modulo provably equivalence in intuitionistic logic) the class of all formulae of the language of arithmetic. In the obvious way, the classes Σ_n^B and Π_n^B give rise to the subtheories $i\Sigma_n^B$ and $i\Pi_n^B$ of Heyting Arithmetic. Note that the theories $I\Pi_n$ and $I\Sigma_n$ are (classically) pairwise equivalent but it is not intuitionistically true for the theories $i\Pi_n^B$ and $i\Sigma_n^B$.

The main aim of this paper is to show how to construct Kripke models for the theories $i\Sigma_n^B$ and $i\Pi_n^B$. As we will see, there is a distinction between properties of models in these two cases. We consider the former theory first.

THEOREM 2. *Let $n \geq 1$ and let $\mathcal{K}$ be a Σ_{n-1}-elementary extension, $I\Sigma_n$-normal Kripke model. Then $\mathcal{K} \Vdash i\Sigma_n^B$.*

Thus, it is easy to construct models of $i\Sigma_n^B$ for each n: to get such a model, we only have to consider a family of (classical) models of $I\Sigma_n$, regarding them as the worlds of a Kripke model and set the accessibility relation as $\prec_{\Sigma_{n-1}}$. Note that as a particular case we get from Theorem 2 that any $I\Sigma_1$-normal model is a model of $i\Sigma_1$.

Now we turn to models of $i\Pi_n^B$. In this case we prove the following theorem.

THEOREM 3. *Let $n \geq 1$ and let $\mathcal{K}$ be Σ_n-elementary extension model of PRA. Then for every $\alpha \in K$: if $\mathfrak{M}_\alpha \models I\Pi_n$ then $\alpha \Vdash i\Pi_n^B$ (and hence $\mathcal{K}_\alpha \Vdash i\Pi_n^B$, where $\mathcal{K}_\alpha$ is a submodel of $\mathcal{K}$ whose domain is $\{\beta : \alpha \preceq \beta\}$).*

So, to obtain a model of $i\Pi_n^B$ it is enough to take an arbitrary (classical) model $\mathfrak{M}$ of $I\Pi_n$, making it the root of the Kripke model being constructed, and consider an arbitrary family $\{\mathfrak{M}_\alpha : \alpha \in J\}$ of Σ_n-elementary extensions of $\mathfrak{M}$. Then the elements of the family $\{\mathfrak{M}_\alpha : \alpha \in J\}$ can be regarded as the worlds of a Kripke model and the relation $\prec_{\Sigma_n}$ induces the desired ordering of the considered model.

Notice that we did not assume that the model of Theorem 3 is $I\Pi_n$-normal. Observe, however, that for an arbitrary Σ_n-elementary extension model, if $\mathfrak{M}_\alpha \models I\Pi_n$ for some $\alpha \in K$, then $\mathfrak{M}_\beta \models I\Pi_n$ for all $\beta \succeq \alpha$. So, in particular, if the root of a Σ_n-elementary extension model $\mathcal{K}$ is a model of $I\Pi_n$, then $\mathcal{K}$ is in fact $I\Pi_n$-normal.

This is still an open problem whether the converse to Theorems 3 and 2 can be proved. However we can prove the following result:

THEOREM 4. *Let $n \geq 0$ and let $\mathcal{K}$ be a Σ_{n+1}-elementary-extension model. Then*

1. *If $\mathcal{K} \Vdash i\Sigma_n^B$ then $\mathcal{K}$ is $I\Sigma_n$-normal.*
2. *If $\mathcal{K} \Vdash i\Pi_n^B$ then $\mathcal{K}$ is $I\Pi_n$-normal.*

We should note that the assumptions about ordering of the model $\mathcal{K}$ by $\prec_{\Sigma_n}$ instead of $\prec_{\Delta_0}$ are not inessential for the theory of the model $\mathcal{K}$. Namely, it turns out that in any Σ_n-elementary extension model of PRA validates the Markov's Principle for Σ_n formulae (the case for $n = 1$ was considered in [3]):

THEOREM 5. *Let $\mathcal{K}$ be a Σ_n-elementary extension model of* PRA. *Then for every $Ax \in \Sigma_n$:*

$$\mathcal{K} \Vdash \neg\neg\exists x Ax \rightarrow \exists x Ax.$$

Recall that the Markov's Principle is not provable in HA even for formulae in Σ_1. Hence, if $\mathcal{K}$ is a Σ_n-elementary extension model of HA, for some $n \geq 1$, then the theory of $\mathcal{K}$ strictly contains Heyting Arithmetic.

Closing this section, we mention a rather surprising feature of cofinal extension models of PRA.

THEOREM 6. *Let $\mathcal{K}$ be a cofinal-extension model of* PRA. *Then $\mathcal{K}$ is a Σ_1-elementary extension model.*

So, in particular, every cofinal-extension model of PRA validates the Markov's Principle for Σ_1 formulae.

§6. **Conclusion.** Let us conclude with posing some open problems which follow from the considerations presented in this paper.

First, from Theorem 5 we know that all Σ_n-elementary extension Kripke models of PRA validate the Markov's Principle for Σ_n formulae. It is however an open question, what exactly arithmetical theory we get if we consider the class of all Σ_n-elementary extension models of PRA.

Second, it would be also interesting to know whether Theorem 4 is optimal. More precisely, we ask whether every Σ_n-elementary extension, $I\Pi_n$-normal Kripke model is a model of $i\Sigma_n^B$ or $i\Pi_n^B$. The positive answer to this question would imply that $I\Sigma_n$-normal, $I\Pi_n$-elementary extension models were exactly the models of $i\Pi_n^B$.

Afterword. The results presented in this paper were obtained mainly in 1998 and presented at Logic Colloquium'99; the proofs of main theorems are published in [4]. Since then some improvements have been made. In particular, we know the answers to the open questions posed in §6: we know an axiomatization of Σ_n-elementary extension models of PRA — it involves the restricted version of Principle of Excluded Middle — and we can prove that Theorem 4 is optimal in the sense explained above. Moreover, we know that for every $n \geq 1$ we can find a model of Σ_n^B which is not $I\Sigma_n$-normal and hence, in particular, none of the theories $i\Sigma_n^B$ can be axiomatized by semi-positive formulae. Of course, this is implies the negative answer to the question concerning semi-positive axiomatizability of $i\Sigma_1$ posed in §3. Let us also mention that our result on cofinal-extension models of Peano Arithmetic stated in §4, was obtained independently, in a more general context, by M. Ardeshir, W. Ruitenburg and S. P. Salehi, in their joint paper [1].

REFERENCES

[1] M. Ardeshir, W. Ruitenburg, and S. P. Salehi, *Intuitionistic axiomatization for bounded extension Kripke models*, **Technical report**, Institute for Studies in Theoretical Physics and Mathematics, Teheran, 1999.

[2] S. R. Buss, *Intuitionistic validity in T-normal Kripke structures*, **Annals of Pure and Applied Logic**, vol. 59 (1993), pp. 159–173.

[3] Z. Marković, *On the structure of Kripke models of Heyting arithmetic*, **Mathematical Logic Quarterly**, vol. 39 (1993), pp. 531–538.

[4] T. Połacik, *Induction schemata valid in Kripke models of arithmetical theories*, **Reports on Mathematical Logic**, vol. 33 (1999), pp. 111–125.

[5] C. Smoryński, *Application of Kripke models*, **Metamathematical investigation of intuitionistic arithmetic and analysis** (A. S. Troelstra, editor), Springer, 1973.

[6] A. S. Troelstra and D. van Dalen, **Constructivism in mathematics. An introduction**, Studies in Logic, vol. 121, 123, North-Holland, 1988.

[7] A. Visser, *Submodels of Kripke models*, **Logic Group Preprint Series 189**, Utrecht University, 1998.

[8] A. Visser and D. Zambella, *Some non-HA models, II*, unpublished.

[9] K. Wehmeier, *Fragments of HA based on Σ_1-induction*, **Archive for Mathematical Logic**, vol. 37 (1997), pp. 37–49.

[10] ———, *Constructing Kripke models of certain fragments of Heyting's arithmetic*, **Publications de l'Institute Mathématique, Nouvelle Série**, vol. 63(77) (1998), pp. 1–8.

INSTITUTE OF MATHEMATICS
UNIVERSITY OF SILESIA
BANKOWA 14
40-007 KATOWICE, POLAND
E-mail: polacik@us.edu.pl

FROM BOUNDED STRUCTURAL RULES
TO LINEAR LOGIC MODALITIES

ANDREJA PRIJATELJ

Abstract. In this paper, a baby-example of a modality free classical linear logic enriched with 2-bounded structural rules is presented. The corresponding embedding theorem into classical linear logic is proved yielding also the cut-elimination property of the system considered.

§1. **Introduction.** In this paper we present a modality-free axiomatic system of classical linear logic $\mathbf{CLL}_a^2$, enriched with 2-bounded structural rules. The system is based on two-fold sequents, the idea taken from Girard's LU [2], separating a structural part from a linear part. Solely in the structural part 2-bounded structural rules may act on an even number of copies of formulas while in the the linear part applications of logical rules specified below can be made. The two parts may also communicate with each other by the so called switching rules. It is easy to give a generalization of $\mathbf{CLL}_a^2$ corresponding to any given $n \geq 2$, and hence left to a reader. Moreover, as shown recently, the systems considered can be turned into equivalent usual sequent calculi by applying the respective switching rule acting from a structural to a linear part. However, it is the so-called auxiliary system $\mathbf{CLL}_a^2$ that helped to remedy all the deficiencies of our previous attempts at formalizing logic with bounded structural rules, in particular the lack of the cut-elimination property. In what follows, we show that $\mathbf{CLL}_a^2$ can be faithfully embedded into classical linear logic (**CLL**), as well as into any of its generalized versions. The same holds true for the corresponding cut-free subsystems, and hence we may conclude that also the systems $\mathbf{CLL}_a^n$ enjoy the cut-elimination property. Moreover, relying on $\mathbf{CLL}_a^n$-derivable and provably equivalent formulas, displayed below for $n = 2$, we can specify the modified expressive power of the multiplicative connectives, i.e., tensor ($\otimes$) and, dually par ($\mathfrak{N}$). More precisely, given $n \geq 2$, then for any $2 \leq k < n$, k copies of a formula A linked by $\otimes$ (denoted by A^k) expresses availability of exactly k copies of A simultaneously, A^{nm} for $m \geq 1$ expresses availability of an arbitrary number of copies A (including zero), while A^{nm+l} with $1 \leq l < n$ expresses availability of at least l copies

Logic Colloquium '99
Edited by J. van Eijck, V. van Oostrom, and A. Visser
Lecture Notes in Logic, 17

of A respectively. This means that up to a given n, $\otimes$ behaves normally as in **CLL**, that A^n and its multiples mimic precisely $!A$, and that A^{nm+l} acts as $!A \otimes A^l$ in **CLL**. Later on, it will be seen that this is the effect of n-bounded structural rules as well as properly modified right $\otimes$-rule and dually, left $\mathcal{R}$-rule. To sum up, given $n \geq 2$, the expressive power of the additive connectives in $\mathbf{CLL}_a^n$ remains the same as in **CLL**. On the other hand, by the multiplicative connectives we can express 'arbitrary many' and 'at least one', if $n = 2$, and exactly $2, \ldots$, exactly $n - 1$, arbitrary many (including zero) as well as at least $1, \ldots$, at least $n - 1$, if $n > 2$.

§2. **Auxiliary axiomatic system with bounded structural rules.** The auxiliary axiomatic system $\mathbf{CLL}_a^2$ is based on the following concept of a two-fold sequent: $\Pi|\Gamma \vdash \Delta|\Sigma$, where Π and Σ may run over finite multisets of an even number of copies of formulas (including the empty set) and Γ, Δ denote arbitrary finite multisets of formulas in the language of classical linear logic without the modalities. We shall refer to Π and Σ as structural parts and to Γ and Δ as linear parts of a sequent considered. Following the standard notation, if any of the parts is uninhabited it will simply be denoted by a blank. Moreover, throughout the below, we shall use the following abbreviations: given $n \geq 2$ and a formula A, let A^n and nA denote n copies of A linked by $\otimes$ and $\mathcal{R}$ respectively for any ordering of brackets (for $n = 1$, A^1 and $1A$ being just A). Let further $A^{(n)}$ denote a multiset of n copies of A, and accordingly given a multiset Γ, let $\Gamma^n = \{A^n; A \in \Gamma\}$, $n\Gamma = \{nA; A \in \Gamma\}$ and $\Gamma^{(n)} := \{A^{(n)}; A \in \Gamma\}$.

Axioms

$$|A \vdash A|; \quad |\bot \vdash \ |; \quad | \ \vdash 1|; \quad |\Gamma, 0 \vdash \Delta|; \quad |\Gamma \vdash \top, \Delta|$$

Logical rules

$$\frac{\Pi|\Gamma \vdash \Delta|\Sigma}{\Pi|\Gamma, 1 \vdash \Delta|\Sigma} \qquad \frac{\Pi|\Gamma \vdash \Delta|\Sigma}{\Pi|\Gamma \vdash \Delta, \bot|\Sigma}$$

$$\frac{\Pi|\Gamma \vdash A, \Delta|\Sigma}{\Pi|\Gamma, \neg A \vdash \Delta|\Sigma} \qquad \frac{\Pi|\Gamma, A \vdash \Delta|\Sigma}{\Pi|\Gamma \vdash \neg A, \Delta|\Sigma}$$

$$\frac{\Pi|\Gamma, B \vdash \Delta|\Sigma}{\Pi|\Gamma, B \,\&\, C \vdash \Delta|\Sigma} \qquad \frac{\Pi|\Gamma, C \vdash \Delta|\Sigma}{\Pi|\Gamma, B \,\&\, C \vdash \Delta|\Sigma}$$

$$\frac{\Pi_1|\Gamma \vdash B, \Delta|\Sigma_1 \quad \Pi_2|\Gamma \vdash C, \Delta|\Sigma_2}{\Pi_1, \Pi_2|\Gamma \vdash B \,\&\, C, \Delta|\Sigma_1, \Sigma_2}$$

$$\frac{\Pi|\Gamma \vdash B, \Delta|\Sigma}{\Pi|\Gamma \vdash B \oplus C, \Delta|\Sigma} \qquad \frac{\Pi|\Gamma \vdash C, \Delta|\Sigma}{\Pi|\Gamma \vdash B \oplus C, \Delta|\Sigma}$$

$$\frac{\Pi_1|\Gamma, B \vdash \Delta|\Sigma_1 \qquad \Pi_2|\Gamma, C \vdash \Delta|\Sigma_2}{\Pi_1, \Pi_2|\Gamma, B \oplus C \vdash \Delta|\Sigma_1, \Sigma_2}$$

Left $\otimes$-rule and right $\invamp$-rule:

$$\frac{\Pi|\Gamma, B, C \vdash \Delta|\Sigma}{\Pi|\Gamma, B \otimes C \vdash \Delta|\Sigma} \qquad \frac{\Pi|\Gamma \vdash B, C, \Delta|\Sigma}{\Pi|\Gamma \vdash B \invamp C, \Delta|\Sigma}$$

Restricted version of the right $\otimes$-rule:

$$\frac{\Pi_1|\Gamma_1 \vdash B, \Delta_1|\Sigma_1 \qquad \Pi_2|\Gamma_2 \vdash C, \Delta_2|\Sigma_2}{\Pi_1, \Pi_2|\Gamma_1, \Gamma_2 \vdash B \otimes C, \Delta_1, \Delta_2|\Sigma_1, \Sigma_2}$$

except when B is of the form A^{2k+1} and C is of the form A^{2m+1}
for any $k, m \in N$.

Promoted right $\otimes$-rule:

$$\frac{\Pi_1|\Gamma_1 \vdash A^{2k+1}, \Delta_1|\Sigma_1 \qquad \Pi_2|\Gamma_2 \vdash A^{2m+1}, \Delta_2|\Sigma_2}{\Pi_1, \Pi_2, \Gamma_1^{(2)}, \Gamma_2^{(2)}| \quad \vdash A^{2(k+m+1)}|\Delta_1^{(2)}, \Delta_2^{(2)}, \Sigma_1, \Sigma_2}$$
for any $k, m \in N$.

Restricted version of the left $\invamp$-rule:

$$\frac{\Pi_1|\Gamma_1, B \vdash \Delta_1|\Sigma_1 \qquad \Pi_2|\Gamma_2, C \vdash \Delta_2|\Sigma_2}{\Pi_1, \Pi_2|B \invamp C, \Gamma_1, \Gamma_2 \vdash \Delta_1, \Delta_2|\Sigma_1, \Sigma_2}$$

except when B is of the form $(2k + 1)A$ and C is of the form $(2m + 1)A$
for any $k, m \in \mathbf{N}$.

Promoted left $\invamp$-rule:

$$\frac{\Pi_1|\Gamma_1, (2k + 1)A \vdash \Delta_1|\Sigma_1 \qquad \Pi_2|\Gamma_2, (2m + 1)A \vdash \Delta_2|\Sigma_2}{\Pi_1, \Pi_2, \Gamma_1^{(2)}, \Gamma_2^{(2)}|2(k + m + 1)A \vdash \quad |\Delta_1^{(2)}, \Delta_2^{(2)}, \Sigma_1, \Sigma_2}$$
for any $k, m \in N$.

Structural rules

Left and right contraction rules:

$$\frac{\Pi, A^{(2n)}|\Gamma \to \Delta|\Sigma}{\Pi, A^{(2k)}|\Gamma \to \Delta|\Sigma} \qquad \frac{\Pi|\Gamma \vdash \Delta|A^{(2n)}, \Sigma}{\Pi|\Gamma \vdash \Delta|A^{(2k)}, \Sigma}$$

for any $n > k \geq 1$.

Left and right weakening rules:

$$\frac{\Pi|\Gamma \vdash \Delta|\Sigma}{\Pi, A^{(2n)}|\Gamma \vdash \Delta|\Sigma} \qquad \frac{\Pi|\Gamma \vdash \Delta|\Sigma}{\Pi|\Gamma \vdash \Delta|A^{(2n)}, \Sigma}$$

for any $n \geq 1$.

Left detensoring rule and right deparing rule:

$$\frac{\Pi, (A^n)^{(2)}|\Gamma \vdash \Delta|\Sigma}{\Pi, A^{(2n)}|\Gamma \vdash \Delta|\Sigma} \qquad \frac{\Pi|\Gamma \vdash \Delta|(nA)^{(2)}, \Sigma}{\Pi|\Gamma \vdash \Delta|A^{(2n)}, \Sigma}$$

for any $n > 1$.

Switching rules

Switching from a linear part to a structural part:

$$\frac{\Pi|\Gamma, A \vdash \Delta|\Sigma}{\Pi, A^{(2)}|\Gamma \vdash \Delta|\Sigma} \qquad \frac{\Pi|\Gamma \vdash \Delta, A|\Sigma}{\Pi|\Gamma \vdash \Delta|A^{(2)}, \Sigma}$$

Switching from a structural part to a linear part:

$$\frac{\Pi, A^{(2n)}|\Gamma \vdash \Delta|\Sigma}{\Pi|A^{2n}, \Gamma \vdash \Delta|\Sigma} \qquad \frac{\Pi|\Gamma \vdash \Delta|A^{(2n)}, \Sigma}{\Pi|\Gamma \vdash 2nA, \Delta|\Sigma}$$

Cut rule:

$$\frac{\Pi_1|\Gamma_1 \vdash A, \Delta_1|\Sigma_1 \qquad \Pi_2|\Gamma_2, A \vdash \Delta_2|\Sigma_2}{\Pi_1, \Pi_2|\Gamma_1, \Gamma_2 \vdash \Delta_1, \Delta_2|\Sigma_1, \Sigma_2}$$

To continue, within $\mathbf{CLL}_a^2$ the following are provably equivalent for any $n \geq 1$ and any formula A:

$$A^2 \vdash\dashv A^{2n} \text{ and dually, } 2A \vdash\dashv 2nA$$
$$A^3 \vdash\dashv A^{2n+1} \text{ and dually, } 3A \vdash\dashv (2n+1)A$$

We shall display below some derivations of the first equivalence given. The rest is equally trivial.

$$\frac{\dfrac{|A^2 \vdash A^2| \quad |A^2 \vdash A^2|}{|A^2, A^2 \vdash A^4| \qquad |A^2 \vdash A^2|}}{\dfrac{|(A^2)^{(n)} \vdash A^{2n}|}{\dfrac{(A^2)^{(2n)}| \quad \vdash A^{2n}|}{\dfrac{A^{(4n)}| \quad \vdash A^{2n}|}{\dfrac{A^{(2)}| \quad \vdash A^{2n}|}{|A^2 \vdash A^{2n}|}}}}}$$

where a double line indicates several successive applications of a specific rule of inference.

$$\frac{\dfrac{|A \vdash A| \qquad\qquad |A \vdash A|}{A^{(2)}, A^{(2)} \ | \vdash A^2| \qquad\qquad |A \vdash A|}}{\dfrac{A^{(2)}, A^{(2)}|A \vdash A^3| \qquad\qquad |A \vdash A|}{\dfrac{A^{(2)}, A^{(2)}, A^{(2)}, A^{(2)}| \vdash A^4| \qquad |A \vdash A|}{\dfrac{\cdots \quad \cdots \quad \cdots \quad \cdots}{\dfrac{\cdots \quad \cdots \quad \cdots \quad \cdots}{\dfrac{A^{(4n)}| \quad \vdash A^{2n}|}{\dfrac{A^{(2)}| \quad \vdash A^{2n}|}{|A^2 \vdash A^{2n}|}}}}}}}$$

where the dot lines indicate several successive interchanging applications of restricted and promoted right $\otimes$-rules. Analogously, a derivation for any other ordering of brackets covered by A^{2n} can easily be made.

For the other turnstile direction of the equivalence under consideration, it suffices to make the following derivation:

$$\frac{\dfrac{|A^2 \vdash A^2|}{\dfrac{(A^2)^{(2)}| \quad \vdash A^2|}{\dfrac{A^{(4)}| \quad \vdash A^2|}{\dfrac{A^{(2n)}| \quad \vdash A^2|}{|A^{2n} \vdash A^2|}}}}}$$

In case $n = 2$ there is no application of weakening rule.

Finally, notice also that for any $n \geq 1$ also $|A \otimes A \vdash A^{2n+1}|$, and dually $|(2n+1)A \vdash A \,\invamp\, A|$ are $\mathbf{CLL}_a^2$-derivable.

To sum up, this indeed means that for any $n \geq 1$ the following are $\mathbf{CLL}_a^2$-derivable:

$$|A \otimes A \vdash A^n|, \text{ and dually, } |nA \vdash A \,\invamp\, A|.$$

§3. Embedding of the system with 2-bounded structural rules into classical linear logic. In what follows we shall show that the axiomatic systems $\mathbf{CLL}_a^2$ can faithfully be embedded into **CLL**. For that purpose we shall first define a translation between the sets of formulas of the underlying languages.

DEFINITION 3.1. *A translation* $(.)^t : For_{\mathbf{CLL}_a^2} \to For_{\mathbf{CLL}}$ *is given inductively as follows*:

- $P^t = P$, *for any propositional letter*;
- $1^t = 1$, $\perp^t = \perp$, $\top^t = \top$ *and* $0^t = 0$;
- $(\neg A)^t = \neg A^t$;
- $(A \,\&\, B)^t = A^t \,\&\, B^t$;
- $(A \oplus B)^t = A^t \oplus B^t$;
- $(A \otimes B)^t = \begin{cases} !C^t, & \text{if } A \otimes B \text{ is of the form } C^{2k}, \text{ with } k \geq 1; \\ !C^t \otimes C^t, & \text{if } A \otimes B \text{ is of the form } C^{2k+1}, \text{ with } k \geq 1; \\ A^t \otimes B^t, & \text{otherwise}; \end{cases}$
- $(A \,\invamp\, B)^t = \begin{cases} ?C^t, & \text{if } A \,\invamp\, B \text{ is of the form } 2kC, \text{ with } k \geq 1; \\ ?C^t \,\invamp\, C^t, & \text{if } A \,\invamp\, B \text{ is of the form } (2k+1)C, \text{ with } k \geq 1; \\ A^t \,\invamp\, B^t, & \text{otherwise}. \end{cases}$

In what follows, we shall use some more abbreviations: given a multiset Γ, let $\Gamma^t = \{A^t; A \in \Gamma\}$, i.e., Γ^t is the corresponding multiset of translated copies of formulas in Γ; moreover, given multisets Π and Σ of an even number of copies of formulas, let $!\Pi^t$ and $?\Sigma^t$ be the **sets** of modalized translations of formulas in Π and Σ by "!" and "?" respectively.

We are now ready to state the

THEOREM 3.2 (Embedding Theorem). $\mathbf{CLL}_a^2 \hookrightarrow \mathbf{CLL}$ *in the following sense*:

$$\Pi | \Gamma \vdash_{\mathbf{CLL}_a^2} \Delta | \Sigma \quad \textit{iff} \quad !\Pi^t, \Gamma^t \vdash_{\mathbf{CLL}} \Delta^t, ?\Sigma^t.$$

PROOF. (Soundness): by induction on the length of $\mathbf{CLL}_a^2$-derivations. We shall here display some cases of the induction step considering promoted right $\otimes$ rules, as well as structural and switching rules.

For the promoted right $\otimes$-rule as the last applied rule within a $\mathbf{CLL}_a^2$-derivation we are to distinguish two cases, one of them displayed below:

$$\frac{\Pi_1 | \Gamma_1 \vdash A^{2k+1}, \Delta_1 | \Sigma_1 \qquad \Pi_2 | \Gamma_2 \vdash A^{2m+1}, \Delta_2 | \Sigma_2}{\Pi_1, \Pi_2, \Gamma_1^{(2)}, \Gamma_2^{(2)} | \quad \vdash A^{2(k+m+1)} | \Delta_1^{(2)}, \Delta_2^{(2)}, \Sigma_1, \Sigma_2}$$
$$\text{for } k, m \geq 1.$$

$$\frac{!\Pi_1^t, \Gamma_1^t \vdash !A^t \otimes A^t, \Delta_1^t, ?\Sigma_1^t}{!\Pi_1^t, !\Gamma_1^t \vdash !A^t \otimes A^t, ?\Delta_1^t, ?\Sigma_1^t}$$

$$\frac{!\Pi_1^t, !\Gamma_1^t \vdash !(!A^t \otimes A^t), ?\Delta_1^t, ?\Sigma_1^t \qquad !(!A^t \otimes A^t) \vdash !A^t}{!\Pi_1^t, !\Gamma_1^t \vdash !A^t, ?\Delta_1^t, ?\Sigma_1^t}$$

$$\frac{!\Pi_1^t, !\Gamma_1^t \vdash !A^t, ?\Delta_1^t, ?\Sigma_1^t}{!\Pi_1^t, !\Gamma_1^t \vdash !A^t, ?\Delta_1^t, ?\Sigma_1^t}$$

$$\overline{!\Pi_1^t, !\Pi_2^t, !\Gamma_1^t, !\Gamma_2^t| \quad \vdash !A^t, ?\Delta_1^t, ?\Delta_2^t, ?\Sigma_1^t, ?\Sigma_2^t}$$

Next, assuming that the last applied rule in a $\mathbf{CLL}_a^2$-derivation is one of the structural rules, say contraction or detensoring, while a reader may try his/her skill with weakening:

Left contraction rule:

$$\frac{\Pi, A^{(2n)}|\Gamma \vdash \Delta|\Sigma}{\Pi, A^{(2k)}|\Gamma \vdash \Delta|\Sigma}$$

for any $n > k \geq 1$.

$$\frac{!\Pi^t, !A^t, \Gamma^t \vdash \Delta^t, ?\Sigma^t}{\Pi, !A^t, \Gamma^t \vdash \Delta^t, ?\Sigma^t}$$

Left detensoring rule:

$$\frac{\Pi, (A^n)^{(2)}|\Gamma \vdash \Delta|\Sigma}{\Pi, A^{(2n)}|\Gamma \vdash \Delta|\Sigma}$$

for any $n > 1$.

$$\frac{!\Pi^t, !(A^n)^t, \Gamma^t \vdash \Delta^t, ?\Sigma^t}{!\Pi^t, !A^t, \Gamma^t \vdash \Delta^t, ?\Sigma^t}$$

To see that the rule just obtained is indeed **CLL**-derivable under the induction hypothesis that its premise is **CLL**-derivable, a reader is to spell out the following two cases:
$n = 2k; n = 2k + 1$.

Switching from a linear part to a structural part:

$$\frac{\Pi|\Gamma, A \vdash \Delta|\Sigma}{\Pi, A^{(2)}|\Gamma \vdash \Delta|\Sigma}$$

$$\frac{!\Pi^t, \Gamma^t, A^t \vdash \Delta^t, ?\Sigma^t}{!\Pi^t, !A^t, \Gamma^t \vdash \Delta^t, ?\Sigma^t}$$

Switching from a structural part to a linear part:

$$\frac{\Pi, A^{(2n)}|\Gamma \vdash \Delta|\Sigma}{\Pi|A^{2n}, \Gamma \vdash \Delta|\Sigma}$$

$$\frac{!\Pi^t, !A^t, \Gamma^t \vdash \Delta^t, ?\Sigma^t}{!\Pi^t, !A^t, \Gamma^t \vdash \Delta^t, ?\Sigma^t}$$

(Faithfulness): Assume now that $!\Pi^t, \Gamma^t \vdash \Delta^t, ?\Sigma^t$ is **CLL**-derivable. Then we want to show that the sequent $\Pi|\Gamma \vdash \Delta|\Sigma$ is $\mathbf{CLL}_a^2$-derivable or equivalently, as can easily be seen, that the sequent $|\Pi^2, \Gamma \vdash \Delta, 2\Sigma|$ is $\mathbf{CLL}_a^2$-derivable. The proof goes by induction on the length of a **CLL**-derivation of $\Pi^2, \Gamma \vdash \Delta, 2\Sigma$ obtained when replacing all modalized occurrences of formulas $!A$ and $?A$ by $A \otimes A$ and $A \,\mathscr{R}\, A$ respectively while keeping all the connectives unchanged as well as the corresponding rules of inference.

We shall here consider just some cases of the induction step involving the modality $!$. The rest is trivial.

Assume the last applied rule of a **CLL**-derivation is $!$-weakening:

$$\frac{!\Pi^t, \Gamma^t \vdash \Delta^t, ?\Sigma^t}{!\Pi^t, \Gamma^t, !A^t \vdash \Delta^t, ?\Sigma^t}$$

covered by applications of $\mathbf{CLL}_a^2$-switching rule from structural to linear part and weakening rule:

$$\frac{\dfrac{|\Pi^2, \Gamma \vdash \Delta, 2\Sigma|}{A^{(2)}|\Pi^2, \Gamma \vdash \Delta, 2\Sigma|}}{|\Pi^2, A \otimes A, \Gamma \vdash \Delta, 2\Sigma|}$$

Assuming the last applied rule of a **CLL**-derivation is $!$-contraction:

$$\frac{!\Pi^t, \Gamma^t, !A^t, !A^t \vdash \Delta^t, ?\Sigma^t}{!\Pi^t, \Gamma^t, !A^t \vdash \Delta^t, ?\Sigma^t}$$

covered by applications of $\mathbf{CLL}^2$-switching rule, left contraction rule, detensoring rules and successive applications of the other switching rule, as shown below:

$$\frac{\dfrac{\dfrac{|\Pi^2, \Gamma, A \otimes A, A \otimes A, \Delta, 2\Sigma|}{(A^2)^{(4)}|\Pi^2, \Gamma \vdash \Delta, 2\Sigma|}}{A^{(8)}|\Pi^2, \Gamma \vdash \Delta, 2\Sigma|}}{\dfrac{A^{(2)}|\Pi^2, \Gamma \vdash \Delta, 2\Sigma|}{|\Pi^2, \Gamma, A \otimes A \vdash \Delta, 2\Sigma|}}$$

Assume the last applied rule of a **CLL**-derivation of is left dereliction:

$$\frac{!\Pi^t, \Gamma^t, A^t \vdash \Delta^t, ?\Sigma^t}{!\Pi^t, \Gamma^t, !A^t \vdash \Delta^t, ?\Sigma^t}$$

covered exactly by applications of both $\mathbf{CLL}^2$-switching rules, as witnessed below:

$$\frac{\dfrac{|\Pi^2, \Gamma, A \vdash \Delta, \Sigma|}{A^{(2)}|\Pi^2, \Gamma \vdash \Delta, 2\Sigma|}}{|\Pi^2, \Gamma, A \otimes A \vdash \Delta, 2\Sigma|}$$

Assume the last applied rule of a **CLL**-derivation is right promotion:

$$\frac{!\Gamma^t \vdash A^t, ?\Delta^t}{!\Gamma^t, \vdash !A^t, ?\Delta^t}$$

covered by applications of $\mathbf{CLL}_a^2$-switching rules, left and right contraction rules, applications of detensoring rule, and an application of promoted right $\otimes$ rule with the same premises, as displayed below:

$$\frac{\dfrac{|\Gamma^2 \vdash A, 2\Delta| \qquad |\Gamma^2 \vdash A, 2\Delta|}{(\Gamma^2)^{(4)}| \quad \vdash A \otimes A|(2\Delta^{(4)}}}{\dfrac{\Gamma^{(8)}| \quad \vdash A \otimes A|\Delta^{(8)}}{\dfrac{\Gamma^{(2)}| \quad \vdash A \otimes A|\Delta^{(2)}}{|\Gamma^2 \vdash A \otimes A, \Delta^2|}}} \qquad \Diamond$$

As an important consequence of the proof of faithfulness of the embedding $\mathbf{CLL}_a^2 \hookrightarrow \mathbf{CLL}$ and the fact that **CLL** enjoys the cut-elimination property we end up with the following

COROLLARY 3.3. *The cut-free system of* $\mathbf{CLL}_a^2$ *can faithfully be embedded into* **CLL**.

COROLLARY 3.4. *The system* $\mathbf{CLL}_a^2$ *enjoys the cut-elimination property.*

REFERENCES

[1] J.-Y. GIRARD, *Linear logic*, **Theoretical Computer Science**, vol. 50 (1987), pp. 1–102.

[2] ———, *Linear logic: a survey*, **Logic and algebra of specification** (Bauer, Brauer, and Schwichtenberg, editors), Heidelberg, 1993, pp. 63–112.

[3] ———, *Linear logic: its syntax and semantics*, **Advances in linear logic** (J.-Y. Girard, Y. Lafont, and L. Reigner, editors), London Mathematical Society Lecture Note Series, vol. 222, Cambridge University Press, 1995.

[4] M. OKADA, *An introduction to linear logic: Expressiveness and phase semantics*, **Theories of types and proofs**, MSJ Memoirs, vol. 2, Tokyo, 1998, pp. 255–295.

[5] A. PRIJATELJ, *Free ordered algebraic structures towards proof theory*, **The Journal of Symbolic Logic**, vol. 66 (2001), no. 2, pp. 597–608.

[6] A. S. TROELSTRA, **Lectures on linear logic**, CSLI Lecture Notes, vol. 29, Stanford, 1992.

DEPARTMENT OF MATHEMATICS
UNIVERSITY OF LJUBLJANA
SLOVENIA
E-mail: Andreja.Prijatelj@fmf.uni-lj.si

A DESCRIPTION OF THE NON-SEQUENTIAL EXECUTION OF PETRI NETS IN PARTIALLY COMMUTATIVE LINEAR LOGIC

CHRISTIAN RETORÉ

Abstract. We encode the execution of Petri nets in Partially Commutative Linear Logic, an intuitionistic logic introduced by Ph. de Groote which contains both commutative and non commutative connectives. We are thus able to faithfully represent the concurrent firing of Petri nets as long as it can be depicted by a series-parallel order. This coding is inspired from the description of context-free languages by Lambek grammars.

§1. Presentation. Since the early days of linear logic, various representations of Petri nets have been proposed — [27, 9, 12] include good surveys; nevertheless these codings are not fully satisfactory. On the logical side the use of proper axioms is not pleasant, and to avoid it the only solution is to include modalities while it is clearly a multiplicative phenomenon; in particular as Petri net accessibility is decidable, it should be encoded in a decidable fragment of linear logic, and it is not yet known whether linear logic with modalities is decidable. On the concurrency side the absence of any record of the execution in the sequent which is proved leaves out a number of interesting questions, like Petri net synthesis or the search of efficient executions. For instance, even the sophisticated treatment of Gehlot [10] only takes into account structural parallelism, and fails to find an efficient execution due to the presence of the marking as shown by Künzle [14, 21].

Here we propose a rather different representation which focuses on events and executions. This is made possible by using the partially commutative calculus, here denoted by PCLL, introduced by Philippe de Groote in 96 [7] — an extension of the published version, to be precise. In this intuitionistic calculus one both have non commutative connectives of the Lambek calculus [15] and the usual commutative connectives of multiplicative linear logic [11]. This kind of calculus has then been extended to a classical setting by Paul Ruet [25], and further studied by Michele Abrusci and Paul Ruet [1], Akim Demaille [8].

2000 *Mathematics Subject Classification.* 03B47,03B60, 03B70, 03F05, 03F52, 68Q85.
Key words and phrases. Petri nets; linear logic; Lambek calculus; categorial grammars.

Logic Colloquium '99
Edited by J. van Eijck, V. van Oostrom, and A. Visser
Lecture Notes in Logic, 17

Roughly speaking it is possible to combine the commutative and non commutative logical connectives by handling structured contexts. These contexts are series-parallel partial orders, i.e., are generated by two kinds of commas: the commutative comma, denoted by $\{\ldots,\ldots\}$ corresponding to the disjoint sum of two contexts, introducing no order between its components, and the non commutative comma $(\ldots;\ldots)$ introducing order between its components: every formula in the first component is before every formula in its second component. This structure allows to take into account the relationship between the commutative and non commutative products, $\otimes$ and $\odot$ respectively corresponding to context operations $\{\ldots,\ldots\}$ and $(\ldots;\ldots)$. This relationship is simply the inclusion of series-parallel orders axiomatized in [3]. It should be observed that this relation is more general than the possibility to replace a commutative product by a non commutative one. Indeed, inclusion of series-parallel orders is basically the distributive law of concurrency viewed as a reduction and not as an equality: $\{(a;b),(a';b')\} \rightarrow (\{a,a'\};\{b,b'\})$ that is if one has to perform a before b and a' before b' without any further constraint, one can in particular perform a and a' simultaneously and perform thereafter b and b' simultaneously. This law cannot be derived by replacing commutative product(s) with non commutative product(s). From a logical perspective it is worth noticing that this relationship is only possible in an intuitionistic calculus as shown in [8, Chapter 6]. For instance the classical calculus considered by [25, 1] only allows for the replacement of commutative products with non commutative products but not for the distributive law. It is also worth noticing that the logical system allows for either relationship: a weaker order entails a stronger order, or the converse; of course for a concurrency interpretation only the first system is relevant.

As usual a marking with $n(P)$ tokens in the place P will be denoted by the formula $\bigotimes_P P^{n(P)}$ where $\otimes$ is the *commutative product* and P^k a short hand for $P \otimes \ldots \otimes P$ k-times. An event e consuming the marking $\mathsf{Pre}[e]$ and producing the marking $\mathsf{Post}[e]$ will be denoted by a formula $\mathsf{Pre}[e] \setminus \mathsf{Post}[e]$ with $\setminus$ being a non commutative implication: so it is not that far from the usual translation $\mathsf{Pre}[e] \multimap \mathsf{Post}[e]$, except that one has $(\mathsf{Pre}[e];\mathsf{Pre}[e] \setminus \mathsf{Post}[e]) \vdash \mathsf{Post}[e]$, but not $(\mathsf{Pre}[e] \setminus \mathsf{Post}[e];\mathsf{Pre}[e]) \vdash \mathsf{Post}[e]$.

We are given a partially ordered multiset of events ϕ, and we can fire simultaneously any subset of minimal events, until all events in ϕ are fired. We will prove that such an execution is possible from an initial marking M and yields the end marking N if and only if the calculus PCLL proves $(M;\phi) \vdash N$.

So in fact we are turning a universal statement into an existential one: *every sequence of step transitions of ϕ is possible from M and yields to N* is shown to be equivalent with *there exists a proof of $(M;\phi) \vdash N$*. To be a bit more precise, our coding and the proof of its correctness handle *executive* sequents $\Gamma \vdash C$ that are the meaningful sequents:

- C is a marking
- formulae of Γ are either markings ($\otimes$-only formulae) or events ($M \setminus N$ with M and N markings)
- the formulae of Γ are endowed with a series-parallel partial order

The provability of an executive sequent means that the concurrent execution of the events in Γ (endowed with the order induced by Γ) leads from the sum of the markings in Γ to the marking C.

Provability corresponds to the possibility of executing the corresponding step transitions, but what do proofs correspond to? They allow to trace every produced or consumed token, which are distinguished in a given proof. Of course the ideal representation of proofs would be proof nets, which exactly identify all proofs depicting the same consumption/production, but up to now proof nets for this calculus do not exist (there is not a yet a sound and complete correctness criterion for recognizing proofs from incorrect proof structures).

But as far as provability is concerned, tokens are not distinguished. Therefore complete models as in [8] may be used instead of proofs to observe the behaviour of Petri nets, since making a distinction between tokens is an artifact — that is nevertheless useful as in [13, 4].

The kind of Petri net execution that we take into account is step transition, where the steps are lower-closed subsets of the partial order of events which expresses the causal constraints in the execution. Step transition are studied in [19] but they are just multisets of events: they are not assumed to be the lower closed subsets of an order on events. On the other hand, the occurrence nets of [13, 4] unfold the causality of a Petri net into a partial order (acyclic graph) where vertices are alternatively events and places. The partial orders of events we consider here are sub-orders of this general order, restricted to events. So our approach is mixed: we consider step transitions, but these steps are lower closed subsets of this general partial order. We only can deal with series-parallel orders, and in this case we are able to replace a complicated statement into the existence of a proof in a multiplicative system. At first sight this work also share some ingredients with the algebraic approach of [17] which describe Petri nets by monoidal categories, for instance the distributive law which for us is a reduction and for them an equality, or transitions between markings. This not so surprising, as models of multiplicative calculi are monoidal categories, but in fact the connection, if any, is far from obvious: we superimpose a commutative and a non commutative structure, our objects are not only markings but all executions, and our morphisms are proofs expressing that if an execution is possible, so is another, while their morphisms are executions leading from a marking to a marking.

We first recall basic definition and results regarding step transition for Petri nets, using the logical notation. The essential result we need is a substitution property for step transition: although not difficult, it is bit tedious, so the proof is postponed to an appendix. Next we introduce the PCLL calculus, and the properties needed for our coding. After a small example we establish the faithfulness of the encoding in both directions. We end with future prospects: indeed our logical description naturally leads to high order Petri nets (mobile nets, where events can produce and consume events, etc.), to Petri nets with credits (where one can assume that some tokens are present provided they are consumed afterwards) and also suggest a new approach to ordinary Petri net synthesis because of the connection between formal grammar and Lambek calculus.

A more thorough report including complete proofs is available from my web page [23].

Acknowledgments. This work owes a lot to Philippe Darondeau; to improve a shallow first connection, he suggested to consider step transition with steps as lower-closed subsets, and answered numerous questions on Petri nets.

Thanks to Marek Bednarczyk, Brigitte Pradin-Chézalviel, Yannick Le Nir, Robert Valette for their helpful comments.

Thanks to the anonymous reviewer for his detailed report which helped to improve this article.

§2. Petri nets and their concurrent firing.

Set theoretic notation. $A - B = \{x \in A \mid x \notin B\}$ is only used when $A \supseteq B$. $A \uplus B$ denotes either multiset union or the union of two disjoint sets.

2.1. Petri nets: definition and notation. A Petri net is defined by a set $\mathcal{P}$ of *places* and a set $\mathcal{E}^{\otimes \backslash \otimes}$ of *events*. In our view the initial marking of a Petri net is not part of its definition.

2.1.1. *Places and markings.* Usually, a marking is defined as a function from $\mathcal{P}$ to N which expresses how many tokens there are in each place. In order to get as close as possible to our coding, let us define a marking as an element of the free commutative monoid over the places: it is clearly an equivalent definition. Composition will be denoted by $\otimes$, and the unit by 1 (the empty marking). As usual, A^n with $n \in N$ and $A \in \mathcal{P}$ is defined by: $A^0 = 1$ and $A^{n+1} = A \otimes A^n$. Due to the equations holding in the free commutative monoid over $\mathcal{P}$, there are many expressions denoting the same marking; for instance $1 \otimes ((A \otimes B) \otimes A)$, $A^2 \otimes B$, $A \otimes (A \otimes B)$ and $(B \otimes A) \otimes A$ all stand for the marking containing two tokens in the place A and one in the place B. Let us denote by $\Pi^{\otimes}$ such marking expressions:

$$\Pi^{\otimes} ::= \mathcal{P} \mid 1 \mid \Pi^{\otimes} \otimes \Pi^{\otimes}$$

Given an expression M in $\Pi^{\otimes}$ and a place A of $\mathcal{P}$ the expression M_A denotes the number of occurrences of A in M, that is the number of tokens of M in A. This number only depends on the marking and not on the expression, so it is definable for a marking.

A marking M' is said to contain a marking M, whenever for each place A, $M'_A \geq M_A$; in this case we write $M' \sqsupseteq M$. If so, then there exists another unique marking X such that $X \otimes M = M'$; this marking X will be denoted by $M' \oslash M$.

2.1.2. *Events.* Each $e \in \mathcal{E}^{\otimes \backslash \otimes}$ is associated with a behavior, that is a pair of maps Pre[e] and Post[e] from places to N. These two maps may be viewed as a pair of markings, indicating how many tokens, for each place A are taken off and put in A by the event e; according to the previous notation, these two numbers will be denoted by Pre[e]$_A$ and Post[e]$_A$. The behavior of an event e will be denoted by Pre[e] \ Post[e] where expressions of $\Pi^\otimes$ are allowed for Pre[e] and Post[e].

Let us mention a classical distinction among Petri nets. A Petri net is said to be unlabeled whenever no two events have the same behavior:

$$\forall e_1, e_2 \in \mathcal{E}^{\otimes \backslash \otimes} \, (\mathsf{Pre}[e_1] = \mathsf{Pre}[e_2] \wedge \mathsf{Post}[e_1] = \mathsf{Post}[e_2]) \Rightarrow e_1 = e_2$$

When a behavior makes use of expressions in $\Pi^\otimes$ equality is the equality of the underlying markings, for instance: $(A \otimes B) \setminus (1 \otimes A) = (B \otimes A) \setminus A$. As it stands, this work does not apply to labeled Petri nets, and this allows to identify an event with its behavior.[1]

Another classical property of Petri nets is purity: for each event e_i and for each place A one has $\mathsf{Pre}[e_i]_A.\mathsf{Post}[e_i]_A = 0$; that is an event never puts any token in a place where he takes some. This distinction is not relevant to our study which works no matter whether Petri nets are pure or not.

2.2. Firing a Petri net according to a partially ordered multiset of events. Given a Petri net R which will remain implicit, we are to define a relation

$$M \xrightarrow{\phi}$$

where

- ϕ is a partially ordered multiset of events of R (the definition follows)
- M is a marking R

The meaning of this relation is that the Petri net R, provided with the initial marking M allows the execution of the partially ordered multiset of events ϕ. Such an execution according to ϕ consists in firing simultaneously any set of minimal events until all events are fired. Let us define this precisely.

2.2.1. *Partially ordered multisets (pomsets) and substitution.* An *ordered enumeration of elements* of E is a triple $(X, <, f)$ where X is a set, $<$ is a partial strict order on X and f a map from X to E. A *partially ordered multiset*

---

[1]Nevertheless our coding could be adapted by using a lexicon as in a Lambek grammar (events are provided with a finite number of formulae by the lexicon). The case of rigid Lambek grammars (one formula per event), would correspond to labeled Petri nets (there can be two events with the same behavior, but each event has a single behavior) while general Lambek grammars would correspond to the extended case in which an event may have several behaviors.

(*pomset*) of elements of E is an equivalence class of ordered enumerations where the equivalence $(X, <, f) \sim (X', <', f')$ is defined by: there exists an order isomorphism h from $(X, <)$ to $(X', <')$ such that $f'(h(x)) = f(x)$ for all $x \in X$. We only consider finite pomsets over E, i.e., X is always finite. When $<$ is a linear order a pomset over E is a finite sequence over E and when $<$ is empty a pomset over E is a multiset over E. We often say "let $(X, <, f)$ be a pomset over E", as our definitions or constructions do not depend on the ordered enumeration which represents the pomset.

Let $(X, <)$ and $(Y, \prec)$ be two partially ordered sets, with $a \in X$ and $X \cap Y = \emptyset$. The order $X[a := Y]$ is the order $\lhd$ over $(X - \{a\}) \uplus Y$ defined by:

- $\lhd|_{X - \{a\}} = <|_{X - \{a\}}$,
- $\lhd|_Y = \prec$,
- $\forall x \in X - \{a\} \; \forall y \in Y \quad x \lhd y \Leftrightarrow x < a$,
- $\forall x \in X - \{a\} \; \forall y \in Y \quad y \lhd x \Leftrightarrow a < x$.

Let $\phi = (X, <, f)$ and $\psi = (Y, \prec, g)$ be two partially ordered enumerations with $a \in X$, $X \cap Y = \emptyset$. The *substitution* $\phi[a := \psi]$ of a by ψ in ϕ is the ordered enumeration $((X - \{a\}) \uplus Y, \lhd, h)$ where $\lhd$ is defined as above and and where $h(x) = f(x)$ for x in $(X - \{a\})$ and $h(y) = g(y)$ for $y \in Y$.

Given two ordered enumerations $(X, <, f) \sim (X', <', f')$, the order isomorphism being $h : X \xrightarrow{\sim} X'$ and two ordered enumerations $(Y, \prec, g) \sim (Y', \prec', g')$ we have $(X, <, f)[a := (Y, \prec, g)] \sim (X', <', f')[h(a) := (Y', \prec', g')]$. Consequently when it is clear which occurrence of $e \in E$ is substituted we will write the abusive notation $\phi[e := \psi]$.[2]

2.2.2. *Concurrent execution of a partially ordered multiset of events.* Let $\phi = (X, <, f)$ be an ordered enumeration representing a partially ordered multiset of events of a Petri net R.

Let Y be a subset of X; let us extend $\mathsf{Pre}[\,]$ and $\mathsf{Post}[\,]$ to a set of events in the simplest way: $\mathsf{Pre}[Y] = \otimes_{y \in Y} \mathsf{Pre}[f(y)]$ and $\mathsf{Post}[Y] = \otimes_{y \in Y} \mathsf{Post}[f(y)]$

DEFINITION 1. The relation $M \xrightarrow{\phi}$ holds whenever for all lower-closed[3] subsets Y of $(X, <)$ one has:

$$(M \otimes \mathsf{Post}[Y]) \oslash \mathsf{Pre}[Y] \sqsupseteq \mathsf{Pre}[\mathcal{F}_\phi(Y)]$$

where $\mathcal{F}_\phi(Y)$ is the *frontier* of Y defined by

$$\mathcal{F}_\phi(Y) = \{x \in X - Y \mid \forall z \in X \; z < x \Rightarrow z \in Y\}$$

Let us explain the intuitive meaning of this definition. The partial order depicts time constraints on the firing of the occurrences of events in X. As

---

[2] In case there are several twin occurrences of e in ϕ, the corresponding partially ordered multiset does not depend on which occurrence of e has been substituted by ψ. [Two elements x and y of an order are said to be twins whenever they cannot be compared and $x < z \Leftrightarrow y < z$ and $z < x \Leftrightarrow z < y$ for all $z \notin \{x, y\}$.]

[3] Y is said to be a lower-closed subset of X whenever $\forall z \in X \; (\exists y \in Y \; z \leq y) \Rightarrow z \in Y$.

the firing respects the time constraints, any set of events that have been fired is a lower-closed subset of X. The events in $\mathcal{F}_\phi(Y)$ are the minimal elements of the complement of $\overline{Y}$ in X (the order on $\overline{Y}$ being $< |_{\overline{Y}}$), hence the events in the frontier of Y are the ones that can be fired next. Consequently the above definition simply says that whatever possible (i.e., lower-closed) part Y of X has been fired there are enough tokens left to fire simultaneously all the events that come next.

Firstly let us observe that this definition makes sense for pomsets of events:

PROPOSITION 2. *If ϕ and ψ are two partially ordered enumerations describing the same partially ordered multiset of events i.e., $\phi \sim \psi$; then $M \xrightarrow{\phi}$ if and only if $M \xrightarrow{\psi}$: so we can speak of the execution of a partially ordered multiset of events.*

Here are some remarks on this definition:

- The notation $\oslash$ assumes that $(M \otimes \mathsf{Post}[Y]) \sqsupseteq \mathsf{Pre}[Y]$; if this did not hold, then there would exist a smaller Y' for which this expression would be meaningful, such that the $\sqsupseteq$ fails.
- One can also define $M \xrightarrow{\phi}$ by: for each anti-chain U letting $T = \{x \mid \forall u \in U \; \neg(x \geq u)\}$, one has: $M \otimes \mathsf{Post}[T] \oslash \mathsf{Pre}[T] \sqsupseteq \mathsf{Pre}[U]$
- Self concurrency is allowed: indeed the frontier of a lower-closed subset Y may contain several occurrences of the same event.

Whenever $M \xrightarrow{\phi}$ the end-marking is uniquely determined:

PROPOSITION 3. *If $M \xrightarrow{\phi}$, then there is a unique marking N such that the final marking obtained by firing ϕ from the initial marking M is N. This marking is defined by $N = M \otimes \mathsf{Post}[X] \oslash \mathsf{Pre}[X]$ — the $\oslash$ makes sense: take $Y = X$ in the definition. In this case we write*:

$$M \xrightarrow{\phi} N$$

PROPOSITION 4. *Let $\phi = (X, <, f)$ be a partially ordered enumeration of events.*

1. *If $M \xrightarrow{\phi} M'$ and if $\psi = (X, \prec, f)$ is an order extension of ϕ ($\forall x, x' \in X \; x < x' \Rightarrow x \prec x'$) then $M \xrightarrow{\psi} M'$.*
2. *In particular when $M \xrightarrow{\phi} M'$ holds, so does $M \xrightarrow{\psi} M'$ for each linearization ψ of ϕ.*

REMARK 5. The converse of this latter point fails, as can be observed from the Petri net R defined by:

- One place A
- Two events: $a = A \setminus A$ and $b = A \otimes A \setminus A \otimes A$

Consider the (multi)set with a single occurrence of each event a and b, and let us consider the three possible pomsets of events on this (multi)set of events:

- $\{a, b\}$ (empty strict order)

- $(a;b)$ (first linearization, $a < b$)
- $(b;a)$ (second linearization, $b < a$)

Since $A \otimes A \not\sqsupseteq (A \otimes A) \otimes A$, one does *not* have $A \otimes A \xrightarrow{\{a,b\}}$ but both the two possible linearizations of this partially ordered set of events can be fired: $A \otimes A \xrightarrow{(a;b)} A \otimes A$ and $A \otimes A \xrightarrow{(b;a)} A \otimes A$.

Let us define the minimum marking M^ϕ of a partially ordered multiset of events ϕ. Letting $L(\phi)$ be the set of the lower-closed subsets of ϕ it is defined by:

$$M^\phi{}_A = \max_{Y \in L(\phi)} \text{Pre}[\mathcal{F}(Y)]_A - \text{Post}[Y]_A + \text{Pre}[Y]_A$$

Observe that $M^\phi{}_A \geq 0$, because of the case $Y = \emptyset$. The name minimum marking is justified by the following fairly obvious proposition:

PROPOSITION 6. *One has* $M^\phi \xrightarrow{\phi}$, *and if* $M \xrightarrow{\phi}$ *then* $M \sqsupseteq M^\phi$.

2.2.3. *Substitution in an execution.*

PROPOSITION 7 (substitution property). *Let*

- $\phi = (X, <, f)$ *be a partially ordered enumeration of events containing an occurrence x of $P \setminus Q$ (i.e., $f(x) = P \setminus Q$)*
- $\psi = (Y, \prec, g)$ *be a partially ordered enumeration of events, such that $Y \cap X = \emptyset$ and such that $P \xrightarrow{\psi} Q$*

then one has:

1.

$$(a): \left(M_0 \xrightarrow{\phi}\right) \implies \left(M_0 \xrightarrow{\phi[x:=\psi]}\right) : (b)$$

2. *If moreover P is the minimum marking for ψ, then the converse also holds*:

$$(b): \left(M_0 \xrightarrow{\phi[x:=\psi]}\right) \implies \left(M_0 \xrightarrow{\phi}\right) : (a)$$

In both cases when (a) and (b) hold, the final markings are equal.

PROOF. The equality of the final markings is obvious. The proof of the main part is rather tedious and postponed to the appendix 9. ⊣

Here is a rather obvious proposition which will allow for the logic to mix events and markings; indeed in the logical model, events are constructed out of markings, and this proposition will be needed for events to appear at the right places in the order on events.

PROPOSITION 8. *Let $\psi[\mathbf{1} \setminus M] = (X, <, f)$ be a pomset of events containing an occurrence x of the event $\mathbf{1} \setminus M$ (i.e., $f(x) = \mathbf{1} \setminus M$) and let $\psi[\ ] = (X - \{x\}, <', f')$ be the partially ordered multiset obtained by suppressing this occurrence of $\mathbf{1} \setminus M$, and taking the induced order on the remaining occurrences of events ($<' = < |_{X - \{x\}}$ and $f' = f|_{X - \{x\}}$).*

If $X \xrightarrow{\psi[\mathbf{1} \setminus M]} Y$ then $X \otimes M \xrightarrow{\psi[\ ]} Y$.

2.2.4. *Series or parallel composition of executions.* Although we shall come back more precisely on these notions for the logical calculus (in paragraph 3.2) we need a few properties of these operations on partial orders.

Given two partially ordered enumerations $(X, <, f)$ and $(Y, \prec, g)$ with $X \cap Y = \emptyset$ we define:[4]

- their *parallel-composition* $(X \uplus Y, < \uplus \prec, f \uplus g)$
- their *series-composition* $(X \uplus Y, < \uplus \prec \uplus (X \times Y), f \uplus g)$

Observe that these operations are well defined for any two partially ordered multisets. Let us denote respectively by $\{\phi, \psi\}$ and $(\phi; \psi)$ the parallel and series composition of two partially ordered multisets of events ϕ and ψ.

The following facts are mere computations and are intuitively obvious — the ones concerning parallel composition derive from the easy equation: $M^{\{\phi_1, \phi_2\}} = M^{\phi_1} \otimes M^{\phi_2}$.

PROPOSITION 9 (step transitions and series or parallel composition).

1. *If* $M_1 \xrightarrow{\phi_1} N_1$ *and* $M_2 \xrightarrow{\phi_2} N_2$ *then* $M_1 \otimes M_2 \xrightarrow{\{\phi_1, \phi_2\}} N_1 \otimes N_2$. *In particular, letting* ϕ_2 *be the empty partially ordered multiset of events (so* $M_2 = N_2$*) which is the unit for parallel composition we have: if* $M \xrightarrow{\phi} N$ *then* $M \otimes P \xrightarrow{\phi} N \otimes P$, *for any marking* P.

2. *If* $M \xrightarrow{\{\phi_1, \phi_2\}} N$ *then there exist markings* M_1, M_2, N_1, N_2, P *such that:*
   - $M = M_1 \otimes M_2 \otimes P$
   - $N = N_1 \otimes N_2 \otimes P$
   - $M_1 \xrightarrow{\phi_1} N_1$
   - $M_2 \xrightarrow{\phi_2} N_2$

3. *If* $M \xrightarrow{(\phi_1; \phi_2)} N$ *then there exists a marking* Q *such that:*
   - $M \xrightarrow{\phi_1} Q$
   - $Q \xrightarrow{\phi_2} N$

§3. The PCLL calculus: sequent calculus. In this section we present the calculus PCLL. Actually our version slightly extends the published version, [7] but clearly was the author's project. Indeed, when his paper was printed he did not yet know the rules axiomatizing the inclusion of series-parallel partial orders [3] but this calculus was designed to incorporate such rules.

3.1. Formulae. Given a set of atomic formulae or propositional variables $\mathcal{P}$, that correspond to places from the Petri net viewpoint, formulae are defined by:

$$\mathcal{F} ::= \mathcal{P} \mid \mathbf{1} \mid \mathcal{F} \otimes \mathcal{F} \mid \mathcal{F} \odot \mathcal{F} \mid \mathcal{F} \multimap \mathcal{F} \mid \mathcal{F} \setminus \mathcal{F} \mid \mathcal{F} / \mathcal{F}$$

So this calculus contains the following connectives:

---

[4] The partial orders $<$ and $\prec$ are viewed as subset of $X \times X$ and $Y \times Y$, and the maps f and g are viewed as subsets of $X \times E$ and $Y \times E$.

- two multiplicative conjunctions:
  - ⊙ the non commutative conjunction of the Lambek calculus.
  - ⊗ the commutative multiplicative conjunction of linear logic.
- the associated implications:
  - \ associated with ⊙ by: $A \odot (A \setminus B) \vdash B$.
  - / associated with ⊙ by $(B / A) \odot A \vdash B$.
  - ⊸ associated with ⊗ by $A \otimes (A \multimap B) \vdash B$ and $(A \multimap B) \otimes A \vdash B$.

As we are going to see, either one can chose that the commutative conjunction entails the non commutative conjunction or the converse: of course this is determined by the structural rules managing the context. Here we use the version according to which $A \odot B \vdash A \otimes B$.

3.2. Contexts. Contexts, that are usually multisets or sequences of formulae, are here structured by series-parallel (partial) orders: they are partially ordered multisets of formulae (in the sense of section 2.2). The need for structured contexts is easily explained: the comma on the left-hand side of a sequent is an implicit conjunction. If we wish to have two kinds of conjunctions then we also need two kinds of "commas".

3.2.1. *Reminder on series-parallel partial orders* (sp-*orders*). *Series-parallel orders*, sp-orders for short are the smallest class of finite strict partial orders containing all (empty) orders over a single point, and closed under series and parallel compositions already used in paragraph 2.2.4.

These binary operations are defined for any two orders ϕ and ψ with respective domains X and Y with $X \cap Y = \emptyset$, and both yield an order with domain $X \uplus Y$.

- *series composition or ordinal sum*

$$(\phi; \psi) = \phi \uplus \psi \uplus (X \times Y) \qquad \subseteq (X \uplus Y) \times (X \uplus Y)$$

- *parallel composition or disjoint sum*

$$\{\phi, \psi\} = \phi \uplus \psi \qquad \subseteq (X \uplus Y) \times (X \uplus Y)$$

The reader interested in sp-orders will find much more details in [18]. Here we just recall the basic properties that we need:

PROPOSITION 10. *Two terms correspond to the same* sp-*order if and only if they are equal up to the algebraic properties of series and parallel compositions*:

$\{\dots, \dots\}$ *is associative and commutative*
$(\dots; \dots)$ *is associative*

Let us mention their famous characterization [28, 22]:

PROPOSITION 11. *A finite order is* sp *if and only if its restriction to four points* a, b, c *and* d *never is* $\{(a, b), (c, b), (c, d)\}$.

which clearly entails:

PROPOSITION 12. *Let ϕ be an SP-order of domain X and let X' be a subset of X. The restriction $\phi' = \phi \cap (X' \times X')$ is an SP-order as well. If t is an SP-term denoting ϕ one obtains a term denoting ϕ' from t by replacing each $x \in (X - X')$ by ε and reducing the term by applying the following equalities:* $(t; \varepsilon) = (\varepsilon; t) = \{t, \varepsilon\}.$[5]

Finally we have found in [3] a complete axiomatization for the inclusion of SP-orders as a rewriting system over SP-terms:

PROPOSITION 13. *Let ϕ and ϕ' be two SP-orders with the same domain, and let s and s' two SP-terms denoting them. One has $\phi \subseteq \phi'$ if and only if $s \longrightarrow^* s'$ where $\longrightarrow^*$ is the reflexive and transitive closure of the following rewriting rules, where $s[w]$ denotes an SP-term containing an occurrence of the subterm w:*

SP-order inclusion

$$s[\{(t; u), (t'; u')\}] \longrightarrow s[(\{t, t'\}; \{u, u'\})]$$
$$s[\{(t; u), u'\}] \longrightarrow s[(t; \{u, u'\})]$$
$$s[\{t, (t'; u')\}] \longrightarrow s[(\{t, t'\}; u')]$$
$$s[\{t, u\}] \longrightarrow s[(t; u)]$$

SP-order equalities

(...;...) associativity

$$s[(t; (u; v))] \longrightarrow s[((t; u); v)]$$
$$s[((t; u); v)] \longrightarrow s[(t; (u; v))]$$

{......} associativity

$$s[\{t, \{u, v\}\}] \longrightarrow s[\{\{t, u\}, v\}]$$
$$s[\{\{t, u\}, v\}] \longrightarrow s[\{t, \{u, v\}\}]$$

{......} commutativity

$$s[\{t, u\}] \longrightarrow s[\{u, t\}]$$

3.2.2. *Contexts:* SP-*terms or* SP-*pomsets of formulae.* Series composition, denoted by $(\ldots; \ldots)$ corresponds to the non commutative conjunction $\odot$, while parallel composition, denoted by $\{\ldots, \ldots\}$ corresponds to the commutative conjunction $\otimes$.

There are two ways to describe contexts depending on the precision we want, e.g. for proof search. Either they are viewed as SP-terms over formulae, or as SP-pomsets of formulae. The first description is better suited for writing down them and for implementing proof search, while the second is more abstract and consists in working with equivalence classes of terms.

Contexts as terms are defined as the following set of expressions in which $\mathcal{F}$ is the set of formulae:

$$\mathcal{C} ::= \mathcal{F} \mid \{\mathcal{C}, \mathcal{C}\} \mid (\mathcal{C}; \mathcal{C})$$

---

[5]Actually ε is the order on an empty domain, which is usually excluded from the class of SP-orders, although it is the unit for both series and parallel composition.

Contexts as SP-*orders* on multisets of formulae are described by the SP-terms denoting them which are the easiest way to write them down, but because of proposition 10 they are considered as equal exactly when they only differ up to the commutativity of $\{\ldots,\ldots\}$ and to the associativity of $\{\ldots,\ldots\}$ and $(\ldots;\ldots)$.

Denoting by $\uplus$ multiset union, the domain of a context is the multiset defined by:

$$|F| = \{F\} \quad |\{\Gamma,\Delta\}| = |(\Gamma;\Delta)| = |\Gamma| \uplus |\Delta|$$

The SP partial order Γ^{SP} associated with a context Γ is the subset of $|\Gamma| \times |\Gamma|$ defined by:

$$F^{\mathrm{SP}} = \emptyset \quad \{\Gamma,\Delta\}^{\mathrm{SP}} = \Gamma^{\mathrm{SP}} \uplus \Delta^{\mathrm{SP}} \quad (\Gamma;\Delta)^{\mathrm{SP}} = \Gamma^{\mathrm{SP}} \uplus \Delta^{\mathrm{SP}} \uplus (|\Gamma| \times |\Delta|)$$

3.3. The rules of the calculus PCLL.

3.3.1. *Axioms.* Axioms are identities:

$$\frac{}{A \vdash A}\ ax.$$

As usual, axioms can be limited to the case where A is a propositional variable.

3.3.2. *Structural rules: which variant?* There are two variants of this calculus depending on whether we want the commutative conjunction to imply the non commutative one, or the converse, and both equally work from a formal point of view. This is set in the choice of the structural rules. Here, regarding that we have in mind executions of Petri nets, we chose to have the *augmenting context* variant which entails $A \odot B \vdash A \otimes B$. The corresponding structural rule is:[6]

$$\frac{\Gamma \vdash C}{\Gamma' \vdash C}\ aug(mentation) \qquad \boxed{\text{if } |\Gamma| = |\Gamma'| \text{ and } \Gamma'^{\mathrm{SP}} \supseteq \Gamma^{\mathrm{SP}}}$$

This rule is not as non deterministic as it may seem. Indeed, because of the rewrite rules axiomatizing the inclusion of series-parallel orders given in proposition 13, *aug.* is equivalent to the following rules on contexts as SP-terms, where $\Psi[\Phi]$ stands for a context containing Φ as a sub-context:

---

[6]The variant *diminishing context* which entails $A \otimes B \vdash A \odot B$ correspond to the structural rule:

$$\frac{\Gamma \vdash C}{\Gamma' \vdash C}\ entropy \qquad \boxed{\text{if } |\Gamma| = |\Gamma'| \text{ and } \Gamma'^{\mathrm{SP}} \subseteq \Gamma^{\mathrm{SP}}}$$

<table>
<tr><th>sp-equalities</th><th>sp-inclusions</th></tr>
</table>

$$\frac{\Psi[((\Gamma;\Delta);\Theta)] \vdash C}{\Psi[(\Gamma;(\Delta;\Theta))] \vdash C} \ (asso;r)$$

$$\frac{\Psi[\{(\Gamma;\Gamma'),(\Delta;\Delta')\}] \vdash C}{\Psi[(\{\Gamma,\Delta\};\{\Gamma',\Delta'\})] \vdash C} \ (rew4)$$

$$\frac{\Psi[(\Gamma;(\Delta;\Theta))] \vdash C}{\Psi[((\Gamma;\Delta);\Theta)] \vdash C} \ (asso;l)$$

$$\frac{\Psi[\{\Gamma,(\Delta;\Delta')\}] \vdash C}{\Psi[(\{\Gamma,\Delta\};\Delta')] \vdash C} \ (rew3l)$$

$$\frac{\Psi[\{\{\Gamma,\Delta\},\Theta\}] \vdash C}{\Psi[\{\Gamma,\{\Delta,\Theta\}\}] \vdash C} \ (asso,r)$$

$$\frac{\Psi[\{(\Gamma;\Gamma'),\Delta\}] \vdash C}{\Psi[(\Gamma;\{\Delta,\Gamma'\})] \vdash C} \ (rew3r)$$

$$\frac{\Psi[\{\Gamma,\{\Delta,\Theta\}\}] \vdash C}{\Psi[\{\{\Gamma,\Delta\},\Theta\}] \vdash C} \ (asso,l)$$

$$\frac{\Psi[\{\Gamma,\Delta\}] \vdash C}{\Psi[(\Gamma;\Delta)] \vdash C} \ (rew2)$$

$$\frac{\Psi[\{\Gamma,\Delta\}] \vdash C}{\Psi[\{\Delta,\Gamma\}] \vdash C} \ (com,)$$

3.3.3. *Other rules: connective introductions, and cut.*

Implication rules	
$$\frac{\Gamma[B] \vdash C \quad \Delta \vdash A}{\Gamma[\{\Delta, A \multimap B\}] \vdash C} \ {\multimap}_h$$	$$\frac{\{A,\Gamma\} \vdash C}{\Gamma \vdash A \multimap C} \ {\multimap}_i$$
$$\frac{\Gamma[B] \vdash C \quad \Delta \vdash A}{\Gamma[(\Delta; A \backslash B)] \vdash C} \ \backslash_h$$	$$\frac{(A;\Gamma) \vdash C}{\Gamma \vdash A \backslash C} \ \backslash_i$$
$$\frac{\Gamma[B] \vdash C \quad \Delta \vdash A}{\Gamma[(B / A; \Delta)] \vdash C} \ /_h$$	$$\frac{(\Gamma;A) \vdash C}{\Gamma \vdash C / A} \ /_i$$

Product/conjunction rules	
$$\frac{\Gamma[(A;B)] \vdash C}{\Gamma[A \odot B] \vdash C} \ {\odot}_h$$	$$\frac{\Delta \vdash A \quad \Gamma \vdash B}{(\Delta;\Gamma) \vdash A \odot B} \ {\odot}_i$$
$$\frac{\Gamma[\{A,B\}] \vdash C}{\Gamma[A \otimes B] \vdash C} \ {\otimes}_h$$	$$\frac{\Delta \vdash A \quad \Gamma \vdash B}{\{\Delta,\Gamma\} \vdash A \otimes B} \ {\otimes}_i$$

Unit rules		Cut rule
$\dfrac{\Gamma \vdash C}{\{\Gamma, \mathbf{1}\} \vdash C}\mathbf{1}_h$	$\dfrac{}{\vdash \mathbf{1}}\mathbf{1}_i$	$\dfrac{\Gamma \vdash A \quad \Delta[A] \vdash B}{\Delta[\Gamma] \vdash B}\ cut$

3.3.4. *Several remarks on* **PCLL** *calculus.*

REMARK 14 (modus ponens). The modus ponens provided by the calculus without the *aug*. rule are: $(B/A; A) \vdash B$, $(A; A \backslash B) \vdash B$ and $\{A, A \multimap B\} \vdash B$, and $\{A \multimap B, A\} \vdash B$. But in this augmenting version of the calculus, where orders are allowed to augment, in addition to these expected modus ponens, we also have $(A; A \multimap B) \vdash B$ and $(A \multimap B; A) \vdash B$ — but neither $\{A, A \backslash B\} \vdash B$ nor $\{A, B / A\} \vdash B$.

REMARK 15 ($\mathbf{1}$: unit for $\odot$ and $\otimes$). The rules for $\mathbf{1}$ show that $\mathbf{1}$ is a unit for $\odot$ and $\otimes$ and the corresponding operations on contexts, respectively $(\ldots; \ldots)$ and $\{\ldots, \ldots\}$. In the rules $\mathbf{1}_h$ we could have decided to insert $\mathbf{1}$ anywhere in the context. However this alternative rule is not needed, since it is derivable using the rule which augments the context.

Here is an obvious proposition which is useful to the main result:

PROPOSITION 16. *Let* $M \in \Pi^{\otimes}$; *then* PCLL *proves* $\Gamma[M] \vdash C$ *if and only if* PCLL *proves* $\Gamma[\mathbf{1} \backslash M] \vdash C$.

The following essential property will be needed as well:

THEOREM 17 (Cut-elimination and subformula property). *The cut-rule is redundant, and in a cut-free proof every formula of every sequent is a subformula of some formula of the conclusion sequent.*

PROOF. A semantical proof can be found in [8] and a syntactic one in [23]. The proof is absolutely standard, the only novelty w.r.t. multiplicative linear logic or Lambek calculus being the commutation of *cut* and *aug*. It results from the trivial monotonicity of order substitution w.r.t. inclusion:

$$\Delta' \subseteq \Delta \Rightarrow \Gamma[\Delta'] \subseteq \Gamma[\Delta] \text{ and } \Gamma'[X] \subseteq \Gamma[X] \Rightarrow \Gamma'[\Delta] \subseteq \Gamma[\Delta]. \qquad \dashv$$

The property below allows us to freely denote a formula by one of its equivalent formulations:

PROPOSITION 18 (Algebraic properties of the connectives). *If* $\Gamma \vdash C$ *is a provable sequent, and if one replaces each formula with an equivalent formula up to the commutativity and associativity of* $\otimes$ *and the associativity of* $\odot$, *one obtains again a provable sequent the proof of which is essentially similar.*

§4. **Encoding Petri nets.** Although it is always easy to criticize previous work, let us nevertheless point out some drawbacks of previous coding of Petri nets into linear logic — for these previous codings, the reader is referred to the surveys [27, 9, 12]. There are objections both from the logic and concurrency viewpoints.

Events and initial state are encoded by *proper* axioms, which are logically not well behaved: cut-elimination and the subformula properties are not as pleasant as in a plain logical calculus — although the *standard* derivations of [27], which do correspond to the usual encoding of Petri nets in linear logic have such properties. Nevertheless still the coding suffers from the following mismatch: proper axioms (events) are reusable, so while Petri nets are a multiplicative phenomenon, we are not in the multiplicative calculus (I)MLL but in the multiplicative-exponential calculus (I)MELL, a system the decidability of which is yet unknown.[7]

What is more worrying from a concurrency viewpoint is the absence of the events from the sequent to be proved. Their occurrences during the firing is encoded by the proof, as well as their order of execution. The absence of some traces of events in the conclusion sequents prevents to study questions like the language of a net or net synthesis. Moreover even the sophisticated work of [10], also dealing with series-parallel executions via a subtle notion of normal proof, does not capture maximally concurrent execution as soon as parallelism is not only due to the events but also to the marking as shown in [14] — counter-example taken up again in [21, 23].

Here we propose a coding which is inspired by the coding of context-free languages by Lambek grammars, see e.g. [15, 5]. In this well-known approach, there is a lexicon mapping terminals to formulae, and the provability of a sequent $T_1, \ldots, T_n \vdash U$ in logical system (Lambek calculus) means that any sequences of terminals $a_1, \ldots, a_n$ whose respective types are $T_1, \ldots, T_n$ is produced by the non-terminal U. Our coding of Petri nets makes use of three kinds of formulae; we of course find again the notations introduced in the section 2.

Places: Propositional variables, elements of $\mathcal{P}$.

Markings: Formulae with $\otimes$ as only connective. A marking with $n(p)$ tokens in each place p of a set Q of places, is denoted by the formula: $\bigotimes_{p \in Q} p^{n(p)}$ — regarding this part of the coding, there is no difference with previous work.

Events: An event denoted by $\mathrm{Pre}[x] \setminus \mathrm{Post}[x]$ in section 2 is represented by this expression viewed as a formula of PCLL. Thus, the set of formulae denoting events is $\mathcal{E}^{\otimes \setminus \otimes} = \Pi^{\otimes} \setminus \Pi^{\otimes}$.

---

[7]In any case, proper axioms are not closed under the substitution of a propositional variable with a formula, so there is no possibility to move to second order, if, for example, quantification over places is needed for specifying the Petri-net behavior.

We can now state our main result precisely:

THEOREM 19. *Given an* SP-*pomset ϕ of events (or of formulae in $\mathcal{E}^{\otimes\backslash\otimes}$) and two markings M and N (or formulae in $\Pi^{\otimes}$) the following propositions are equivalent*:

1. PCLL *proves* $(M;\phi) \vdash N$
2. $M \xrightarrow{\phi} N$

The (2) $\Rightarrow$ (1) is precisely proposition 22, which follows in section 6.

The (1) $\Rightarrow$ (2) results from a slightly more general result, proposition 25 of section 7, which concerns executive sequents — the ones that make sense w.r.t. Petri nets:

DEFINITION 20. A sequent $\Gamma \vdash C$ is said to be *executive* whenever:

- all formulae of Γ are either markings ($\otimes$-only formulae) or events ($M \backslash N$ with M and N markings).
- C is a marking.

The proposition 25 shows that whenever an executive sequent $\Gamma \vdash C$ is provable, one has $M \xrightarrow{\phi} C$, where

- ϕ is the restriction of the SP-order Γ to the events of Γ,[8]
- M is the sum ($\otimes$) of all the markings in Γ.

One of the key points is that executive sequents are well behaved w.r.t. provability (proposition 23): normal proofs of executive sequents only contain executive sequents. The other is the substitution property on Petri net execution, proposition 7.

Before proving this, let us consider a small example.

§5. A small example.

Let us a consider a Petri net with two places a and b, and two events $x = a \backslash b$ and $y = b \backslash a$ — each of them moving one token from one place to the other. If the initial marking is one token in each place, there are two different ways to execute x and y: either simultaneously and in this case the token used by y (resp. x) cannot be the one produced by x (resp. y), or sequentially and in this latter case, the token consumed by the second event can either be the one produced by the first event or the one that was already there.

Intuitively, how does our model take this difference into account? A given proof completely describes an execution: one can actually trace the consumption and production of tokens. Tokens are introduced by pairs in axioms, one being positive and the other negative — the usual notion of polarity, see any logic text book. They can be followed in each rule of a cut free proof, so in a sequent one can see by which events a token is produced and consumed. We

---

[8]As seen in proposition 12 the restriction of an SP-order is also an SP-order.

indicate this by labeling the propositional variables with the number of the axiom they come from.

A proof net representation would provide a much clearer representation, since the tracing of the tokens corresponds to paths in proof nets. Unfortunately up to now there does not exists a proof net formalism for this calculus. Indeed, we could present proofs as proof structures of [1] but up to now there does not exists a sound and complete correctness criterion for recognizing proofs among them when the augmenting rule is allowed (their notion of proof net only allows for a $\otimes$ to be replaced by a $\odot$.)

Series composition corresponds to sequential composition of two executions, while parallel composition corresponds to the concurrent composition of two executions, so let us analyze the possible proofs of the sequents corresponding respectively to both ways to execute the Petri net:

$\{a, b\}; (a \setminus b; b \setminus a) \vdash a \otimes b$ There are two essentially different proofs. In the first proof the second event consumes the token produced by the first event, and in the other proof, the second event consumes the token that was there since the initial marking.

$$
\cfrac{
  \cfrac{
    \cfrac{
      \cfrac{
        \cfrac{
          \cfrac{
            \cfrac{
              \cfrac{b_1 \vdash b_1 \quad a_2 \vdash a_2}{(b_1; b_1 \setminus a_2) \vdash a_2} \, \backslash h \quad a_3 \vdash a_3
            }{((a_3; a_3 \setminus b_1); b_1 \setminus a_2) \vdash a_2} \, \backslash h \quad b_4 \vdash b_4
          }{\{b_4, ((a_3; a_3 \setminus b_1); b_1 \setminus a_2)\} \vdash a_2 \otimes b_4} \, \otimes_i
        }{\{b_4, (a_3; (a_3 \setminus b_1; b_1 \setminus a_2))\} \vdash a_2 \otimes b_4} \, asso; r
      }{\{\{b_4, a_3\}, (a_3 \setminus b_1; b_1 \setminus a_2))\} \vdash a_2 \otimes b_4} \, aug.(rew3l)
    }{(\{a_3, b_4\}; (a_3 \setminus b_1; b_1 \setminus a_2)) \vdash a_2 \otimes b_4} \, aug.(rew2)
  }{(\{a_3 \otimes b_4\}; (a_3 \setminus b_1; b_1 \setminus a_2)) \vdash a_2 \otimes b_4} \, \otimes h
$$

$$
\cfrac{
  \cfrac{
    \cfrac{
      \cfrac{
        \cfrac{b_1 \vdash b_1 \quad a_2 \vdash a_2}{(b_1; b_1 \setminus a_2) \vdash a_2} \, \backslash h \quad \cfrac{a_3 \vdash a_3 \quad b_4 \vdash b_4}{(a_3; a_3 \setminus b_4) \vdash b_4} \, \backslash h
      }{\{(b_1; b_1 \setminus a_2), (a_3; a_3 \setminus b_4)\} \vdash a_2 \otimes b_4} \, \otimes_i
    }{(\{a_3, b_1\}; \{a_3 \setminus b_4, b_1 \setminus a_2\}) \vdash a_2 \otimes b_4} \, aug.(rew4)
  }{(\{a_3, b_1\}; (a_3 \setminus b_4; b_1 \setminus a_2)) \vdash a_2 \otimes b_4} \, aug.(rew2)
}{(a_3 \otimes b_1; (a_3 \setminus b_4; b_1 \setminus a_2)) \vdash a_2 \otimes b_4} \, \otimes h
$$

$\{a, b\}; \{a \setminus b, b \setminus a\} \vdash a \otimes b$ In this case there is a proof as well, similar to the second one above (skipping the final $aug.(rew2)$ rule); but there is no proof which would correspond to the first proof. It is indeed impossible (both intuitively and formally) that the token consumed by $b \setminus a$ is the one produced by $a \setminus b$, since these two events take place simultane-ously. Using the fact that only executive sequents appear in a proof of an executive sequent (proposition 23) an easy induction on cut-free proofs in PCLL shows the following property which clearly prohibits a proof yielding $\{a_3, b_4\}; \{a_3 \setminus b_1, b_1 \setminus a_2\} \vdash a_2 \otimes b_4$.

PROPOSITION 21. *If the two occurrences x_i of a propositional variable intro-duced by the same axiom occur in the formulae $M \otimes x_i$ and $F \otimes x_i \setminus G$ then one always has $(M \otimes x_i) < (F \otimes x_i \setminus G)$ — in the SP-order over formulae.*

§6. From Petri nets to proofs in PCLL.

PROPOSITION 22. *Let ϕ be an SP-pomset of events of a Petri net, which can also be viewed as a context whose formulae are events.*

If $M \xrightarrow{\phi} N$ then PCLL proves $(M; \phi) \vdash N$.

PROOF. We proceed by induction on the SP-term ϕ.

$\phi = P \setminus Q$ (ϕ is a single event) Since $M \xrightarrow{P \setminus Q} N$, there exist markings P, Q, R such that $M = P \otimes R$ and $N = Q \otimes R$.

$$\frac{\dfrac{\dfrac{P \vdash P \quad Q \vdash Q}{(P; P \setminus Q) \vdash Q} \setminus_h \quad R \vdash R}{\dfrac{\{(P; P \setminus Q), R\} \vdash Q \otimes R}{\dfrac{(\{P, R\}; P \setminus Q) \vdash Q \otimes R}{(P \otimes R; P \setminus Q) \vdash Q \otimes R} \otimes_h} aug.} \otimes_i}{}$$

$\phi = \{\phi_1, \phi_2\}$ We know from 2 of proposition 9 that there exist markings M_1, M_2, N_1, N_2 and R such that:

- $M = M_1 \otimes M_2 \otimes R$
- $N = N_1 \otimes N_2 \otimes R$
- $M_1 \xrightarrow{\phi_1} N_1$
- $M_2 \xrightarrow{\phi_2} N_2$

By induction hypothesis we know that PCLL proves $(M_1; \phi_1) \vdash N_1$ and $(M_2; \phi_2) \vdash N_2$.

$$\frac{\dfrac{\dfrac{\dfrac{\dfrac{(M_1; \phi_1) \vdash N_1 \quad (M_2; \phi_2) \vdash N_2}{\{(M_1; \phi_1), (M_2; \phi_2)\} \vdash N_1 \otimes N_2} \otimes_i}{(\{M_1, M_2\}; \{\phi_1, \phi_2\}) \vdash N_1 \otimes N_2} aug. \quad R \vdash R}{\{(\{M_1, M_2\}; \{\phi_1, \phi_2\}), R\} \vdash N_1 \otimes N_2 \otimes R} \otimes_i}{(\{M_1, M_2, R\}; \{\phi_1, \phi_2\}) \vdash N_1 \otimes N_2 \otimes R} aug.}{\dfrac{(\{M_1, M_2 \otimes R\}; \{\phi_1, \phi_2\}) \vdash N_1 \otimes N_2 \otimes R}{(M_1 \otimes M_2 \otimes R; \{\phi_1, \phi_2\}) \vdash N_1 \otimes N_2 \otimes R} \otimes_h} \otimes_h$$

$\phi = (\phi_1; \phi_2)$ We know from 3 of proposition 9 that there exists a marking Q such that $M \xrightarrow{\phi_1} Q$ and $Q \xrightarrow{\phi_2} N$ therefore, by induction hypothesis PCLL proves $M; \phi_1 \vdash Q$ and $Q; \phi_2 \vdash N$.

$$\frac{\dfrac{(M; \phi_1) \vdash Q \quad (Q; \phi_2) \vdash N}{((M; \phi_1); \phi_2) \vdash N} \ cut}{(M; (\phi_1; \phi_2)) \vdash N} \ \text{``;'' associative}$$

This yields a non cut-free proof but when the whole proof is built, we can eliminate them, and thus obtain a cut-free proof. $\dashv$

§7. From proofs in PCLL to Petri net execution.

7.1. A property of the PCLL calculus on executive sequents. Recall from definition 20 that executive sequents of PCLL are the ones whose right-hand side is a marking, and whose left hand-side only consists in markings or events. Although it is more than a mere language restriction, executive sequents are closed under provability in PCLL calculus in the following sense:

PROPOSITION 23. *Let δ be a cut free proof of an executive sequent; then each sequent in δ is itself an executive sequent. Consequently the only rules of δ are*

$$aug., \mathbf{1}_h, \backslash_h, \otimes_i, \otimes_h$$

and its axioms are either $\vdash \mathbf{1}$ ($\mathbf{1}_i$) or $M \vdash M$ (ax.) with $M \in \Pi^{\otimes}$.

PROOF. Because of the subformula property (proposition 17) only formulae of $\mathcal{E}^{\otimes \backslash \otimes} \uplus \Pi^{\otimes}$ can appear in δ. So δ does not contain any of the rules $\odot_h$, $\odot_i$, $/_h$, $/_i$, $\multimap_h$ or $\multimap_i$ since all these rules introduce a connective which does not appear in the conclusion sequent. To complete the proof, let us show that if the right-hand side of a sequent of δ contains a formula of $\mathcal{E}^{\otimes \backslash \otimes}$, then so does the right hand side of the sequent below it.

So assume there is a formula H containing the symbol $\backslash$ in the right-hand side of a sequent of δ — a cut free proof of an executive sequent $\Gamma \vdash C$. Now let us see that whatever the rule having this sequent as one of its premises we obtain an $\backslash$ symbol in the right-hand side of the conclusion sequent of the rule.

<u>$aug., \mathbf{1}_h$</u> Assume there is a formula of $\mathcal{E}^{\otimes \backslash \otimes}$ in the right hand-side of the premise of either of these two rules; then there is one in the right hand side of the conclusion sequent of either rule. <u>$\backslash_i$</u> This rule would create in the right-hand side of the conclusion sequent a formula with two symbols $\backslash$'s, that is a formula which is not a subformula of the conclusion sequent of δ — this rule is not used below the problematic sequent.

<u>$\backslash_h$</u>

- either H is kept in the right hand-side of the sequent, so there is a symbol $\backslash$ in the right-hand side of the conclusion sequent of the rule,

- or a formula $H \setminus X$ is created in the left-hand side of the conclusion sequent; this formula has at least two symbols $\setminus$, so it is a formula which is not a subformula of the conclusion sequent of δ, this subcase is impossible.

$\otimes_i$ This would create a formula $H \otimes U$ which is not a subformula of the conclusion sequent of δ, this case is impossible.

$\otimes_h$ H is kept in the right-hand side of the conclusion sequent.

The presence of a $\setminus$ symbol in the right hand-side of a sequent of δ would entail the presence of such a symbol in C, the right hand-side of the sequent $\Gamma \vdash C$ proved by δ, and this conflicts with $\Gamma \vdash C$ being an executive sequent.

As we have shown that δ only contains executive sequents, that is contains no $\setminus$ symbol in the right-hand side of any sequent, it is clear that the rule $\setminus_i$ is not used: indeed, it introduces a $\setminus$ symbol in the right-hand side of its conclusion sequent.

Since the proof only contains executive sequents the axioms can only be $M \vdash M$ with $M \in \Pi^\otimes$ or $\vdash \mathbf{1}$. $\dashv$

7.2. From PCLL proofs to concurrent executions. Let $\Gamma \vdash C$ be an executive sequent. We denote by $\overline{\Gamma}^{I \setminus \mathcal{M}}$ the context obtained by replacing each marking M of Γ by the event $\mathbf{1} \setminus M$. Notice that from proposition 16 we know that PCLL proves $\Gamma \vdash C$ if and only if PCLL proves $\overline{\Gamma}^{I \setminus \mathcal{M}} \vdash C$.

PROPOSITION 24. *If PCLL proves an executive sequent* $\Gamma \vdash C$ *then* $\mathbf{1} \xrightarrow{\overline{\Gamma}^{I \setminus \mathcal{M}}} C$.

PROOF. We proceed by induction on the height of a normal (cut-free) proof of $\Gamma \vdash C$. Because of 23, we know the only possible rules are $\setminus_h$, $\mathbf{1}_h$, $\otimes_h$, $\otimes_i$, *aug.* and the only possible axioms are $\vdash \mathbf{1}$ $(\mathbf{1}_i)$ or $M \vdash M$ $(ax.)$ with $M \in \Pi^\otimes$. The last rule is an axiom so it is either $\vdash \mathbf{1}$, or $M \vdash M$ with $M \in \Pi^\otimes$. Nothing to say: $\mathbf{1} \xrightarrow{\emptyset} \mathbf{1}$ and $\mathbf{1} \xrightarrow{I \setminus M} M$ hold.

The last rule is $\mathbf{1}_h$

$$\frac{\Gamma \vdash C}{\{\Gamma, \mathbf{1}\} \vdash C} \, \mathbf{1}_h$$

By induction hypothesis we have $\mathbf{1} \xrightarrow{\overline{\Gamma}^{I \setminus \mathcal{M}}} C$, and $\mathbf{1} \xrightarrow{I \setminus I} \mathbf{1}$ by proposition 9.1 we have $\mathbf{1} = \mathbf{1} \otimes \mathbf{1} \xrightarrow{\{\overline{\Gamma}^{I \setminus \mathcal{M}}, I \setminus I\}} C \otimes \mathbf{1} = C$ and $\{\overline{\Gamma}^{I \setminus \mathcal{M}}, \mathbf{1} \setminus \mathbf{1}\} = \overline{\{\Gamma, \mathbf{1}\}}^{I \setminus \mathcal{M}}$.

The last rule is *aug.*

$$\frac{\Gamma \vdash C}{\Gamma' \vdash C} \, aug\,(mentation) \qquad \boxed{\text{if } |\Gamma| = |\Gamma'| \text{ and } \Gamma'^{\text{SP}} \supseteq \Gamma^{\text{SP}}}$$

By induction hypothesis we have $1 \xrightarrow{\overline{\Gamma}^{I \backslash \mathcal{M}}} C$ and since $\overline{\Gamma}^{I \backslash \mathcal{M}}$ is a suborder of $\overline{\Gamma'}^{I \backslash \mathcal{M}}$ we have $1 \xrightarrow{\overline{\Gamma'}^{I \backslash \mathcal{M}}} C$ by proposition 4.

The last rule is $\otimes_i$

$$\frac{\Delta \vdash A \quad \Gamma \vdash B}{\{\Delta, \Gamma\} \vdash A \otimes B} \otimes_i$$

By induction hypothesis we have both $1 \xrightarrow{\overline{\Delta}^{I \backslash \mathcal{M}}} A$ and $1 \xrightarrow{\overline{\Gamma}^{I \backslash \mathcal{M}}} B$. Thus by 1 of proposition 9 we have

$$1 = 1 \otimes 1 \xrightarrow{\overline{\{\Delta, \Gamma\}}^{I \backslash \mathcal{M}} = \{\overline{\Delta}^{I \backslash \mathcal{M}}, \overline{\Gamma}^{I \backslash \mathcal{M}}\}} A \otimes B$$

The last rule is $\otimes_h$

$$\frac{\Gamma[\{A, B\}] \vdash C}{\Gamma[A \otimes B] \vdash C} \otimes_h$$

By induction hypothesis we have: $1 \xrightarrow{\overline{\Gamma}^{I \backslash \mathcal{M}}[\{I \backslash A, I \backslash B\}]} C$.

Since $1 \xrightarrow{\{I \backslash A, I \backslash B\}} A \otimes B$ we can use the proposition 7.2 with $\psi = \{1 \backslash A, 1 \backslash B\}$ and $x = 1 \backslash A \otimes B$ — indeed 1 is the minimum marking of $\{1 \backslash A, 1 \backslash B\}$. We thus obtain $1 \xrightarrow{\overline{\Gamma}^{I \backslash \mathcal{M}}[I \backslash A \otimes B]} C$ that is: $1 \xrightarrow{\overline{\Gamma[A \otimes B]}^{I \backslash \mathcal{M}}} C$.

The last rule is $\backslash_h$

$$\frac{\Gamma[B] \vdash C \quad \Delta \vdash A}{\Gamma[(\Delta; A \backslash B)] \vdash C} \backslash_h$$

By induction hypothesis we have $1 \xrightarrow{\overline{\Gamma}^{I \backslash \mathcal{M}}[I \backslash B]} C$ and we know that $1 \xrightarrow{I \backslash A; A \backslash B} B$ holds.

So we can apply proposition 7.1 with $x = 1 \backslash B$ and $\psi = (1 \backslash A; A \backslash B)$. This yields: $1 \xrightarrow{\overline{\Gamma}^{I \backslash \mathcal{M}}[(I \backslash A; A \backslash B)]} C$.

Since $1 \xrightarrow{\overline{\Delta}^{I \backslash \mathcal{M}}} A$, we can again apply proposition 7.1 with $x = 1 \backslash A$ and $\psi = \overline{\Delta}^{I \backslash \mathcal{M}}$. We thus obtain $1 \xrightarrow{\overline{\Gamma}^{I \backslash \mathcal{M}}[(\overline{\Delta}^{I \backslash \mathcal{M}}; A \backslash B)]} C$, that is: $1 \xrightarrow{\overline{\Gamma[(\Delta; A \backslash B)]}^{I \backslash \mathcal{M}}} C$. $\dashv$

PROPOSITION 25. *If* PCLL *proves* $(M; \phi) \vdash C$ *where* ϕ *is an* SP-*pomset of events, then* $M \xrightarrow{\phi} C$.

PROOF. From proposition 24 we have $1 \xrightarrow{\overline{\Gamma}^{I \backslash \mathcal{M}}} C$. Because ϕ is an SP-pomset of events, $\overline{\Gamma}^{I \backslash \mathcal{M}} = (1 \backslash M; \phi)$, and thus $1 \xrightarrow{\overline{(I \backslash M; \phi)}^{I \backslash \mathcal{M}}} C$. Applying proposition 8 with $\psi[1 \backslash M] = (1 \backslash M; \phi)$ we obtain $M \xrightarrow{\phi} C$. $\dashv$

§8. Future prospects.

8.1. Petri net synthesis. The synthesis of Petri nets from formal languages is the following question:

> *Given a set of events A and a language $L \subseteq A^*$, does there exists a Petri net **R** with a marking M such that the sequences of events of the possible firings of **R** with M as its initial marking are precisely the words in L? If so, how can **R** be constructed from L?*

This question has been solved for deterministic context-free languages by Darondeau in [6]. Our encoding allows a logical formulation of this question — which has not yet been investigated, and possibly leads nowhere. Our logical approach can be undertaken because the kind of logical systems we are using can be viewed as formal grammars, describing context-free languages.

Introduced by Lambek in his pioneering article [15] for natural language analysis via categorial grammars, Lambek-grammars reduce parsing to provability in a non commutative logic known as Lambek calculus. This calculus is exactly the non commutative part of PCLL: connectives are restricted to $\backslash, /, \odot$ and context only allows $(\ldots;\ldots)$, and the only structural rule is associativity — contexts are sequences of formulae. A lexicon associates each terminal a in A with a finite sets of formulae $\mathcal{L}(a)$. A word $a_1 \ldots a_n$ of A^* belongs to the language generated by the Lambek grammars (i.e., the lexicon, they are lexicalized grammars) whenever for each i there exists a formula $t_i \in \mathcal{L}(a_i)$ such that the Lambek calculus proves

$$t_1, \ldots, t_n \vdash S$$

Lambek grammars describe all context-free languages [2, 5] (even if only $\backslash$ is allowed) [9] and only them [20, 5].

Assume the context-free language that we want to obtain as the language of a Petri net is defined by a Lambek grammar with only $\backslash$. We wish to obtain a Petri net whose sequences of events are the $a_1 \ldots a_n$'s such that $t_1; \ldots; t_n \vdash S$ with $a_i \in \mathcal{L}(a_i)$. So the question is to find $M = M_1 \otimes \ldots \otimes M_m$ and to associate with each a_i one formula $\overline{a}_i$ of the shape $P_1 \otimes \ldots P_p \backslash Q_1 \otimes \ldots Q_q$ such that (1) and (2) are equivalent:

1. There exists N of the shape $N_1 \otimes N_2 \otimes \ldots \otimes N_n$ such that PCLL proves

$$M; \overline{a}_1; \overline{a}_2; \ldots; \overline{a}_k \vdash N$$

2. for all i there exists $t_i \in \mathcal{L}(a_i)$ such that

$$t_1; \ldots; t_n \vdash S$$

---

[9]Given a context-free grammar G, put it into Greibach normal form G'. For each non terminal a, the lexicon $\mathcal{L}$ is defined by $X_1 \backslash X_2 \backslash \ldots X_n \backslash X \in \mathcal{L}(a)$ whenever G' contains the rule $X \rightarrow X_n \ldots X_1 a$. The Lambek grammar generates the same language as G' and so the same language as G.

We are rather optimistic for this approach: indeed, there is a strong similarity between the terminal-type (the formula associated with a_i according to the lexicon) and the event-type (the formula corresponding to the event a_i).

$$
\begin{array}{rcl}
t_i &=& X_i^1 \odot \quad \ldots \quad \odot X_i^{k_i} \quad \backslash \qquad Y_i \\
\overline{a}_i &=& P_1 \otimes \quad \ldots \quad \otimes P_p \quad \backslash \quad Q_1 \otimes \ldots \otimes Q_q
\end{array}
$$

This suggests to use formula unification for solving net-synthesis questions.

8.2. Petri nets with credits. Our coding does not use the backward implication $/$. The meaning of such a connective is interesting from a computational viewpoint. Intuitively, an event M / N consumes a marking that will appear later on: $(M / N; N) \vdash M$. This should correspond to the possibility to have a credit N that ought to be consumed later on.

This first application that come to the idea is to use this for protocols: one can specify that a token has to be received by an event.

The second one is computational linguistics, since the diminishing context version of this calculus restricted to first order formulae $M \backslash N / P$ is the one we used in [16] to describe minimalist grammars of Stabler [26] which describe mildly context-sensitive languages—see [24] for more details on the relation between linear logic and computational linguistics. Although the connection is yet unclear, our hope is to extract a Petri net model for parsing, as for instance pushdown automata correspond to context-free languages.

8.3. High order Petri nets. Our coding of Petri nets only makes use of first order implication. The PCLL calculus naturally enables the definition of high order Petri nets, where events could consume and produce markings or events (second order nets), or even higher order events. From the logical view point this is quite natural and should cause no trouble. For instance a higher order event $(R \otimes (N \backslash M)) \backslash ((P \backslash Q) \otimes S)$ consumes the marking R and the event $N \backslash M$ and produces a new event $P \backslash Q$ and a marking S. Clearly it is mandatory to bound the order (implication nesting) of formulae (e.g. to order at most 2, where an event can be consumed); indeed the whole PCLL logic leads to hardly interpretable or at least irrealistic mobile systems, too far away from actual computational processes. Most properties are preserved since the subformula property 17 guarantees that no formula of order more than p is needed for proving formulae of order p. So this approach suggests a neat treatment of mobile processes. This could also be combined with the notion of credit of the previous paragraph.

§9. Appendix: proof of the substitution property in concurrent executions (proposition 7). In this section we prove the two parts of proposition 7. It should be observed that this proposition holds for any pomset of events, and not just for SP-pomsets of events.

Before proving it we need to know, how the lower closed subsets and their frontiers behave with respect to substitution.

9.1. Lower closed subsets, frontiers and substitution.

NOTATION 26. We are given a pomset ϕ with an occurrence of x, another pomset ψ and a lower closed subset Y of $\phi[x := \psi]$:

- We consider multisets as sets, that is we index the elements, and no two elements have the same index.
- Recall we only use $A - B$ when $A \supset B$.
- $C = A \uplus B$ means $C = A \cup B$ and $A \cap B = \emptyset$.
- X the domain of ψ.
- $\min(X)$ the set of the minimal elements of ψ.
- $\Phi = \phi[x := \psi]$.
- W the domain of Φ.
- W' the domain of ϕ so $W' = (W - X) \uplus \{x\}$.
- $S(x)$ is the set of all the immediate successors of x in ϕ, which is also the set of all the immediate successors of any of the maximal elements of ψ in Φ.
- $P(x)$ is the set of all the immediate predecessors of x in ϕ, which is also the set of all the immediate predecessors of any of the minimal elements of ψ in Φ.

PROPOSITION 27. *Let Z be a lower closed subset of Φ, and let $X' = Z \cap X$. Then exactly one of the following cases holds*:
1. $X' = X$
   In this case $\mathcal{F}_\Phi(Z) = \mathcal{F}_\phi(Z')$ with $Z' = (Z - X) \uplus \{x\}$.
2. $\emptyset \subsetneq X' \subsetneq X$
   In this case $\mathcal{F}_\Phi(Z) = \mathcal{F}_\psi(X') \uplus (\mathcal{F}_\phi(Z') - \{x\})$ with $Z' = Z - X'$. Observe that $\mathcal{F}_\Phi(Z)$ can contain elements of $\min(X)$, when $\mathcal{F}_\psi(X')$ does.
3. $X' = \emptyset$ *and* $Z \supset P(x)$
   In this case $\mathcal{F}_\Phi(Z) = (\mathcal{F}_\phi(Z) - \{x\}) \uplus \min(X)$.
4. $X' = \emptyset$ *and* $Z \not\supset P(X)$
   In this case $\mathcal{F}_\Phi(Z) = \mathcal{F}_\phi(Z)$.

PROOF. The list is clearly an exhaustive one and all cases are disjoint. We have to check that the equalities for the frontiers hold.

1. $X' = X$. Let $Z' = (Z - X) \uplus \{x\}$. We have $\Phi|_{W-Z} = \phi|_{W'-Z'}$. Therefore $\mathcal{F}_\Phi(Z) = \mathcal{F}_\phi(Z')$.
2. $\emptyset \subsetneq X' \subsetneq X$. Let $Z' = Z - X'$.
   - $\mathcal{F}_\phi(Z')$ *is well defined.* We have to show that Z' is a lower closed subset of Φ without elements of X, hence a lower closed subset of ϕ. Observe that as soon as an element $z \notin X$ is above one element of X then it is above every element of X. Consequently if there would exists an $x \in X$ below an element z of Z' then all elements of X

would be below z, and this would conflicts with $X' \neq X$ (that is $X \not\subseteq Z$).

- $Z' \supseteq P(X)$. Indeed Z is lower closed and contains an element of X while any element of $P(X)$ is below any element of X.

- $x \in \mathcal{F}_\phi(Z')$. Let z be an element such that $z < x$ in ϕ. Then there exists $p \in P(X)$ such that $z \leq p$, hence $z \in Z'$, because Z' is lower closed and contains $P(X)$ (previous item).

- If $z \in \mathcal{F}_\Phi(Z)$ and $\exists w \in X'$ $w \leq z$ then $z \in \mathcal{F}_\psi(X')$. Firstly let us show that $z \in X$. If $z \notin X$ since $\exists w \in X'$ $w \leq z$, we would have $z \geq u$ for all $u \in X$, conflicting with $X' \neq X$. Secondly, consider $z' \in X$ such that $z' < z$ w.r.t. ψ. Then we have $z' \in Z$, hence $z' \in X' = Z \cap X$. Consequently, $z \in \mathcal{F}_\psi(X')$.

- If $z \in \mathcal{F}_\Phi(Z)$ and $\neg\exists x \in X'$ $x \leq z$ then $z \in \mathcal{F}_\phi(Z') - \{x\}$. Observe that $z \neq x$ since $x \notin W$. Consider $u < z$ w.r.t. ϕ. As $u \neq x$ and $z \neq x$ this amounts to $u < z$ in Φ, hence $u \in Z$ and as $u \notin X$ we have $u \in Z'$. Hence $z \in \mathcal{F}_\phi(Z') - \{x\}$.

- If $z \in \mathcal{F}_\phi(Z') - \{x\}$ then $z \in \mathcal{F}_\Phi(Z)$. Notice that $z \in W - X$. Consider $u < z$ w.r.t. Φ. If $u \in X$ then for every element t of X we would have $t < z$ w.r.t. Φ, and $x < z$ w.r.t. ϕ; with $z \in \mathcal{F}_\phi(Z')$ this would entail $x \in Z'$, and this is impossible because $x \in \mathcal{F}_\phi(Z')$. Since $u \notin X$ and $z \neq x$ the relation $u < z$ w.r.t. Φ means $u < z$ w.r.t. ϕ, and therefore $u \in Z'$ hence $u \in Z$.

- If $z \in \mathcal{F}_\psi(X')$ then $z \in \mathcal{F}_\Phi(Z)$. Let $u < z$ w.r.t. Φ. If $u \in X$, then $u \in X' \subset Z$. If $u \notin X$, as $z \in X$, then $\exists p \in P(X)$ $u \leq p$. As $Z \supset P(X)$ and Z is lower closed, $Z \ni u$.

- Consequently, $\mathcal{F}_\Phi(Z) = \mathcal{F}_\psi(X') \uplus (\mathcal{F}_\phi(Z') - \{x\})$.

3. $X' = \emptyset$ and $Z \supset P(x)$.

- $x \in \mathcal{F}_\phi(Z)$. Indeed $x \notin Z$ and $\forall u < x$ $\exists p \in P(X)$ $u \leq p$, and as $p \in Z$ and Z is lower closed we have $u \in Z$.

- $\min(X) \subset \mathcal{F}_\Phi(Z)$. Indeed let $m \in \min(X)$ then $\forall u < m$ $\exists p \in P(X)$ $u \leq p$, and as $m \in Z$ and Z is lower closed we have $u \in Z$.

- If $z \notin X$ and $z \in \mathcal{F}_\Phi(Z)$ then $z \in \mathcal{F}_\phi(Z)$. Let $z \in \mathcal{F}_\Phi(Z)$, $z \notin X$, so $z \in W'$. Let $u < z$ w.r.t. ϕ. Then $u \in Z$, and therefore $u \neq x$. Hence $z \in \mathcal{F}_\phi(Z)$.

- If $z \neq x$ and $z \in \mathcal{F}_\phi(Z)$ then $z \in \mathcal{F}_\Phi(Z)$. Let $z \in \mathcal{F}_\phi(Z)$, $z \neq x$, so $z \in W$. Let $u < z$ w.r.t. Φ. Then $u \in Z$, and therefore $u \notin X$. Hence $z \in \mathcal{F}_\Phi(Z)$.

- Consequently $(\mathcal{F}_\Phi(Z) - \min(X)) = (\mathcal{F}_\phi(Z) - \{x\})$.

4. $X' = \emptyset$ and $Z \not\supset P(X)$.

- If $z \in \mathcal{F}_\Phi(Z)$ then $z \in \mathcal{F}_\phi(Z)$. We have $z \notin X$; otherwise as $\forall p \in P(X)$ $p < z$, one would have $P(X) \subset Z$. Therefore $z \in W'$.

Let $u < z$ w.r.t. ϕ; then $u \neq x$ (otherwise $P(X) \subset Z$). So $u \in W$ and $u \in Z$, hence $z \in \mathcal{F}_\phi(Z)$.

- If $z \in \mathcal{F}_\phi(Z)$ then $z \in \mathcal{F}_\Phi(Z)$. We have $z \neq x$; otherwise as $\forall p \in P(X) \ p < z$, one would have $P(X) \subset Z$. Let $u < z$ w.r.t. Φ then $u \notin X$, $u \in W$, and thus $u < z$ w.r.t. ϕ; so $u \in Z$. Hence $z \in \mathcal{F}_\Phi(Z)$. $\dashv$

9.2. Proof of proposition 7. To facilitate the computation we will extend the operation on markings to elements of the free abelian group over places and this corresponds to a negative number of tokens in a place.[10] It is harmless to compute markings using such expressions provided the result is a real a marking (all places have a positive or null exponent).

For simplification, we drop the $\otimes$ product. As we deal with element of the free abelian group over places, the expression $M \oslash N$ is always defined and means MN^{-1}.

The partial order $\sqsupseteq$ extends to elements of the free abelian group over places: $M \sqsupseteq N$ if for every place A the exponent of A in M is bigger than the exponent of A in N.

Observe that

$$\mathsf{Pre}[A \uplus B] = \mathsf{Pre}[A]\mathsf{Pre}[B] \text{ and } \mathsf{Post}[A \uplus B] = \mathsf{Post}[A]\mathsf{Post}[B] \qquad (\diamond)$$

and that $\mathsf{Pre}[A - B] = \mathsf{Pre}[A]\mathsf{Pre}[B]^{-1}$ and $\mathsf{Post}[A - B] = \mathsf{Post}[A]\mathsf{Post}[B]^{-1}$ since $A - B$ presupposes that $A \supseteq B$.

PROPOSITION 28 (substitution property — 7.1 expansion). *Let*

- $\phi = (X, <, f)$ *be a partially ordered enumeration of events containing an occurrence x of $P \setminus Q$ (i.e., $f(x) = P \setminus Q$)*
- $\psi = (Y, \prec, g)$ *be a partially ordered enumeration of events, such that $Y \cap X = \emptyset$ and such that $P \overset{\psi}{\longrightarrow} Q$*

then one has:

$$(a) : \left(M_0 \overset{\phi}{\longrightarrow} \right) \quad \Longrightarrow \quad \left(M_0 \overset{\phi[x:=\psi]}{\longrightarrow} \right) : (b)$$

PROOF. Observe that $\mathsf{Post}[x] \ (\mathsf{Pre}[x])^{-1} = Q \ P^{-1} = \mathsf{Post}[X] \ \mathsf{Pre}[X]^{-1} \ (\diamond)$.

We use the notation 26. Let Z be a lower closed subset of Φ, and let $X' = X \cap Z$, we have to show that $M_0 \ \mathsf{Post}[Z]\mathsf{Pre}[Z]^{-1} \sqsupseteq \mathcal{F}_\Phi(Z)$. We follow the cases of proposition 27.

---

[10]Linear logic notation, $\otimes$, oblige us to a "multiplicative" notation, while an additive one would be more intuitive. We would have vectors of integers, indicating how many tokens are present or missing in each place. It is nevertheless absolutely equivalent, since a free abelian group is the same as a $\mathbb{Z}$-module.

1. $X' = X$

   In this case $\mathcal{F}_\Phi(Z) = \mathcal{F}_\phi(Z')$ with $Z' = (Z - X) \uplus \{x\}$.

   $$\text{Pre}[Z] = \text{Pre}[Z'] \;\; P \;\; \text{Pre}[X]^{-1}$$

   $$\text{Post}[Z] = \text{Post}[Z'] \; Q \;\; \text{Post}[X]^{-1}$$

   Therefore one has the following equalities where the last one is due to $(\diamond)$:

   $$M_0 \;\; \text{Post}[Z]\text{Pre}[Z]^{-1}$$
   $$= M_0 \;\; \text{Post}[Z'] \;\; Q \;\; \text{Post}[X]^{-1} \;\; (\text{Pre}[Z'] \;\; P \;\; \text{Pre}[X]^{-1})^{-1}$$
   $$= M_0 \;\; \text{Post}[Z'] \;\; \text{Pre}[Z']^{-1} \;\; Q \;\; P^{-1} \;\; (\text{Post}[X] \;\; \text{Pre}[X]^{-1})^{-1}$$
   $$= M_0 \;\; \text{Post}[Z'] \;\; \text{Pre}[Z']^{-1}$$

   Because $M_0 \overset{\phi}{\longrightarrow}$ and Z' is lower closed subset of ϕ, we know that

   $$M_0 \;\; \text{Post}[Z']\text{Pre}[Z']^{-1} \sqsupseteq \text{Pre}[\mathcal{F}_\phi(Z')] = \text{Pre}[\mathcal{F}_\Phi(Z)]$$

2. $\emptyset \subsetneq X' \subsetneq X$

   In this case $\mathcal{F}_\Phi(Z) = \mathcal{F}_\psi(X') \uplus (\mathcal{F}_\phi(Z') - \{x\})$ with $Z' = Z - X'$. Let us call $U = \mathcal{F}_\phi(Z') - \{x\}$.

   Because Z' is a lower closed subset of ϕ and $M_0 \overset{\phi}{\longrightarrow}$ we have:

   $$M_0 \;\; \text{Post}[Z'] \;\; \text{Pre}[Z']^{-1} \sqsupseteq \text{Pre}[\mathcal{F}_\phi(Z')] = \text{Pre}[U] \;\; \text{Pre}[x] = \text{Pre}[U] \;\; P \quad (*)$$

   Because $P \overset{\psi}{\longrightarrow}$ and X' is a lower closed subset of ψ we have:

   $$P \;\; \text{Post}[X'] \;\; \text{Pre}[X']^{-1} \sqsupseteq \text{Pre}[\mathcal{F}_\psi(X')] \qquad\qquad (**)$$

   $$M_0 \;\; \text{Post}[Z] \;\; \text{Pre}[Z]^{-1}$$
   $$= M_0 \;\; \text{Post}[Z'] \;\; \text{Post}[X'] \;\; \text{Pre}[X']^{-1} \;\; \text{Pre}[Z']^{-1}$$
   $$\sqsupseteq \text{Pre}[U] \;\; P \;\; \text{Post}[X'] \;\; \text{Pre}[X']^{-1} \qquad\text{because of } (*)$$
   $$\sqsupseteq \text{Pre}[U] \;\; \text{Pre}[\mathcal{F}_\psi(X')] \qquad\qquad\qquad\text{because of } (**)$$
   $$= \text{Pre}[\mathcal{F}_\Phi(Z)]$$

3. $X' = \emptyset$ and $Z \supset P(x)$

   In this case $\mathcal{F}_\Phi(Z) = (\mathcal{F}_\phi(Z) - \{x\}) \uplus \min(X)$ Let U be the multiset of events such that $\mathcal{F}_\phi(Z) = \{x\} \uplus U$ and $\mathcal{F}_\Phi(Z) = U \uplus (\min(X))$.

   Because $M_0 \overset{\phi}{\longrightarrow}$ and Z is lower closed subset of ϕ, we know that

   $$M_0 \;\; \text{Post}[Z] \;\; \text{Pre}[Z]^{-1} \sqsupseteq \text{Pre}[\mathcal{F}_\phi(Z)] = \text{Pre}[x] \;\; \text{Pre}[U]$$

   Because $P \overset{\psi}{\longrightarrow}$, considering the $\emptyset$ which is lower closed and whose frontier in ψ is $\min(X)$ we have we have $P \sqsupseteq \text{Pre}[\min(X)]$.

   Therefore

   $$M_0 \;\; \text{Post}[Z] \;\; \text{Pre}[Z]^{-1} \sqsupseteq \text{Pre}[x] \;\; \text{Pre}[U]$$
   $$\sqsupseteq \text{Pre}[\min(X)] \;\; \text{Pre}[U] = \text{Pre}[\mathcal{F}_\Phi(Z)]$$

4. $X' = \emptyset$ and $Z \not\sqsupseteq P(X)$

In this case $\mathcal{F}_\Phi(Z) = \mathcal{F}_\phi(Z)$. Because $M_0 \xrightarrow{\phi}$ and Z is lower closed subset of ϕ, we know that

$$M_0 \ \text{Post}[Z] \ \text{Pre}[Z]^{-1} \sqsupseteq \text{Pre}[\mathcal{F}_\phi(Z)] = \text{Pre}[\mathcal{F}_\Phi(Z)] \qquad \dashv$$

PROPOSITION 29 (substitution property — 7.2 contraction). *Let*

- $\phi = (X, <, f)$ *be a partially ordered enumeration of events containing an occurrence x of $P \setminus Q$ (i.e., $f(x) = P \setminus Q$)*
- $\psi = (Y, \prec, g)$ *be a partially ordered enumeration of events, such that $Y \cap X = \emptyset$ and such that $P \xrightarrow{\psi} Q$ with P being the minimum marking of ψ.*

then one has:

$$(b): \left(M_0 \xrightarrow{\phi[x:=\psi]}\right) \implies \left(M_0 \xrightarrow{\phi}\right): (a)$$

PROOF. We still use notation 26. Let Z' be a lower closed subset of ϕ. We have to show that $M_0 \ \text{Pre}[Z']^{-1} \ \text{Post}[Z'] \sqsupseteq \text{Pre}[\mathcal{F}_\phi(Z')]$.

1. $x \notin \mathcal{F}_\phi(Z')$

Let Z be the lower closed subset of Φ defined by $Z = Z'$ if $x \notin Z'$ and by $Z = (Z' - \{x\}) \uplus X$ if $x \in Z'$. Observe that

$$\text{Post}[Z]\text{Pre}[Z]^{-1} = \text{Post}[Z']\text{Pre}[Z']^{-1} \qquad (*)$$

We have $\mathcal{F}_\phi(Z') = \mathcal{F}_\Phi(Z)$. Because of $(*)$ we have:

$$M_0 \ \text{Post}[Z'] \ \text{Pre}[Z']^{-1} = M_0 \ \text{Post}[Z] \ \text{Pre}[Z]^{-1}$$

and since $M_0 \xrightarrow{\Phi}$ we have

$$M_0 \ \text{Post}[Z] \ \text{Pre}[Z]^{-1} \sqsupseteq \text{Pre}[\mathcal{F}_\Phi(Z)] = \text{Pre}[\mathcal{F}_\phi(Z)]$$

2. $x \in \mathcal{F}_\phi(Z')$

Let us call $U = \mathcal{F}_\phi(Z) - \{x\}$. Let X' be any lower closed subset of ψ. Then $Z = Z' \uplus X'$ is a lower closed subset of Φ and $\mathcal{F}_\Phi(Z) = U \uplus \mathcal{F}_\psi(X')$. Since $M_0 \xrightarrow{\Phi}$, we have:

$$M_0 \ \text{Post}[Z'] \ \text{Post}[X'] \ \text{Pre}[Z']^{-1} \ \text{Pre}[X']^{-1} \sqsupseteq \text{Pre}[U] \ \text{Pre}[\mathcal{F}_\psi(X')]$$

that is:

$$(M_0 \ \text{Pre}[U]^{-1} \ \text{Post}[Z'] \ \text{Pre}[Z']^{-1}) \ \text{Post}[X'] \ \text{Pre}[X']^{-1} \sqsupseteq \text{Pre}[\mathcal{F}_\psi(X')]$$

Because this holds for any lower closed subset X' of ψ, and because P is the minimum marking of ψ, we have

$$(M_0 \ \text{Pre}[U]^{-1} \ \text{Post}[Z'] \ \text{Pre}[Z']^{-1}) \sqsupseteq P$$

and therefore

$$(M_0 \ \text{Post}[Z'] \ \text{Pre}[Z']^{-1}) \sqsupseteq \text{Pre}[U] \ P = \text{Pre}[U]\text{Pre}[x] = \text{Pre}[\mathcal{F}_\phi(Z')] \qquad \dashv$$

REFERENCES

[1] MICHELE V. ABRUSCI and PAUL RUET, *Non-commutative logic I: the multiplicative fragment*, **Annals of Pure and Applied Logic**, vol. 101 (1999), no. 1, pp. 29–64.

[2] Y. BAR-HILLEL, C. GAIFMAN, and E. SHAMIR, *On categorial and phrase-structure grammars*, **Bulletin of the Research Council of Israel**, vol. F (1960), no. 9, pp. 1–16.

[3] DENIS BECHET, PHILIPPE DE GROOTE, and CHRISTIAN RETORÉ, *A complete axiomatisation of the inclusion of series-parallel partial orders*, **Rewriting Techniques and Applications, RTA'97** (H. Comon, editor), LNCS, vol. 1232, Springer Verlag, 1997, pp. 230–240.

[4] EIKE BEST and RAYMOND DEVILLERS, *Sequential and concurrent behaviour in Petri net theory*, **Theoretical Computer Science**, vol. 55 (1987), pp. 87–136.

[5] WOJCIECH BUSZKOWSKI, *Mathematical linguistics and proof theory*, **Handbook of logic and language** (J. van Benthem and A. ter Meulen, editors), North-Holland Elsevier, Amsterdam, 1997, pp. 683–736.

[6] PHILLIPE DARONDEAU, *Deriving unbounded Petri nets from formal languages,*, **Rapport de Recherche 3365**, INRIA, fevrier 1998, http://www.inria.fr/.

[7] PHILIPPE DE GROOTE, *Partially commutative linear logic: sequent calculus and phase semantics*, **Third Roma workshop: Proofs and linguistic categories — applications of logic to the analysis and implementation of natural language** (Michele Abrusci and Claudia Casadio, editors), Bologna: CLUEB, 1996, pp. 199–208.

[8] AKIM DEMAILLE, *Logiques linéaires hybrides et leurs modalités*, **Thése de doctorat, spécialité informatique**, Ecole Nationale Supérieure des Télécommunications de Paris, juin 1999.

[9] UFFE ENGEBERG and GLYNN WINSKEL, *Completeness results for linear logic on Petri nets*, **Annals of Pure and Applied Logic**, vol. 86 (1997), no. 2, pp. 101–135.

[10] VIJAY GEHLOT, *A proof-theoretic approach to semantics of concurrency*, **Ph.D. thesis**, University of Pennsylvania, 1992.

[11] JEAN-YVES GIRARD, *Linear logic*, **Theoretical Computer Science**, vol. 50 (1987), no. 1, pp. 1–102.

[12] FRANÇOIS GIRAULT, *Formalisation en logique linéaire du fonctionnement des réseaux de Petri*, **Thése de doctorat, spécialité informatique industrielle**, Université Paul Sabatier, Toulouse, décembre 1997.

[13] URSULA GOLTZ and WOLFGANG REISIG, *The non sequential behaviour of Petri nets*, **Information and Computation**, vol. 57 (1983), pp. 125–147.

[14] LUIS ALLAN KÜNZLE, *Raisonnement temporel basé sur les réseaux de Petri pour des systèmes manipulant des ressources*, **Thése de doctorat, spécialité informatique industrielle**, Université Paul Sabatier, Toulouse, septembre 1997.

[15] JOACHIM LAMBEK, *The mathematics of sentence structure*, **American Mathematical Monthly**, vol. 65 (1958), pp. 154–169.

[16] ALAIN LECOMTE and CHRISTIAN RETORÉ, *Extending Lambek grammars: a logical account of minimalist grammars*, **Proceedings of the 39th annual meeting of the Association for Computational Linguistics, ACL 2001** (Toulouse), vol. 354–361, ACL, July 2001.

[17] JOSÉ MESEGUER and UGO MONTANARI, *Petri nets are monoids*, **Information and Computation**, vol. 88 (1990), pp. 105–155.

[18] ROLF H. MÖHRING, *Computationally tractable classes of ordered sets*, **Algorithms and order** (I. Rival, editor), NATO ASI series C, vol. 255, Kluwer, 1989, pp. 105–194.

[19] MADHAVAN MUKUND, *Petri nets and step transition systems*, **International Journal of Foundations of Computer Science**, vol. 3 (1992), no. 4, pp. 443–478.

[20] MATI PENTUS, *Lambek grammars are context-free*, **Logic In Computer Science, LICS'93**, IEEE Computer Society Press, 1993.

[21] BRIGITTE PRADIN-CHÉZALVIEL, LUIS ALLAN KÜNZLE, FRANÇOIS GIRAULT, and ROBERT VALETTE, *Calculating duration of concurrent scenarios in time Petri nets*, **APII - Journal Européen des Systèmes Automatisés**, vol. 33 (1999), no. 8–9, pp. 943–958.

[22] CHRISTIAN RETORÉ, *Réseaux et séquents ordonnés*, **Thése de Doctorat, spécialité Mathématiques**, Université Paris 7, février 1993.

[23] ———, *Petri-nets step-transitions and proofs in partially commutative linear logic*, **Technical report, 4288**, INRIA, 2001, http://www.inria.fr/.

[24] ———, **Logique linéaire et syntaxe des langues, Habilitation à diriger des recherches**, Université de Nantes, 2002, 196 pp., http://perso.wanadoo.fr/christian.retore/.

[25] PAUL RUET, *Logique non-commutative et programmation concurrente*, **Thése de doctorat, spécialité logique et fondements de l'informatique**, Université Paris 7, 1997.

[26] EDWARD STABLER, *Derivational minimalism*, **Logical Aspects of Computational Linquistics, LACL 96** (Christian Retoré, editor), LNCS/LNAI, vol. 1328, Springer-Verlag, 1997, pp. 68–95.

[27] ANNE SJERP TROELSTRA, **Lectures in linear logic**, CSLI Lecture Notes, vol. 29, CSLI, 1992, (distributed by Cambridge University Press).

[28] J. VALDES, R. E. TARJAN, and E. L. LAWLER, *The recognition of Series-Parallel digraphs*, **SIAM Journal on Computing**, vol. 11 (1982), no. 2, pp. 298–313.

IRISA (INRIA-RENNES)
    CAMPUS UNIVERSITAIRE DE BEAULIEU
        F-35042 RENNES CEDEX, FRANCE
E-mail: Christian.Retore@inria.fr
URL: http://perso.wanadoo.fr/retore/christian/

A VERY SLOW GROWING HIERARCHY FOR Γ_0

ANDREAS WEIERMANN

Abstract. We investigate systems of fundamental sequences for the limits less than Γ_0 and show that there is a (natural) Bachmann system of fundamental sequences such that the resulting slow growing hierarchy $(G_\alpha)_{\alpha<\Gamma_0}$ does not majorize the elementary functions.

§1. Introduction. This article is part of a general program of classifying systems of fundamental sequences and their induced subrecursive hierarchies. Such hierarchies have been used so far to approach the classification problem of the general recursive functions from below and to classify provably recursive functions of several subsystems of analysis and set theory. Some years ago it has been shown e.g., in a joint paper with Buchholz and Cichon [3] that the concept of a normed Bachmann system forms a suitable basis to develop a smooth and satisfying theory of fast growing hierarchies like the Hardy hierarchy.

Since by now the pointwise or so called slow growing hierarchy has been established as a very natural and genuine subrecursive hierarchy (cf., e.g., [5, 9, 10]) it is an immediate and obvious problem to investigate whether it is as well possible to develop a general theory of the pointwise hierarchy on the basis of normed Bachmann systems. (Recall that the pointwise hierarchy is defined as follows: $G_0(x) := 0$; $G_{\alpha+1}(x) := G_\alpha(x) + 1$; $G_\lambda(x) := G_{\lambda[x]}(x)$ if λ is a limit and $\sup_{x<\omega} \lambda[x] = \lambda$.) It turned out that this problem is more complex than in the case of the Hardy hierarchy, since the pointwise hierarchy is very sensitive with respect to innocently looking modifications of the underlying system of fundamental sequences [12, 14].

These investigations lead naturally to the growth rate classification problem for the pointwise hierarchy. Assume that we have given a proof-theoretic ordinal τ which is represented by a set of unique ordinal notations. Assume

---

This paper presents a recent result which was indicated at the end of the author's conference lecture. Recent surveys on the comparison between the slow and fast growing hierarchy are given e.g., in [10, 18]. Further applications of hierarchies to proof theory and recursion theory can be found, for example, in [2, 6, 17, 11, 13]

The author is supported by a Heisenberg grant of the Deutsche Forschungsgemeinschaft.

Logic Colloquium '99
Edited by J. van Eijck, V. van Oostrom, and A. Visser
Lecture Notes in Logic, 17

further that we have given a natural norm on τ. For $\alpha < \tau$, $N\alpha$ is typically defined as the number of function symbols (which are neither the constant 0 nor the addition function symbol $+$) occurring in the term which represents α. Is it possible to classify the resulting $(G_\alpha)_{\alpha<\tau}$ when it is defined with respect to a Bachmann system which is normed with N? In the case of the segment of ordinals less than ε_0 it turned out that there are natural examples in which the induced pointwise hierarchy either consists of elementary functions only or of a hierarchy of functions which classifies the PA-provably recursive functions. Surprisingly, for the segment of ordinals below $\varphi\varepsilon_0 0$ it turns out that there are natural examples of induced pointwise hierarchies which may either consist entirely of functions which are elementary recursive in H_{ω^ω} ([5, 8]), or of functions which are elementary recursive in F_α for some $\alpha < \varepsilon_0$ ([16]), or of functions which are elementary recursive in F_α for some $\alpha < \varphi\varepsilon_0 0$ ([12]). These investigations indicate that for standard ordinal notation systems like the one for $\varphi\varepsilon_0 0$ natural variations of normed Bachmann systems yield pointwise hierarchies which are somehow densely distributed in between the standard pointwise and the fast growing hierarchy. These normed Bachmann system can be chosen *naturally*, i.e., in such a way that the growth rate of the resulting Hardy hierarchy is always equivalent to the growth rate of the standard Hardy hierarchy.

Since in all these examples the induced pointwise hierarchies grow at least as fast as the standard pointwise hierarchy (which is defined with respect to the standard assignment of fundamental sequences) it is quite natural to ask whether this property would give an invariant of the standard assignment. In this article we show that this is not the case.

The paper starts with a general investigation on normed Bachmann systems for Γ_0 and then we investigate the growth rate behavior of the pointwise hierarchy for specific normed Bachmann systems which induce rather slow growing pointwise hierarchies. We show that there exists a natural Bachmann system for Γ_0 such that the induced pointwise hierarchy consists entirely of elementary recursive functions. The proof is based on lengthy but elementary numerical calculations with the functions involved. At the end we indicate that there are various essentially different pointwise hierarchies which grow slower than the standard pointwise hierarchy.

To us it seems that it might be an interesting question whether it is always (depending on the choice of the underlying Bachmann system of fundamental sequences) possible to classify the PA-provably recursive function in terms of a pointwise growing hierarchy which is induced by a normed Bachmann system along a proof-theoretic ordinal. Moreover it seems interesting to ask whether there are always proof-theoretic ordinals τ at which the pointwise hierarchy and the Hardy hierarchy match up. Our investigations indicate that both questions may have a negative answer.

Anyway, our investigations yield that given a proof-theoretic ordinal τ and its associated (natural) norm function $N : \tau \to \omega$ there is always a (natural) Bachmann system $\cdot[\cdot]$ which induces a pointwise hierarchy of a minimal possible growth rate. It is defined via

$$\alpha[x] := \max\{\beta < \alpha : N\beta \leq N\alpha + x - 1\}.$$

We assume familiarity with the basic theory of the ordinals below Γ_0 (as for example developed in [7]) and sufficient familiarity with [3].

§2. Basic definitions.

DEFINITION 1 (cf. [7]). *Recursive definition of a set T of terms, a subset $P \subseteq T$, and a binary relation $<$ on T. (P corresponds to the set of additive principal numbers in T. The function $\lambda\xi.\overline{\varphi}\alpha\xi$ denotes the function which enumerates the maximal α-critical numbers.)*

1. $0 \in T$;
2. $\alpha_1, \ldots, \alpha_m \in P$ & $\alpha_1 \geq \cdots \geq \alpha_m$ & $m \geq 2 \Rightarrow \alpha_1 + \cdots + \alpha_m \in T$;
3. $\alpha, \beta \in T \Rightarrow \overline{\varphi}\alpha\beta \in T$ & $\overline{\varphi}\alpha\beta \in P$;
4. $0 \neq \beta \Rightarrow 0 < \beta$;
5. $\alpha_1, \ldots, \alpha_m \in P$ & $m \geq 2$ & $\alpha_1, \ldots, \alpha_m < \beta$ & $\beta \in P$
$$\Rightarrow \alpha_1 + \cdots + \alpha_m < \beta;$$
6. $\alpha = \alpha_1 + \cdots + \alpha_m$ & $\beta = \beta_1 + \cdots + \beta_n$:
   (a) $m < n$ & $\forall i \leq m(\alpha_i = \beta_i) \Rightarrow \alpha < \beta$;
   (b) $\exists i \leq \min\{m, n\}(\alpha_i < \beta_i$ & $\forall j < i(\alpha_j = \beta_j)) \Rightarrow \alpha < \beta$;
7. $\alpha = \overline{\varphi}\alpha_1\alpha_2$ & $\beta = \overline{\varphi}\beta_1\beta_2$
   (a) $\alpha_1 < \beta_1$ & $\alpha_2 < \beta \Rightarrow \alpha < \beta$;
   (b) $\alpha_1 = \beta_1$ & $\alpha_2 < \beta_2 \Rightarrow \alpha < \beta$;
   (c) $\beta_1 < \alpha_1$ & $\alpha \leq \beta_2 \Rightarrow \alpha < \beta$.

LEMMA 1.

1. $\langle T, < \rangle$ *is a well-order.*
2. $otype(T, <) = \Gamma_0$.

Conventions. In the sequel small Greek letters range over elements of T. We denote the natural number 1 by $\overline{\varphi}00$ and define as usual the T-numerals (natural numbers) with 0, 1 and $+$. In the sequel small Latin letters range over natural numbers. We further identify $\overline{\varphi}01$ with ω. Let Lim be the set of elements in T which are neither 0 nor of the form $\alpha + 1$. In the sequel λ, λ' range over elements of Lim. For $\alpha > 0$ we write $\alpha =_{NF} \overline{\varphi}\beta\gamma + \delta$ if $\alpha = \overline{\varphi}\beta\gamma + \delta$ and $\delta < \overline{\varphi}\beta(\gamma + 1)$. For $\alpha \in T$ let $\overline{\varphi}_\alpha : T \to T$ be defined by $\overline{\varphi}_\alpha(\beta) := \overline{\varphi}\alpha\beta$. For $m < \omega$ we define $\overline{\varphi}_\alpha^m : T \to T$ recursively via $\overline{\varphi}_\alpha^0(\beta) := \beta$ and $\overline{\varphi}_\alpha^{m+1}(\beta) := \overline{\varphi}_\alpha(\overline{\varphi}_\alpha^m(\beta))$. Let $Fix_\alpha := \{\lambda \in T : \exists\beta\exists\gamma(\alpha < \beta$ & $\lambda = \overline{\varphi}\beta\gamma)\}$. The elements of Fix_α are fixed points of the unary Veblen function φ_α (cf. [7]).

LEMMA 2.

1. *If* $0 < \gamma < \overline{\varphi}\alpha 0$ *then there exist a unique* $m > 0$ *and unique* $\beta_1, \ldots, \beta_m, \gamma_1,$ $\ldots, \gamma_m$ *with* $\gamma =_{NF} \overline{\varphi}_{\beta_1}(\ldots(\overline{\varphi}_{\beta_m} 0 + \gamma_m)\ldots) + \gamma_1$ *and* $\beta_1, \ldots, \beta_m < \alpha$.

2. *If* $\overline{\varphi}\alpha\beta < \gamma < \overline{\varphi}\alpha(\beta + 1)$ *then there exist a unique* $m > 0$ *and unique* $\beta_1, \ldots, \beta_m, \gamma_1, \ldots, \gamma_m$ *with* $\gamma =_{NF} \overline{\varphi}_{\beta_1}(\ldots(\overline{\varphi}_{\beta_m} \overline{\varphi}\alpha\beta + \gamma_m)\ldots) + \gamma_1$ *and* $\beta_1, \ldots, \beta_m < \alpha$.

3. *Assume that* $\lambda \in Fix_\alpha$ *and* $\lambda < \gamma < \overline{\varphi}\alpha\lambda$. *Then there exist a unique* $m > 0$ *and unique* $\beta_1, \ldots, \beta_m, \gamma_1, \ldots, \gamma_m$ *with* $\gamma =_{NF} \overline{\varphi}_{\beta_1}(\ldots(\overline{\varphi}_{\beta_m}\lambda + \gamma_m)\ldots) + \gamma_1$ *and* $\beta_1, \ldots, \beta_m < \alpha$.

DEFINITION 2 (cf. [3]).

1. $N0 := 0$;
2. $N\alpha := N\alpha_1 + \cdots + N\alpha_m$ *if* $\alpha = \alpha_1 + \cdots + \alpha_m > \alpha_1 \geq \cdots \geq \alpha_m$ *with* $\alpha_1, \ldots, \alpha_m \in P$ & $m \geq 2$;
3. $N\alpha := N\alpha_1 + N\alpha_2 + 1$ *if* $\alpha = \overline{\varphi}\alpha_1\alpha_2$.

DEFINITION 3 (cf. [3]). *Assume that* $\alpha > 0$:

1. $\alpha[x] := \max\{\beta < \alpha : N\beta \leq N\alpha + x\}$;
2. $\alpha[\![x]\!] := \max\{\beta < \alpha : N\beta \leq x\}$.

LEMMA 3.

1. $\alpha \in Lim \Rightarrow N(\alpha[x]) = N\alpha + x$.
2. $\alpha \in Lim \Rightarrow N(\alpha[\![x]\!]) = x$.
3. $\alpha[x] < \beta < \alpha \Rightarrow \alpha[x] \leq \beta[0]$ (*Bachmann property*).

DEFINITION 4. *Let* $\cdot(\cdot) : \Gamma_0 \times \omega \to \Gamma_0$. *We call* $\langle \Gamma_0, \cdot(\cdot), N \rangle$ *a normed Bachmann system if*

1. $(\forall \alpha, \beta, x)\,[\alpha(x) < \beta < \alpha \Rightarrow N\alpha < N\beta]$ *and*
2. $(\forall \alpha)\,N\alpha \leq N\alpha(0) + 1$.

DEFINITION 5 (The slow growing hierarchy (cf. [5, 8])). *Recursive definition of a number-theoretic function* $G_\alpha : \omega \to \omega$ *for* $\alpha < \Gamma_0$ *with respect to the assignment* $\cdot[\cdot]$ *from Definition 3.*

1. $G_0(x) := 0$;
2. $G_{\alpha+1}(x) := G_\alpha(x) + 1$;
3. $G_\lambda(x) := G_{\lambda[x]}(x)$ *if* λ *is a limit.*

The following theorem shows that in some sense the assignment $\cdot[\cdot] : \Gamma_0 \times \omega \to \Gamma_0$ from Definition 3 has (among all normed Bachmann systems) minimal growth rate and it produces (among all normed Bachmann systems) slow growing hierarchies of minimal possible growth rate.

THEOREM 1. *Assume that* $\cdot(\cdot) : \Gamma_0 \times \omega \to \Gamma_0$ *is chosen so that* $\langle \Gamma_0, \cdot(\cdot), N \rangle$ *is a normed Bachmann system. Then* $\alpha(x+1) \geq \alpha[x]$. *Assume that* $(G'_\alpha)_{\alpha < \Gamma_0}$ *is the slow growing hierarchy defined with respect to* $\cdot(\cdot)$. *Then* $G'_\alpha(x+1) \geq G_\alpha(x)$.

PROOF. Put $p(x) := x$ for $x < \omega$ and $p(\lambda + x) := N(\lambda(x))$ for $x < \omega$ and $\lambda \in Lim$. We have $\lambda > \lambda(x) > \lambda(x-1) > \cdots > \lambda(1) > \lambda(0)$.

Since $\langle \Gamma_0, \cdot(\cdot), N \rangle$ is a normed Bachmann system we obtain $N(\lambda(x)) > N(\lambda(x-1)) > \cdots > N(\lambda(1)) > N(\lambda(0)) \geq N\lambda - 1$ hence $p(\lambda + x) \geq N(\lambda(x)) \geq N\lambda + x - 1$. Theorem 5 in [3] yields $\alpha(x+1) = \max\{\beta < \alpha : N\beta \leq p(\alpha + x + 1)\}$ hence $\alpha(x+1) \geq \alpha[x]$. By induction on α we show $G'_\alpha(x+1) \geq G_\alpha(x)$. This is clear for $\alpha = 0$ and follows immediately from the induction hypothesis if α is a successor. Assume that α is a limit. Then the induction hypothesis yields $G'_\alpha(x+1) = G'_{\alpha(x+1)}(x+1) \geq G_{\alpha(x+1)}(x)$. The proof of Lemma 3 in [3] yields $G_{\alpha(x+1)}(x) \geq G_{\alpha[x]}(x) = G_\alpha(x)$. $\dashv$

In the sequel we compute a recursive definition of $\lambda[x]$.

LEMMA 4. *Assume that* $\alpha =_{NF} \overline{\varphi}\beta\gamma + \delta$ *with* $\delta > 0$. *Then* $\alpha[\![x]\!] = (\overline{\varphi}\beta\gamma)[\![x]\!]$ *if* $x < N\beta + N\gamma + 1$ *and* $\alpha[\![x]\!] = \overline{\varphi}\beta\gamma$ *if* $x = N\beta + N\gamma + 1$. *Further* $\alpha[\![x]\!] = \overline{\varphi}\beta\gamma + \delta[\![x - N\beta - N\gamma - 1]\!]$ *if* $x > N\beta + N\gamma + 1$.

LEMMA 5.

1. $(\overline{\varphi}0(\alpha+1))[\![x]\!] = \overline{\varphi}0\alpha \cdot y + (\overline{\varphi}0\alpha)[\![z]\!]$ *where* y *is maximal with* $N(\alpha+1) \cdot y \leq x$ *and* $z \in \{0, \ldots, N\alpha\}$ *satisfies* $N(\alpha+1) \cdot y + z = x$.
2. *If* $\lambda \in Fix_0$ *then* $(\overline{\varphi}0\lambda)[\![x]\!] = \lambda \cdot y + \lambda[\![z]\!]$ *where* y *is maximal with* $N\lambda \cdot y \leq x$ *and* $z \in \{0, \ldots, N\lambda - 1\}$ *satisfies* $N\lambda \cdot y + z = x$.

PROOF. Let $\beta := (\overline{\varphi}0(\alpha+1))[\![x]\!]$. Then $N\beta = x$ and $\beta \geq \overline{\varphi}0\alpha \cdot y + (\overline{\varphi}0\alpha)[\![z]\!]$. Since $\overline{\varphi}0(\alpha+1) = \sup\{\overline{\varphi}0\alpha \cdot y : y < \omega\}$ there exists a unique $k < \omega$ and a unique $\gamma < \overline{\varphi}0\alpha$ such that $\beta = \overline{\varphi}0\alpha \cdot k + \gamma$. If k would not be maximal with $(N\alpha + 1) \cdot k \leq x$ then $N(\alpha+1) \cdot (k+1) \leq x$ and $\overline{\varphi}0\alpha \cdot (k+1)$ would be a better choice than β for $(\overline{\varphi}0(\alpha+1))[\![x]\!]$. Thus $k = y$. Further $z := N\gamma < N\alpha + 1$ by choice of y. Thus $\gamma \leq (\overline{\varphi}0\alpha)[\![z]\!]$ hence $\beta \leq \overline{\varphi}0\alpha \cdot y + (\overline{\varphi}0\alpha)[\![z]\!]$. This shows the first assertion and the second is proved by replacing $\overline{\varphi}0\alpha$ through λ in the previous argument. $\dashv$

For notational convenience we put $\alpha[\![-1]\!] := -1$ and $\overline{\varphi}_{-1}\alpha := \alpha$.

LEMMA 6.

1. $(\overline{\varphi}(\alpha+1)0)[\![x]\!] = 0$ *if* $x = 0$ *and* $(\overline{\varphi}(\alpha+1)0)[\![x]\!] = \overline{\varphi}^y_\alpha \overline{\varphi}_{\alpha[z]}0$ *if* $x > 0$ *where* y *is maximal with* $(N\alpha + 1) \cdot y \leq x$ *and* $(N\alpha + 1) \cdot y + z + 1 = x$ *with* $-1 \leq z < N\alpha$.
2. $(\overline{\varphi}(\alpha+1)(\beta+1))[\![x]\!] = (\overline{\varphi}(\alpha+1)\beta)[\![x]\!]$ *if* $x < N\alpha + N\beta + 2$ *and* $(\overline{\varphi}(\alpha+1)(\beta+1))[\![x]\!] = \overline{\varphi}(\alpha+1)\beta$ *if* $x = N\alpha + N\beta + 2$. *Further* $(\overline{\varphi}(\alpha+1)(\beta+1))[\![x]\!] = \overline{\varphi}^y_\alpha \overline{\varphi}_{\alpha[z]}\overline{\varphi}(\alpha+1)\beta$ *if* $x > N\alpha + N\beta + 2$ *where* y *is maximal with* $(N\alpha + 1) \cdot y \leq x - N\alpha - N\beta - 2$ *and* $(N\alpha + 1) \cdot y + z + N\alpha + N\beta + 3 = x$ *and* $-1 \leq z < N\alpha$.
3. *Assume that* $\lambda \in Fix_{\alpha+1}$. $(\overline{\varphi}(\alpha+1)\lambda)[\![x]\!] = \lambda[\![x]\!]$ *if* $x < N\lambda$ *and* $(\overline{\varphi}(\alpha+1)\lambda)[\![x]\!] = \lambda$ *if* $x = N\lambda$. $(\overline{\varphi}(\alpha+1)\lambda)[\![x]\!] = \overline{\varphi}^y_\alpha \overline{\varphi}_{\alpha[z]}\lambda$ *if* $x > N\lambda$ *where* y *is maximal with* $(N\alpha + 1) \cdot y + N\lambda \leq x$ *and* $(N\alpha + 1) \cdot y + z + N\lambda + 1 = x$ *with* $-1 \leq z < N\alpha$.

PROOF. Let $\beta := (\overline{\varphi}(\alpha+1)0)[\![x]\!]$. Then $N\beta = x$. If $x = 0$ then $\beta = 0$. Assume $x > 0$. Then $\beta > 0$. Let $\beta =_{NF} \overline{\varphi}_{\beta_1}(\overline{\varphi}_{\beta_2}(\ldots(\overline{\varphi}_{\beta_m}0 + \gamma_m)\ldots) + \gamma_2) + \gamma_1$ where $m \geq 1$ and $\overline{\varphi}_{\beta_i}((\ldots) + 1) > \gamma_i$ for $1 \leq i \leq m$. $\beta < \overline{\varphi}(\alpha+1)0$ yields $\beta_i < \alpha + 1$ for $1 \leq i \leq m$. We claim that $\gamma_i = 0$ for $1 \leq i \leq m$. Assume for a contradiction that $\gamma_i > 0$ for some $i \in \{1, \ldots, m\}$. Let β' be a nominal form such that $\beta'[\star := \overline{\varphi}_{\beta_i}(\ldots) + \gamma_i] = \beta$. Then $\overline{\varphi}0(\beta'[\star := \overline{\varphi}_{\beta_i}(\ldots) + 0])$ would be a better choice than β for $(\overline{\varphi}(\alpha+1)0)[\![x]\!]$. Thus $\gamma_i = 0$ for $1 \leq i \leq m$. We claim that $\beta_1 \geq \cdots \geq \beta_m$. Otherwise there exists an $i < m$ such that $\beta_i < \beta_{i+1}$. Choose a permutation π of $\{1, \ldots, m\}$ such that $\beta_{\pi(1)} \geq \cdots \geq \beta_{\pi(m)}$. Let $\beta' := \overline{\varphi}_{\beta_{\pi(1)}} \ldots \overline{\varphi}_{\beta_{\pi(1)}} 0$. Then $\beta' > \beta$, $N\beta' = N\beta$ and β' would be a better choice than β for $(\overline{\varphi}(\alpha+1)0)[\![x]\!]$. Assume now that $\beta_1 < \alpha$. We claim that $m = 1$ in this case. If $m \geq 2$ then $\overline{\varphi}_{\beta_1+1}0$ would be a better choice than β for $(\overline{\varphi}(\alpha+1)0)[\![x]\!]$. Thus $m = 1$. We have $N\beta_1 < N\alpha$ since otherwise $\overline{\varphi}\alpha0$ would be a better choice than β for $(\overline{\varphi}(\alpha+1)0)[\![x]\!]$. Let $z := N\beta_1$. Then $z < N\alpha$ and $z + 1 = x$ thus $\beta_1 = \alpha[\![z]\!]$. Let $y := 0$. Then y is maximal with $(N\alpha+1) \cdot y + z + 1 = x$ and in this case the assertion is shown.

Assume finally that $\beta_1 = \alpha$. Choose i maximal with $\alpha = \beta_1 = \cdots = \beta_i$. If $i + 2 \leq m$ then $\overline{\varphi}_{\beta_1} \ldots \overline{\varphi}_{\beta_i} \overline{\varphi}_{\beta_{i+1}+1}0$ would be a better choice than β for $(\overline{\varphi}(\alpha+1)0)[\![x]\!]$. Assume $i + 1 \leq m$. Then $\beta_{i+1} < \alpha$. Let $z := N\beta_{i+1}$. Then $z < N\alpha$ since otherwise $\overline{\varphi}_{\beta_1} \ldots \overline{\varphi}_{\beta_i} \overline{\varphi}_\alpha 0$ would be a better choice than β for $(\overline{\varphi}(\alpha+1)0)[\![x]\!]$. Further $(N\alpha+1) \cdot i + z + 1 = x$. Let $y := i$. Then y is maximal with $(N\alpha+1) \cdot y \leq x$ and in this case the assertion is shown. Assume $i = m$. Let $y := m$ and $z := -1$. Then y is maximal with $(N\alpha+1) \cdot y \leq x$.

The second and third assertion follow similarly. For the second assertion we employ assertion 2 of Lemma 2: If $\overline{\varphi}\alpha\beta \leq \gamma < \overline{\varphi}\alpha(\beta+1)$ then $\gamma = \overline{\varphi}_{\beta_1}(\overline{\varphi}_{\beta_2} \ldots \overline{\varphi}_{\beta_m}(\overline{\varphi}\alpha\beta + \gamma_m)\ldots) + \gamma_2) + \gamma_1$ with $\beta_1, \ldots, \beta_m < \alpha$ and hence $N\gamma \geq N\overline{\varphi}\alpha\beta$. For the third assertion we employ assertion 3 of Lemma 2: If $\lambda \leq \gamma < \overline{\varphi}\alpha\lambda$ then $\gamma = \overline{\varphi}_{\beta_1}(\overline{\varphi}_{\beta_2} \ldots (\overline{\varphi}_{\beta_m}(\lambda + \gamma_m)\ldots) + \gamma_2) + \gamma_1$ with $\beta_1, \ldots, \beta_m < \alpha$ and hence $N\gamma \geq N\lambda$. $\dashv$

LEMMA 7.

1. $(\overline{\varphi}\lambda0)[\![x]\!] = \begin{cases} 0 & \text{if } x = 0, \\ \overline{\varphi}\lambda[\![x-1]\!]0 & \text{if } x > 0. \end{cases}$

2. $(\overline{\varphi}\lambda(\alpha+1))[\![x]\!] = \begin{cases} (\overline{\varphi}\lambda\alpha)[\![x]\!] & \text{if } x < N\lambda + N\alpha + 1, \\ \overline{\varphi}\lambda\alpha & \text{if } x = N\lambda + N\alpha + 1, \\ \overline{\varphi}_{\lambda[\![x-N\lambda-N\alpha-2]\!]}\overline{\varphi}\lambda\alpha & \text{if } x > N\lambda + N\alpha + 1. \end{cases}$

3. Assume that $\lambda' \in Fix_\lambda$.

$$(\overline{\varphi}\lambda\lambda')[\![x]\!] = \begin{cases} \lambda'[\![x]\!] & \text{if } x < N\lambda', \\ \lambda' & \text{if } x = N\lambda', \\ \overline{\varphi}_{\lambda[\![x-N\lambda'-1]\!]}\lambda' & \text{if } x > N\lambda'. \end{cases}$$

PROOF. Let $\beta := (\overline{\varphi}\lambda 0)[\![x]\!]$. If $x = 0$ then $\beta = 0$. Assume that $\beta > 0$. Let $\beta =_{NF} \overline{\varphi}_{\beta_1}(\ldots(\overline{\varphi}_{\beta_m}0 + \gamma_m)\ldots) + \gamma_1$ where $\beta_1,\ldots,\beta_m < \lambda$. We claim that $m = 1$. Otherwise we would have $m > 1$. Then choose $i \in \{1,\ldots,m\}$ with $\beta_i = \max\{\beta_1,\ldots,\beta_m\}$. Then $\overline{\varphi}_{\beta_i+1}0$ would be a better choice than β for $(\overline{\varphi}\lambda 0)[\![x]\!]$. Thus $m = 1$. We claim that $\gamma_1 = 0$. Otherwise $\overline{\varphi}0\overline{\varphi}_{\beta_1}0$ would be a better choice than β for $(\overline{\varphi}\lambda 0)[\![x]\!]$. Hence $\gamma_1 = 0$ and $N\beta_1 = x - 1$. Thus $\overline{\varphi}\beta_1 0 \leq \overline{\varphi}\lambda[\![x - 1]\!]0$.

The second and third assertion follow similarly. ⊣

LEMMA 8. *If $\lambda \notin Fix_\alpha$ then $\overline{\varphi}\alpha\lambda[x] > \lambda$.*

PROOF. Assume that $\lambda =_{NF} \overline{\varphi}\xi\eta + \delta$.

Case 1: $\delta > 0$. Then $\lambda[x] \geq \overline{\varphi}\xi\eta$ and thus $\overline{\varphi}\alpha\lambda[x] \geq \overline{\varphi}\alpha\overline{\varphi}\xi\eta > \overline{\varphi}\xi\eta, \delta$ hence $\overline{\varphi}\alpha\lambda[x] > \lambda$.

Case 2: $\delta = 0$. Then, by assumption, $\xi \leq \alpha$. $\eta < \overline{\varphi}\xi\eta$ yields $\eta < (\overline{\varphi}\xi\eta)[x]$ hence $\overline{\varphi}\alpha\lambda[x] > \overline{\varphi}\alpha\eta \geq \overline{\varphi}\xi\eta = \lambda$. ⊣

LEMMA 9. *Assume that $\lambda \notin Fix_\alpha$ and $\overline{\varphi}\alpha\lambda[x] \leq \gamma < \overline{\varphi}\alpha\lambda$. Then there exists δ with $\lambda[x] \leq \delta < \lambda$ and there exist $m \geq 0$ and $\beta_1,\ldots,\beta_m, \gamma_0, \gamma_1,\ldots,\gamma_m$ with $\gamma =_{NF} \overline{\varphi}_{\beta_1}(\ldots\overline{\varphi}_{\beta_m}(\overline{\varphi}_\alpha\delta + \gamma_m)\cdots + \gamma_1) + \gamma_0$ and $\beta_1,\ldots,\beta_m < \alpha$.*

PROOF. By induction on $N\gamma$. Assume $\gamma =_{NF} \overline{\varphi}\xi\eta + \rho$. Then $\overline{\varphi}\alpha\lambda[x] \leq \overline{\varphi}\xi\eta < \overline{\varphi}\alpha\lambda$. If $\xi = \alpha$ then we are done. We claim that $\xi > \alpha$ is not possible. Otherwise we would have $\gamma \leq \lambda < \lambda + 1 \leq \overline{\varphi}\alpha(\lambda[x])$. If $\xi < \alpha$ then $\overline{\varphi}\alpha\lambda[x] \leq \overline{\varphi}\xi\eta$ yields $\overline{\varphi}\alpha\lambda[x] \leq \eta < \overline{\varphi}\alpha\lambda$. The assertion follows when we apply the induction hypothesis to η. ⊣

LEMMA 10. *If $\lambda \notin Fix_\alpha$ then $(\overline{\varphi}\alpha\lambda)[x] = \overline{\varphi}\alpha\lambda[x]$.*

PROOF. Let $\beta = (\overline{\varphi}\alpha\lambda)[x]$. Then $\overline{\varphi}\alpha\lambda[x] \leq \beta \leq \overline{\varphi}\alpha\lambda$. Thus there is a δ with $\lambda[x] \leq \delta < \lambda$ and with $\beta =_{NF} \overline{\varphi}_{\beta_1}(\ldots\overline{\varphi}_{\beta_m}(\overline{\varphi}_\alpha\delta + \gamma_m)\cdots + \gamma_1) + \gamma_0$ and $\beta_1,\ldots,\beta_m < \alpha$. We claim that $m = 0$. Otherwise we would have $m \geq 1$. $\beta_1 < \alpha$ yields $\overline{\varphi}_{\beta_1}(\ldots\overline{\varphi}_{\beta_m}(\overline{\varphi}_\alpha\delta) + \gamma_m\ldots) + \gamma_1 < \overline{\varphi}\alpha(\delta + 1)$ and $\overline{\varphi}\alpha(\delta + 1)$ would be a better choice than β for $(\overline{\varphi}\alpha\lambda)[x]$. Thus $m = 0$. A similar argument yields $\gamma_0 = 0$. $N\beta \leq N\overline{\varphi}\alpha\lambda + x$ yields $N\delta \leq N\lambda + x$. Thus $\delta < \lambda$ yields $\beta = \overline{\varphi}\alpha\delta \leq \overline{\varphi}\alpha\lambda[x]$. ⊣

THEOREM 2.

1. $(\alpha_1 + \cdots + \alpha_m)[x] = \alpha_1 + \cdots + \alpha_m[x]$.
2. $(\overline{\varphi}0(\beta + 1))[x] = \overline{\varphi}0\beta + (\overline{\varphi}0(\beta + 1))[\![x + 1]\!]$.
3. $(\overline{\varphi}0\lambda)[x] = \overline{\varphi}0(\lambda[x])$ *if $\lambda \notin Fix_0$*.
4. $(\overline{\varphi}0\lambda)[x] = \lambda + (\overline{\varphi}0\lambda)[\![x + 1]\!]$ *if $\lambda \in Fix_0$*.
5. $(\overline{\varphi}(\alpha + 1)0)[x] = \overline{\varphi}_\alpha^y\overline{\varphi}_{(\alpha+1)[z]}0$ *where y is maximal with $(N\alpha + 1) \cdot (y-1) \leq x+1$ and $z \in \{-1,\ldots,N\alpha\}$ satisfies $(N\alpha+1)\cdot(y-1)+z+1 = x+1$.*

6. $(\overline{\varphi}(\alpha+1)(\beta+1))[x] = \overline{\varphi}_\alpha^y \overline{\varphi}_{(\alpha+1)[z]}(\overline{\varphi}(\alpha+1)\beta)$ where y is maximal with $(N\alpha+1) \cdot y \le x+1$ and $z \in \{-1, \ldots, N\alpha\}$ satisfies $(N\alpha+1) \cdot y + z = x$.

7. $(\overline{\varphi}(\alpha+1)\lambda)[x] = \overline{\varphi}(\alpha+1)(\lambda[x])$ if $\lambda \notin Fix_{\alpha+1}$.

8. $(\overline{\varphi}(\alpha+1)\lambda)[x] = \overline{\varphi}_\alpha^y \overline{\varphi}_{(\alpha+1)[z]}\lambda$ if $\lambda \in Fix_{\alpha+1}$ where y is maximal with $(N\alpha+1) \cdot (y-1) \le x+1$ and $z \in \{-1, \ldots, N\alpha\}$ satisfies $(N\alpha+1) \cdot (y-1) + z + 1 = x+1$.

9. $(\overline{\varphi}\lambda 0)[x] = \overline{\varphi}\lambda[x]0$.

10. $(\overline{\varphi}\lambda(\beta+1))[x] = \overline{\varphi}(\lambda[\![x]\!])(\overline{\varphi}\lambda\beta)$.

11. $(\overline{\varphi}\lambda\lambda')[x] = \overline{\varphi}\lambda(\lambda'[x])$ if $\lambda' \notin Fix_\lambda$.

12. $(\overline{\varphi}\lambda\lambda')[x] = \overline{\varphi}(\lambda[x])\lambda'$ if $\lambda' \in Fix_\lambda$.

PROOF. This follows from Lemmas 4, 5, 6, 7, 10 and $\lambda[x] = \lambda[\![N\lambda + x]\!]$ for any $\lambda \in Lim$. ⊣

DEFINITION 6 (The standard assignment of fundamental sequences). *Recursive definition of $\alpha\{x\}$ for $\alpha \in T$ and $x < \omega$.*

1. $0\{x\} := 0$;

2. $(\alpha_1 + \cdots + \alpha_m)\{x\} := \alpha_1 + \cdots + \alpha_m\{x\}$;

3. $(\overline{\varphi}00)\{x\} := 0$;

4. $(\overline{\varphi}0(\beta+1))\{x\} := \overline{\varphi}0\beta \cdot (x+1)$;

5. $(\overline{\varphi}0\lambda)\{x\} := \overline{\varphi}0(\lambda\{x\})$ if $\lambda \notin Fix_0$;

6. $(\overline{\varphi}0\lambda)\{x\} := \lambda \cdot (x+1)$ if $\lambda \in Fix_0$;

7. $(\overline{\varphi}(\alpha+1)0)\{x\} := \overline{\varphi}_\alpha^{x+1}(0)$;

8. $(\overline{\varphi}(\alpha+1)(\beta+1))\{x\} := \overline{\varphi}_\alpha^{x+1}(\overline{\varphi}(\alpha+1)\beta)$;

9. $(\overline{\varphi}(\alpha+1)\lambda)\{x\} := \overline{\varphi}(\alpha+1)(\lambda\{x\})$ if $\lambda \notin Fix_{\alpha+1}$;

10. $(\overline{\varphi}(\alpha+1)\lambda)\{x\} := \overline{\varphi}_\alpha^{x+1}\lambda$ if $\lambda \in Fix_{\alpha+1}$;

11. $(\overline{\varphi}\lambda 0)\{x\} := \overline{\varphi}\lambda\{x\}0$;

12. $(\overline{\varphi}\lambda(\beta+1))\{x\} := \overline{\varphi}(\lambda\{x\})(\overline{\varphi}\lambda\beta)$;

13. $(\overline{\varphi}\lambda\lambda')\{x\} := \overline{\varphi}\lambda(\lambda'\{x\})$ if $\lambda' \notin Fix_\lambda$;

14. $(\overline{\varphi}\lambda\lambda')\{x\} := \overline{\varphi}(\lambda\{x\})\lambda'$ if $\lambda' \in Fix_\lambda$.

DEFINITION 7. *Recursive definition of a function p*

1. $p(x) := x$;

2. $p(\alpha + x) := N\alpha_1 + \cdots + N\alpha_{m-1} + p(\alpha_m + x)$, if $\alpha = \alpha_1 + \cdots + \alpha_m \in Lim$;

3. $p(\overline{\varphi}0(\beta+1) + x) := N(\beta+1) \cdot (x+1)$;

4. $p(\overline{\varphi}0\lambda + x) := 1 + p(\lambda + x)$ if $\lambda \notin Fix_0$;

5. $p(\overline{\varphi}0\lambda + x) := N\lambda \cdot (x+1)$ if $\lambda \in Fix_0$;

6. $p(\overline{\varphi}(\alpha+1)0 + x) := N(\alpha+1) \cdot (x+1)$;

7. $p(\overline{\varphi}(\alpha+1)(\beta+1) + x) := (N\alpha+1) \cdot (x+1) + N\alpha + N\beta + 2$;

8. $p(\overline{\varphi}(\alpha+1)\lambda + x) := N\alpha + 2 + p(\lambda + x)$ if $\lambda \notin Fix_{\alpha+1}$;

9. $p(\overline{\varphi}(\alpha+1)\lambda + x) := (N\alpha+1) \cdot (x+1) + N\lambda$ if $\lambda \in Fix_{\alpha+1}$;

10. $p(\overline{\varphi}\lambda 0 + x) := 1 + p(\lambda + x)$;

11. $p(\overline{\varphi}\lambda(\beta + 1) + x) := p(\lambda + x) + N\lambda + N\beta + 1;$

12. $p(\overline{\varphi}\lambda\lambda' + x) := N\lambda + 1 + p(\lambda' + x)$ if $\lambda' \notin Fix_\lambda;$

13. $p(\overline{\varphi}\lambda\lambda' + x) := p(\lambda + x) + N\lambda' + 1$ if $\lambda' \in Fix_\lambda.$

LEMMA 11.

1. $\alpha\{x\} = \max\{\beta < \lambda \colon N\beta \leq p(\lambda + x)\}.$
2. $N\alpha \leq p(\alpha) + 1 \leq p(\alpha + 1).$

PROOF. This follows from Lemmas 4, 5, 6, 7, 9 and 10. $\dashv$

COROLLARY 1. $\langle \Gamma_0, N, \cdot\{\cdot\}\rangle$ is a normed Bachmann system.

PROOF. This follows from the previous Lemma and Theorem 4 of [3] since we have $(\forall\alpha)[N0 \leq N\alpha]$ and $(\forall\alpha)[N(\alpha + 1) \leq N\alpha + 1].$ $\dashv$

LEMMA 12.

1. $G_\alpha(x) = G_{\alpha_1}(x) + \cdots + G_{\alpha_m}(x)$ if $\alpha = \alpha_1 + \cdots + \alpha_m > \alpha_1 \geq \cdots \geq \alpha_m$ with $\alpha_1, \ldots, \alpha_m \in P.$
2. $G_{\overline{\varphi}0\alpha}(x) \geq G_\alpha(x) \geq N\alpha.$
3. $G_\alpha(x) < G_\alpha(x + 1)$ if $\alpha \geq \omega.$
4. $\alpha[x] < \beta < \alpha \implies G_{\alpha[x]}(x) \leq G_\beta(x).$
5. $\alpha < \beta \implies (\exists y)(\forall x \geq y)[G_\alpha(x) < G_\beta(x)].$

For measuring the extreme slow growth of the elements of (G_α) we introduce an auxiliary hierarchy (T_α). This hierarchy is needed in the proof of Theorem 3.

DEFINITION 8. Recursive definition of a number-theoretic function $T_\alpha \colon \omega \to \omega$ for $\alpha < \Gamma_0$

1. $T_0(x) := 0;$
2. $T_\alpha(x) := T_{\alpha[x]}(x) + 2$ if $\alpha > 0.$

LEMMA 13.

1. $T_\alpha(x) = 2 \cdot \max\{k \colon (\exists\alpha_0, \ldots, \alpha_{k-1})$
$$[\alpha_0 < \cdots < \alpha_{k-1} \leq \alpha \And (\forall i < k)[N\alpha_i \leq x]]\}.$$
2. $\alpha \leq \beta \implies T_\alpha(x) \leq T_\beta(x).$
3. $\alpha < \beta \And N\alpha \leq x \implies T_\alpha(x) + 1 < T_\beta(x).$
4. $\alpha \geq \omega \implies T_\alpha(x) < T_\alpha(x + 1).$

LEMMA 14. Let $U_x := \{\alpha < \Gamma_0 \colon N\alpha \leq x\}$ and $\#U_x$ be the cardinality of U_x. Then $\#U_x \leq 4^{4^x}.$

PROOF. By induction on x. Obviously $\#U_0 = 1$ and $\#U_{x+1}$ is less than or equal to one plus the cardinality of the Cartesian product $U_x \times U_x \times U_x$. For, if $\alpha =_{NF} \overline{\varphi}\beta\gamma + \delta > 0$ with $\alpha \in U_{x+1}$, then $\beta, \gamma, \delta \in U_x$. $\dashv$

LEMMA 15. $T_\alpha(x) \leq 4^{4^x} \cdot 2.$

PROOF. $T_\alpha(x)$ is twice the cardinality of $\{\beta \in U_x \colon \beta < \alpha\}$. $\dashv$

This bound on T_α is very rough but it suffices for our purposes. In fact one can show that each T_α is exponentially bounded.

DEFINITION 9.

1. $g_0(x) := 0$;
2. $g_\alpha(x) := G_{\alpha[x+1]}(x) + 1$ for $\alpha > 0$.

LEMMA 16.

1. $\alpha < \beta$ & $N\alpha \leq x + 1 \Rightarrow G_\alpha(x) \leq G_\beta(x)$,
2. $\alpha \leq \beta \Rightarrow g_\alpha(x) \leq g_\beta(x)$.

To obtain bounds on $G_\alpha(x)$ we started with some numerical calculations. The results of these calculations formed the basis of the pointwise collapsing operation $\alpha \mapsto C_x(\alpha, g_\alpha(x))$ which we introduce now.

DEFINITION 10. *Let* $x, g < \omega$. *If* $\alpha = 0$ *then* $C_x(\alpha, g) := 0$. *If* $\alpha > 0$ *then* assume that $\alpha =_{NF} \overline{\varphi} \alpha_1 (\ldots \overline{\varphi} \alpha_m (\overline{\varphi} \alpha_{m+1} 0 + \beta_m) \cdots + \beta_1) + \beta_0$. *Assume that* $i \in \{1, \ldots, m+1\}$ *is maximal with* $\alpha_1 \geq \cdots \geq \alpha_i$. *If* $i = m + 1$ *then*

$$C_x(\alpha, g) := G_{\beta_0}(x) + g^{x^{T_{\alpha_1}(x)+1}} \cdot (G_{\alpha_1}(x) + 1)$$
$$+ g^{x^{T_{\alpha_1}(x)}} \cdot G_{\beta_1}(x)$$
$$+ \cdots +$$
$$+ g^{x^{T_{\alpha_1}(x)} + \cdots + x^{T_{\alpha_{m-1}}(x)} + x^{T_{\alpha_m}(x)+1}} \cdot (G_{\alpha_m}(x) + 1)$$
$$+ g^{x^{T_{\alpha_1}(x)} + \cdots + x^{T_{\alpha_m}(x)}} \cdot G_{\beta_m}(x)$$
$$+ g^{x^{T_{\alpha_1}(x)} + \cdots + x^{T_{\alpha_{m+1}}(x)+1}} \cdot (G_{\alpha_{m+1}}(x) + 1).$$

In case that $\alpha_i < \alpha_{i+1}$ *assume that* $j \in \{0, \ldots, i-1\}$ *is maximal with* $\alpha_1 \geq \cdots \geq \alpha_j > \alpha_{i+1}$. *In this situation we define* $C_x(\alpha, g)$ *recursively as follows:*

$$C_x(\alpha, g) := G_{\beta_0}(x) + g^{x^{T_{\alpha_1}(x)+1}} \cdot (G_{\alpha_1}(x) + 1)$$
$$+ g^{x^{T_{\alpha_1}(x)}} \cdot G_{\beta_1}(x)$$
$$+ \cdots +$$
$$+ g^{x^{T_{\alpha_1}(x)} + \cdots + x^{T_{\alpha_{i-1}}(x)} + x^{T_{\alpha_i}(x)+1}} \cdot (G_{\alpha_i}(x) + 1)$$
$$+ g^{x^{T_{\alpha_1}(x)} + \cdots + x^{T_{\alpha_i}(x)}} \cdot G_{\beta_i}(x)$$
$$+ g^{x^{T_{\alpha_1}(x)} + \cdots + x^{T_{\alpha_j}(x)}} \cdot c_x(i)$$

where $c_x(i) := C_x(\overline{\varphi}_{\alpha_{i+1}}(\ldots \overline{\varphi}_{\alpha_m}(\overline{\varphi}_{\alpha_{m+1}} 0 + \beta_m) \cdots + \beta_{i+1}), g)$.

THEOREM 3. *Let* $x \geq 3$ *and* $g := g_\alpha(x)$. *Then* $G_\alpha(x) \leq C_x(\alpha, g)$.

PROOF. By induction on α. Let $x \geq 3$ be fixed during the proof. We abbreviate $G_\alpha(x)$ by G_α and $T_\alpha(x)$ by T_α. If $\alpha = 0$ then the assertion is clear. Now assume that $0 < \alpha =_{NF} \overline{\varphi} \alpha_1 (\ldots \overline{\varphi} \alpha_m (\overline{\varphi} \alpha_{m+1} 0 + \beta_m) \cdots + \beta_1) + \beta_0$. Choose i and j according to Definition 10. If $\beta_0 \in Lim$ or $\beta_0 = \gamma + 1$ then the assertion follows from the induction hypothesis. We assume that $\beta_0 = 0$.

Case 1. There exists an $l \leq i$ with $\beta_l > 0$.

Choose l minimal with this property.

Case 1.1. $\beta_l \in Lim$. We obtain

$$
\begin{aligned}
G_\alpha &= G_{\overline{\varphi}_{\alpha_1}(\dots\overline{\varphi}_{\alpha_l}((\dots)+\beta_l)\dots)} \\
&= G_{\overline{\varphi}_{\alpha_1}(\dots\overline{\varphi}_{\alpha_l}((\dots)+\beta_l[x])\dots)} \\
&\leq \cdots + g^{x^{T\alpha_1}+\cdots+x^{T\alpha_l}+1} \cdot (G_{\alpha_l}+1) + g^{x^{T\alpha_1}+\cdots+x^{T\alpha_l}} \cdot G_{\beta_l[x]} + \cdots \\
&= \cdots + g^{x^{T\alpha_1}+\cdots+x^{T\alpha_l}+1} \cdot (G_{\alpha_l}+1) + g^{x^{T\alpha_1}+\cdots+x^{T\alpha_l}} \cdot G_{\beta_l} + \cdots \\
&= C_x(\alpha, g).
\end{aligned}
$$

Case 1.2. $\beta_l = \gamma + 1$.

Case 1.2.1. $\alpha_l = 0$. We have

$$
\begin{aligned}
G_\alpha &= G_{\overline{\varphi}_{\alpha_1}(\dots\overline{\varphi}_{\alpha_{l-1}}\overline{\varphi}0((\dots)+\gamma+1)\dots)} \\
&= G_{\overline{\varphi}_{\alpha_1}(\dots\overline{\varphi}_{\alpha_{l-1}}(\overline{\varphi}0((\dots)+\gamma)+(\overline{\varphi}0((\dots)+\gamma+1)))[x+1])\dots)} \\
&\leq g^{x^{T\alpha_1}+1} \cdot (G_{\alpha_1}+1) + \cdots + g^{x^{T\alpha_1}+\cdots+x^{T\alpha_{l-1}}+1} \cdot (G_{\alpha_{l-1}}+1) \\
&\quad + g^{x^{T\alpha_1}+\cdots+x^{T\alpha_{l-1}}} \cdot G_{(\overline{\varphi}0((\dots)+\gamma+1))[x+1]} \\
&\quad + g^{x^{T\alpha_1}+\cdots+x^{T\alpha_{l-1}}+x^{T_0}+1} \cdot (G_0+1) \\
&\quad + g^{x^{T\alpha_1}+\cdots+x^{T\alpha_{l-1}}+x^{T_0}} \cdot G_\gamma + \cdots \\
&\leq \cdots + g^{x^{T\alpha_1}+\cdots+x^{T\alpha_{l-1}}+x^{T_0}} \cdot (G_\gamma+1) + \cdots \\
&= C_x(\alpha, g),
\end{aligned}
$$

since $G_{(\overline{\varphi}0((\dots)+\gamma+1))[x+1]} \leq g$ and $x^{T_0} \geq 1$.

Case 1.2.2. $\alpha_l = \delta + 1$.

Case 1.2.2.1. $N\alpha_l > x + 1$. Then $N\delta \geq x + 1$. We obtain

$$
\begin{aligned}
G_\alpha &= G_{\overline{\varphi}_{\alpha_1}(\dots\overline{\varphi}_{\alpha_{l-1}}\overline{\varphi}_{\delta+1}((\dots)+\gamma+1)\dots)} \\
&= G_{\overline{\varphi}_{\alpha_1}(\dots\overline{\varphi}_{\alpha_{l-1}}\overline{\varphi}_{(\delta+1)[x]}\overline{\varphi}_{\delta+1}((\dots)+\gamma)\dots)} \\
&\leq g^{x^{T\alpha_1}+1} \cdot (G_{\alpha_1}+1) + \cdots + g^{x^{T\alpha_1}+\cdots+x^{T\alpha_{l-1}}+1} \cdot (G_{\alpha_{l-1}}+1) \\
&\quad + g^{x^{T\alpha_1}+\cdots+x^{T\alpha_{l-1}}+x^{T(\delta+1)[x]}+1} \cdot (G_{(\delta+1)[x]}+1) \\
&\quad + g^{x^{T\alpha_1}+\cdots+x^{T\alpha_{l-1}}} \cdot (g^{x^{T\delta+1}+1} \cdot (G_{\delta+1}+1) + g^{x^{T\delta+1}} \cdot G_\gamma + \cdots) \\
&\leq \cdots + g^{x^{T\alpha_1}+\cdots+x^{T\alpha_{l-1}}+x^{T\delta+1}+1} \cdot (G_{\delta+1}+1) + \\
&\quad g^{x^{T\alpha_1}+\cdots+x^{T\alpha_{l-1}}+x^{T\delta+1}} \cdot G_\gamma + \cdots \\
&\leq \cdots + g^{x^{T\alpha_1}+\cdots+x^{T\alpha_{l-1}}+x^{T\delta+1}} \cdot (G_\gamma+1) + \cdots \\
&= C_x(\alpha, g).
\end{aligned}
$$

since $x^{T_{(\delta+1)[x]}+1}+1 \leq x^{T_{\delta+1}}$ and $G_{(\delta+1)[x]}+1 \leq g$. Notice that we have applied the induction hypothesis to the configuration $\alpha_1 \geq \cdots \geq \alpha_{l-1} \geq (\delta+1)[\![x]\!] < \delta+1$.

Case 1.2.2.2. $N\alpha_l \leq x+1$.

Then $\overline{\varphi}_{\delta+1}(\cdots+\gamma+1)[x] = \overline{\varphi}_{\delta}{}^y\overline{\varphi}_{(\delta+1)[z]}\overline{\varphi}_{\delta+1}(\cdots+\gamma)$ for some $y, z \leq x$. We obtain

$$
\begin{aligned}
G_\alpha &= G_{\overline{\varphi}_{\alpha_1}(\ldots\overline{\varphi}_{\alpha_{l-1}}\overline{\varphi}_{\delta+1}((\ldots)+\gamma+1)\ldots)} \\
&= G_{\overline{\varphi}_{\alpha_1}(\ldots\overline{\varphi}_{\alpha_{l-1}}\overline{\varphi}_{\delta}{}^y\overline{\varphi}_{(\delta+1)[z]}\overline{\varphi}_{\delta+1}((\ldots)+\gamma)\ldots)} \\
&\leq g^{x^{T_{\alpha_1}+1}} \cdot (G_{\alpha_1}+1) + \cdots + g^{x^{T_{\alpha_1}+\cdots+x^{T_{\alpha_{l-1}}+1}}} \cdot (G_{\alpha_{l-1}}+1) \\
&\quad + g^{x^{T_{\alpha_1}+\cdots+x^{T_\delta+1}}} \cdot (G_\delta+1) + \ldots \\
&\quad + g^{x^{T_{\alpha_1}+\cdots+x^{T_{\alpha_l-1}}+x^{T_\delta}\cdot y+x^{T_{(\delta+1)[z]}+1}}} \cdot (G_{(\delta+1)[z]}+1) \\
&\quad + g^{x^{T_{\alpha_1}+\cdots+x^{T_{\alpha_l-1}}+x^{T_{\delta+1}+1}}} \cdot (G_{\delta+1}+1) \\
&\quad + g^{x^{T_{\alpha_1}+\cdots+x^{T_{\alpha_l-1}}+x^{T_{\delta+1}}}} \cdot G_\gamma + \ldots \\
&\leq \cdots + g^{x^{T_{\alpha_1}+\cdots+x^{T_{\alpha_l-1}}+x^{T_{\delta+1}}}} \cdot (G_\gamma+1) + \ldots \\
&= C_x(\alpha, g).
\end{aligned}
$$

since $G_\delta+1 \leq g$ and $x^{T_\delta}\cdot y+x^{T_{(\delta+1)[z]}+1}+y+2 \leq x^{T_{\delta+1}}$ because $T_\delta+1 < T_{\delta+1}$ holds since $N\delta \leq x$. Notice that we have applied the induction hypothesis to the configuration $\alpha_1 \geq \cdots \geq \alpha_{l-1} \geq \delta \geq \cdots \geq \delta \geq (\delta+1)[\![z]\!] < \delta+1$.

Case 1.2.3. $\alpha_l \in Lim$. We obtain

$$
\begin{aligned}
G_\alpha &= G_{\overline{\varphi}_{\alpha_1}(\ldots\overline{\varphi}_{\alpha_{l-1}}\overline{\varphi}_{\alpha_l}((\ldots)+\gamma+1)\ldots)} \\
&= G_{\overline{\varphi}_{\alpha_1}(\ldots\overline{\varphi}_{\alpha_{l-1}}\overline{\varphi}_{\alpha_l[x]}(\overline{\varphi}_{\alpha_l}(\ldots)+\gamma)\ldots)} \\
&\leq g^{x^{T_{\alpha_1}+1}} \cdot (G_{\alpha_1}+1) + \cdots + g^{x^{T_{\alpha_1}+\cdots+x^{T_{\alpha_{l-1}}+1}}} \cdot (G_{\alpha_{l-1}}+1) \\
&\quad + g^{x^{T_{\alpha_1}+\cdots+x^{T_{\alpha_l[x]}+1}}} \cdot (G_{\alpha_l[x]}+1) \\
&\quad + g^{x^{T_{\alpha_1}+\cdots+x^{T_{\alpha_l-1}}}} \cdot (g^{x^{T_{\alpha_l}+1}} \cdot (G_{\alpha_l}+1) + g^{x^{T_{\alpha_l}}} \cdot G_\gamma + \ldots) \\
&\leq \cdots + g^{x^{T_{\alpha_1}+\cdots+x^{T_{\alpha_l-1}}+x^{T_{\alpha_l}+1}}} \cdot (G_{\alpha_l}+1) \\
&\quad + g^{x^{T_{\alpha_1}+\cdots+x^{T_{\alpha_l-1}}+x^{T_{\alpha_l}}}} \cdot (G_\gamma+1) + \ldots \\
&= C_x(\alpha, g).
\end{aligned}
$$

since $x^{T_{\alpha_l[x]}+1}+1 \leq x^{T_{\alpha_l}}$ and $G_{\alpha_l[x]}+1 \leq g$.

Case 2. $\beta_l = 0$ for $1 \leq l \leq i$.

Case 2.1. $i = m+1$. If $\alpha_{m+1} = 0$ then we can proceed similarly as in case 1.2 since then $\overline{\varphi}_{\alpha_{m+1}}0 = 1$. For example let us consider the case $\alpha_m = 0$.

We obtain

$$
\begin{aligned}
G_\alpha &= G_{\overline{\varphi}_{\alpha_1}(\dots\overline{\varphi}_{\alpha_{l-1}}\overline{\varphi}0\overline{\varphi}00\dots)} \\
&= G_{\overline{\varphi}_{\alpha_1}(\dots\overline{\varphi}_{\alpha_{l-1}}(\overline{\varphi}00+x+1)\dots)} \\
&\le g^{x^{T\alpha_1}+1}\cdot(G_{\alpha_1}+1)+\cdots+g^{x^{T\alpha_1}+\cdots+x^{T\alpha_{l-1}}+1}\cdot(G_{\alpha_{l-1}}+1) \\
&\quad +g^{x^{T\alpha_1}+\cdots+x^{T\alpha_{l-1}}}\cdot(x+1) \\
&\quad +g^{x^{T\alpha_1}+\cdots+x^{T\alpha_{l-1}}+x^{T0+1}}\cdot1 \\
&\le g^{x^{T\alpha_1}+1}\cdot(G_{\alpha_1}+1)+\cdots+g^{x^{T\alpha_1}+\cdots+x^{T\alpha_{l-1}}+1}\cdot(G_{\alpha_{l-1}}+1) \\
&\quad +g^{x^{T\alpha_1}+\cdots+x^{T\alpha_{l-1}}+x^{T0+1}}\cdot1 \\
&\quad +g^{x^{T\alpha_1}+\cdots+x^{T\alpha_{l-1}}+x^{T0}+x^{T0+1}}\cdot1 \\
&= C_x(\alpha,g).
\end{aligned}
$$

Case 2.1.1. Assume that $\alpha_{m+1}\in Lim$. We obtain

$$
\begin{aligned}
G_\alpha &= G_{\overline{\varphi}_{\alpha_1}(\dots\overline{\varphi}_{\alpha_m}(\overline{\varphi}_{\alpha_{m+1}}0)\dots)} \\
&= G_{\overline{\varphi}_{\alpha_1}(\dots\overline{\varphi}_{\alpha_m}(\overline{\varphi}_{\alpha_{m+1}[x]}0)\dots)} \\
&\le \cdots+g^{x^{T\alpha_1}+\cdots+x^{T\alpha_{m+1}[x]}+1}\cdot(G_{\alpha_{m+1}[x]}+1) \\
&\le \cdots+g^{x^{T\alpha_1}+\cdots+x^{T\alpha_{m+1}}+1}\cdot(G_{\alpha_{m+1}}+1) \\
&= C_x(\alpha,g).
\end{aligned}
$$

Case 2.1.2. Assume that $\alpha_{m+1}=\delta+1$.
Case 2.1.2.1. $N\alpha_{m+1}>x+1$. Then $N\delta\ge x+1$. We obtain

$$
\begin{aligned}
G_\alpha &= G_{\overline{\varphi}_{\alpha_1}(\dots\overline{\varphi}_{\alpha_m}\overline{\varphi}_{\delta+1}0\dots)} \\
&= G_{\overline{\varphi}_{\alpha_1}(\dots\overline{\varphi}_{\alpha_m}\overline{\varphi}_\delta(\overline{\varphi}_{(\delta+1)[x]}0)\dots)} \\
&\le \cdots+g^{x^{T\alpha_1}+\cdots+x^{T\alpha_m}+x^{T\delta}+1}\cdot(G_\delta+1) \\
&\quad +g^{x^{T\alpha_1}+\cdots+x^{T\alpha_m}+x^{T\delta}+x^{T(\delta+1)[x]}+1}\cdot(G_{(\delta+1)[x]}+1) \\
&\le \cdots+g^{x^{T\alpha_1}+\cdots+x^{T\alpha_{m-1}}+x^{T\delta+1}+1}\cdot(G_{\delta+1}+1) \\
&= C_x(\alpha,g).
\end{aligned}
$$

since $x^{T\delta}+x^{T(\delta+1)[x]+1}+1\le x^{T\delta+1+1}$ and $G_{(\delta+1)[x]}+1\le g$.

Case 2.1.2.2. $N\alpha_{m+1} \leq x+1$. Then $N\delta \leq x$. We obtain for some $y, z \leq x$

$$
\begin{aligned}
G_\alpha &= G_{\overline{\varphi}_{\alpha_1}(\ldots\overline{\varphi}_{\alpha_m}\overline{\varphi}_{\delta+1}0\ldots)} \\
&= G_{\overline{\varphi}_{\alpha_1}(\ldots\overline{\varphi}_{\alpha_m}\overline{\varphi}_\delta{}^y(\overline{\varphi}_{(\delta+1)[z]}0)\ldots)} \\
&\leq g^{x^{T\alpha_1}+1} \cdot (G_{\alpha_1}+1) + \cdots + g^{x^{T\alpha_1}+\cdots+x^{T\alpha_m}+1} \cdot (G_{\alpha_m}+1) \\
&\quad + g^{x^{T\alpha_1}+\cdots+x^{T\alpha_m}+x^{T\delta}+1} \cdot (G_\delta+1) \\
&\quad + \cdots + g^{x^{T\alpha_1}+\cdots+x^{T\alpha_m}+x^{T\delta}\cdot y+x^{T(\delta+1)[z]}+1} \cdot (G_{(\delta+1)[x+1]}+1) \\
&\leq \cdots + g^{x^{T\alpha_1}+\cdots+x^{T\alpha_m}+x^{T\delta+1}+1} \cdot (G_{\delta+1}+1) \\
&= C_x(\alpha, g),
\end{aligned}
$$

since $x^{T\delta+1} + x^{T(\delta+1)[z]+1} + 1 \leq x^{T\delta+1+1}$ because $T\delta < T\delta+1$ since $N\delta \leq x$.

Case 2.2. $i < m+1$. Recall that $\alpha_1 \geq \cdots \geq \alpha_i < \alpha_{i+1}$ and that j is maximal with $0 \leq j < m$ and $\alpha_1 \geq \cdots \geq \alpha_j \geq \alpha_{i+1}$.

Case 2.2.1. $\alpha_i = 0$.

Case 2.2.1.1. $\alpha_{i-1} < \alpha_{i+1}$. We have

$$
\begin{aligned}
G_\alpha &= G_{\overline{\varphi}_{\alpha_1}(\ldots\overline{\varphi}_{\alpha_{i-1}}\overline{\varphi}0(\overline{\varphi}_{\alpha_{i+1}}(\ldots)\ldots)} \\
&= G_{\overline{\varphi}_{\alpha_1}(\ldots\overline{\varphi}_{\alpha_{i-1}}((\overline{\varphi}_{\alpha_{i+1}}(\ldots)+(\overline{\varphi}0\overline{\varphi}_{\alpha_{i+1}}(\ldots))[x+1])\ldots)} \\
&\leq g^{x^{T\alpha_1}+1} \cdot (G_{\alpha_1}+1) + \cdots + g^{x^{T\alpha_1}+\cdots+x^{T\alpha_{i-1}}+1} \cdot (G_{\alpha_{i-1}}+1) \\
&\quad + g^{x^{T\alpha_1}+\cdots+x^{T\alpha_{i-1}}} \cdot G_{\overline{\varphi}0(\overline{\varphi}_{\alpha_{i+1}}(\ldots))[x+1]} \\
&\quad + g^{x^{T\alpha_1}+\cdots+x^{T\alpha_j}} \cdot (g^{x^{T\alpha_{i+1}}+1} \cdot (G_{\alpha_{i+1}}+1)+\ldots) \\
&\leq \cdots + g^{x^{T\alpha_1}+\cdots+x^{T\alpha_{i-1}}+1} \cdot (G_{\alpha_{i-1}}+1) \\
&\quad + g^{x^{T\alpha_1}+\cdots+x^{T\alpha_{i-1}}+x^1} \cdot (G_0+1) + \ldots \\
&= C_x(\alpha, g)
\end{aligned}
$$

since $G_{\overline{\varphi}0\overline{\varphi}_{\alpha_{i+1}}(\ldots))[x+1]} + 1 \leq g$ and because in both of the configurations $\langle \alpha_1, \ldots, \alpha_{i-1}, 0, \alpha_{i+1}\rangle, \langle \alpha_1, \ldots, \alpha_{i-1}, \alpha_{i+1}\rangle$ which are to consider j is maximal with $\alpha_1 \geq \cdots \geq \alpha_j \geq \alpha_{i+1}$.

Case 2.2.1.2. $\alpha_{i-1} \geq \alpha_{i+1}$. In this critical case we have $j = i-1$. Assume that l is maximal with $\alpha_1 \geq \cdots \geq \alpha_{i-1} \geq \alpha_{i+1} \geq \cdots \geq \alpha_l$. If $\alpha_l < \alpha_{l+1}$ choose k maximal with $\alpha_1 \geq \cdots \geq \alpha_{i-1} \geq \alpha_{i+1} \geq \cdots \geq \alpha_k \geq \alpha_{l+1}$. We have

$$
\begin{aligned}
G_\alpha &= G_{\overline{\varphi}_{\alpha_1}(\ldots\overline{\varphi}_{\alpha_{i-1}}\overline{\varphi}0\overline{\varphi}_{\alpha_{i+1}}(\ldots)\ldots)} \\
&= G_{\overline{\varphi}_{\alpha_1}(\ldots\overline{\varphi}_{\alpha_{i-1}}(\overline{\varphi}_{\alpha_{i+1}}(\ldots\overline{\varphi}_{\alpha_l}\overline{\varphi}_{\alpha_{l+1}}\ldots)+(\overline{\varphi}0\overline{\varphi}_{\alpha_{i+1}}(\ldots))[x+1])\ldots)} \\
&\leq g^{x^{T\alpha_1}+1} \cdot (G_{\alpha_1}+1) + \cdots + g^{x^{T\alpha_1}+\cdots+x^{T\alpha_{i-1}}+1} \cdot (G_{\alpha_{i-1}}+1) \\
&\quad + g^{x^{T\alpha_1}+\cdots+x^{T\alpha_{i-1}}} \cdot G_{\overline{\varphi}0(\overline{\varphi}_{\alpha_{i+1}}(\ldots))[x+1]}
\end{aligned}
$$

$$+ \cdots + g^{x^{T_{\alpha_1}} + \cdots + x^{T_{\alpha_k}}} \cdot (g^{x^{T_{\alpha_{l+1}}+1}} \cdot (G_{\alpha_{l+1}} + 1) + \dots)$$

$$\leq \cdots + g^{x^{T_{\alpha_1}} + \cdots + x^{T_{\alpha_i-1}} + x^{T_0}} \cdot (G_0 + 1)$$

$$+ g^{x^{T_{\alpha_1}} + \cdots + x^{T_{\alpha_i-1}}} \cdot (g^{x^{T_{\alpha_i-1}+1}} \cdot (G_{\alpha_{i+1}} + 1) + \dots)$$

$$= C_x(\alpha, g)$$

Here we have to make the following case distinction: assume that n is maximal with $\alpha_{i+1} \geq \cdots \geq \alpha_n \geq \alpha_{l+1}$. If $n > i$ then n is the maximal k with $\alpha_1 \geq \cdots \geq \alpha_{i-1} \geq \alpha_{i+1} \geq \cdots \geq \alpha_k \geq \alpha_{l+1}$. If $n = i$ then $\alpha_{i+1} < \alpha_{l+1}$. Then $j = i - 1$ is not smaller than the maximal k with $\alpha_1 \geq \cdots \geq \alpha_k \geq \alpha_{l+1}$. In both cases the definition of $C_x(\alpha, g)$ yields the desired estimates.

Case 2.2.2. $\alpha_i = \delta + 1$.

Case 2.2.2.1. $N\alpha_i > x + 1$. Then $N\delta \geq x + 1$. We obtain

$$G_\alpha = G_{\overline{\varphi}_{\alpha_1}(\dots\overline{\varphi}_{\alpha_{i-1}}\overline{\varphi}_{\delta+1}\overline{\varphi}_{\alpha_{i+1}}(\dots)\dots)}$$

$$= G_{\overline{\varphi}_{\alpha_1}(\dots\overline{\varphi}_{\alpha_{i-1}}\overline{\varphi}_{\delta}\overline{\varphi}_{(\delta+1)[x+1]}\overline{\varphi}_{\alpha_{i+1}}(\dots)\dots)}$$

$$\leq g^{x^{T_{\alpha_1}+1}} \cdot (G_{\alpha_1} + 1) + \cdots + g^{x^{T_{\alpha_1}} + \cdots + x^{T_\delta+1}} \cdot (G_\delta + 1)$$

$$+ g^{x^{T_{\alpha_1}} + \cdots + x^{T_\delta} + x^{T_{(\delta+1)[x]}+1}} \cdot (G_{(\delta+1)[x]} + 1)$$

$$+ g^{x^{T_{\alpha_1}} + \cdots + x^{T_{\alpha_j}}} \cdot (g^{x^{T_{\alpha_{i+1}}+1}} \cdot (G_{\alpha_{i+1}} + 1) + \dots)$$

$$\leq \cdots + g^{x^{T_{\alpha_1}} + \cdots + x^{T_{\alpha_i-1}} + x^{T_\delta+1}+1} \cdot (G_{\delta+1} + 1) + \dots$$

$$= C_x(\alpha, g)$$

since $G_{(\delta+1)[x]} + 1 \leq g$ and $x^{T_\delta} + x^{T_{(\delta+1)[x]}+1} + 1 \leq x^{T_\delta+1}+1$.

Case 2.2.2.2. $N\alpha_i \leq x + 1$.

Then $\overline{\varphi}_{\delta+1}\overline{\varphi}_{\alpha_{i+1}}(\dots)[x] = \overline{\varphi}_\delta{}^y\overline{\varphi}_{(\delta+1)[z]}\overline{\varphi}_{\alpha_{i+1}}(\dots)$ for some $y, z \leq x$. We obtain

$$G_\alpha = G_{\overline{\varphi}_{\alpha_1}(\dots\overline{\varphi}_{\alpha_{i-1}}\overline{\varphi}_{\delta+1}\overline{\varphi}_{\alpha_{i+1}}(\dots)\dots)}$$

$$= G_{\overline{\varphi}_{\alpha_1}(\dots\overline{\varphi}_{\alpha_{i-1}}\overline{\varphi}_\delta{}^y\overline{\varphi}_{(\delta+1)[z]}\overline{\varphi}_{\alpha_{i+1}}(\dots)\dots)}$$

$$\leq g^{x^{T_{\alpha_1}+1}} \cdot (G_{\alpha_1} + 1) + \cdots + g^{x^{T_{\alpha_1}} + \cdots + x^{T_{\alpha_i-1}+1}} \cdot (G_{\alpha_i-1} + 1)$$

$$+ g^{x^{T_{\alpha_1}} + \cdots + x^{T_\delta+1}} \cdot (G_\delta + 1) + \dots$$

$$+ g^{x^{T_{\alpha_1}} + \cdots + x^{T_{\alpha_i-1}} + x^{T_\delta} \cdot y + x^{T_{(\delta+1)[z]}+1}} \cdot (G_{(\delta+1)[z]} + 1)$$

$$+ g^{x^{T_{\alpha_1}} + \cdots + x^{T_{\alpha_j}}} \cdot (g^{x^{T_{\alpha_{i+1}}+1}} \cdot (G_{\alpha_{i+1}+1} + 1) + \dots)$$

$$\leq \cdots + g^{x^{T_{\alpha_1}} + \cdots + x^{T_{\alpha_i-1}} + x^{T_\delta+1}+1} \cdot (G_{\delta+1} + 1) + \dots$$

$$= C_x(\alpha, g)$$

since $G_{(\delta+1)[z]} + 1 \leq g$ and $x^{T_\delta} \cdot y + x^{T_{(\delta+1)[z]}+1} + y + 2 \leq x^{T_\delta+1}+1$ because $T_\delta < T_{\delta+1}$ holds since $N\delta \leq x$.

Case 2.2.3. Assume that $\alpha_i \in Lim$. We obtain

$$
\begin{aligned}
G_\alpha &= G_{\overline{\varphi}_{\alpha_1}(\ldots \overline{\varphi}_{\alpha_i-1} \overline{\varphi} \alpha_i \overline{\varphi}_{\alpha_{i+1}} (\ldots) \ldots)} \\
&= G_{\overline{\varphi}_{\alpha_1}(\ldots \overline{\varphi}_{\alpha_i-1} \overline{\varphi}_{\alpha_i[x]} \overline{\varphi}_{\alpha_{i+1}} (\ldots) \ldots)} \\
&\leq \cdots + g^{x^{T_{\alpha_1}} + \cdots + x^{T_{\alpha_i[x]} + 1}} \cdot (G_{\alpha_i[x]} + 1) + \ldots \\
&\leq \cdots + g^{x^{T_{\alpha_1}} + \cdots + x^{T_{\alpha_i} + 1}} \cdot (G_{\alpha_i} + 1) + \ldots \\
&= C_x(\alpha, g). \qquad \dashv
\end{aligned}
$$

It is now quite simple to obtain elementary recursive bounds on G_α for $\alpha < \Gamma_0$. The bounds obtained here are rather rough and presumably they can be sharpened considerably. Nevertheless they suffice to yield the desired result.

LEMMA 17. $N\alpha \leq x \geq 5 \Rightarrow G_\alpha(x) \leq x^{x^{x^{T_\alpha(x+1)}}}$.

PROOF. By induction on α.

Assume that $\alpha =_{NF} \overline{\varphi} \alpha_1(\ldots \overline{\varphi} \alpha_m(\overline{\varphi} \alpha_{m+1} 0 \mid \beta_m) \cdots + \beta_1) + \beta_0$. Let $g := g_\alpha(x)$. Then

$$
\begin{aligned}
G_\alpha(x) &\leq G_{\beta_0}(x) + g^{x^{T_{\alpha_1}(x)+1}} \cdot (G_{\alpha_1}(x) + 1) \\
&\quad + g^{x^{T_{\alpha_1}(x)}} \cdot G_{\beta_1}(x) \\
&\quad + \cdots + \\
&\quad + g^{x^{T_{\alpha_1}(x)} + \cdots + x^{T_{\alpha_m}(x)+1}} \cdot (G_{\alpha_m}(x) + 1) \\
&\quad + g^{x^{T_{\alpha_1}(x)} + \cdots + x^{T_{\alpha_m}(x)}} \cdot G_{\beta_m}(x) \\
&\quad + g^{x^{T_{\alpha_1}(x)} + \cdots + x^{T_{\alpha_{m+1}}(x)+1}} \cdot (G_{\alpha_{m+1}}(x) + 1).
\end{aligned}
$$

The induction hypothesis yields $\max\{G_{\alpha_i}(x), G_{\beta_i}(x)\} \leq x^{x^{x^{T_{\alpha[x+1]}(x+1)}}}$ for $1 \leq i \leq m + 1$. Assertion 4 of Lemma 13 yields $T_{\alpha_i}(x) + 1 \leq T_{\alpha[x+1]}(x + 1)$. Therefore $x^{T_{\alpha_1}(x)} + \cdots + x^{T_{\alpha_i}(x)+1} \leq x^{T_{\alpha[x+1]}(x+1)+1}$ holds for $1 \leq i \leq m + 1$ since $m \leq x$. Putting things together the assertion follows. $\qquad \dashv$

THEOREM 4. $N\alpha \leq x \geq 5 \Rightarrow G_\alpha(x) \leq x^{x^{x^{4^{4^{x+1} \cdot 2}}}}$.

COROLLARY 2. $(G_\alpha)_{\alpha < \Gamma_0}$ *does not majorize the elementary recursive functions.*

Remarks. 1. Let $(G_\alpha^{st})_{\alpha < \Gamma_0}$ denote the slow growing hierarchy which is defined with respect to the standard assignment $\cdot \{\cdot\}$ from Definition 6. It has been shown for example in [17] that every function which is elementary recursive in the Ackermann function is eventually dominated by a function G_α^{st} for some $\alpha < \Gamma_0$. For variants of $\cdot \{\cdot\}$ similar results have already been proved e.g., in [4].

2. Let $\cdot \{\cdot\}_1$ be defined recursively as in the cases 1–3 and 5–14 of Definition 6 and via $(\overline{\varphi} 0(\beta + 1))\{x\}_1 := \overline{\varphi} 0\beta + (\overline{\varphi} 0(\beta + 1))[\![x + 1]\!]$. Assume that $(G_\alpha^1)_{\alpha < \Gamma_0}$ be defined with respect to $\cdot \{\cdot\}_1$. Then $(G_\alpha^1)_{\alpha < \overline{\varphi}_2 0}$ classifies the elementary

recursive functions and $\overline{\varphi}\omega 0$ is the least ordinal ξ such that there is a match between $(G_\alpha^1)_{\alpha<\xi}$ and $(G_\alpha^{st})_{\alpha<\xi}$.

3. Let $\cdot\{\cdot\}_2$ be defined recursively as in the cases 1–3, 5 and 7–14 of Definition 6 and via $(\overline{\varphi}0(\beta+1))\{x\}_2 := \overline{\varphi}0\beta + (\overline{\varphi}0(\beta+1))[\![x+1]\!]$ as well as $\overline{\varphi}0\lambda\{x\}_2 := \lambda + \overline{\varphi}0\lambda[\![x+1]\!]$ if $\lambda \in Fix_0$. Let $(G_\alpha^2)_{\alpha<\Gamma_0}$ be defined with respect to $\cdot\{\cdot\}_2$. Then Γ_0 is the least η such that $(G_\alpha^2)_{\alpha<\eta}$ classifies the elementary recursive functions.

4. Let $\cdot\{\cdot\}_3$ be the canonical extension of N to the ψ-based Buchholz notation system for the Howard Bachmann ordinal $\psi\varepsilon_{\Omega+1}$. Further let $(G_\alpha^3)_{\alpha<\psi\varepsilon_{\Omega+1}}$ be defined with respect to $\cdot\{\cdot\}_3$. Then $(G_\alpha^3)_{\alpha<\psi\varepsilon_{\Omega+1}}$ does not majorize the elementary recursive functions. A proof of this result has recently been given in [15]. The proof is very technical and will thus not be included in this paper. In this system the ordinal Γ_0 appears as $\psi\Omega^\Omega$. We conjecture that $(G_\alpha^3)_{\alpha<\psi\Omega^\Omega}$ is very close to the hierarchy considered in this paper but we have not verified this till now.

Open problems. 1. Assume the obvious extension of Definition 3, the definition of the fundamental sequences, to larger proof-theoretic ordinals. Are there ξ and η such that $(G_\alpha)_{\alpha<\xi}$ majorizes the provably recursive functions of PA and such that $(G_\alpha)_{\alpha<\eta}$ matches $(F_\alpha)_{\alpha<\eta}$? If these ordinals exist, then determine ξ and η. The results of this paper and [15] indicate that such ordinals are either rather large or they do not exist.

2. Do the results of this paper have applications in proof theory, e.g., in Bounded Arithmetic (cf., e.g., [1]) or in combinatorics, for example in Ramsey Theory?

3. Is it possible to get a detailed asymptotics of G_α for $\alpha < \Gamma_0$?

4. Is it possible to get a detailed asymptotics of T_α for $\alpha < \Gamma_0$?

Acknowledgments. I would like to thank Ingo Lepper for careful proof-reading. Also I would like to thank the referee for some helpful remarks.

REFERENCES

[1] A. BECKMANN, S. BUSS, and C. POLLETT, *Ordinal notations and well-orderings in bounded arithmetic*, **Annals of Pure and Applied Logic**, vol. 120, no. 1–3.

[2] BENJAMIN BLANKERTZ and ANDREAS WEIERMANN, *How to characterize provably total functions by the Buchholz operator method*, **Gödel '96 (Brno, 1996)** (Petr Hájek, editor), Lecture Notes in Logic, vol. 6, Springer, Berlin, 1996, pp. 205–213.

[3] WILFRIED BUCHHOLZ, E. ADAM CICHON, and ANDREAS WEIERMANN, *A uniform approach to fundamental sequences and hierarchies*, **Mathematical Logic Quarterly**, vol. 40 (1994), pp. 273–286.

[4] E. ADAM CICHON and STANLEY SCOTT WAINER, *The slow-growing and the Grzegorczyk Hierarchies*, **The Journal of Symbolic Logic**, vol. 48 (1983), pp. 399–408.

[5] JEAN-YVES GIRARD, Π_2^1-*logic. I. Dilators*, **Annals of Mathematical Logic**, vol. 21 (1981), no. 2–3, pp. 75–219.

[6] HARVEY E. ROSE, *Subrecursion*, Clarendon Press, Oxford, 1984.

[7] KURT SCHÜTTE, *Proof theory*, Springer, Berlin, 1977.

[8] HELMUT VOGEL, *Ausgezeichnete Folgen für prädikative Ordinalzahlen und prädikativ-rekursive Funktionen*, **Zeitschrift für Mathematische Logik und Grundlagen der Mathematik**, vol. 23 (1977), no. 5, pp. 435–438.

[9] STANLEY SCOTT WAINER, *Slow growing versus fast growing*, **The Journal of Symbolic Logic**, vol. 54 (1989), no. 2, pp. 608–614.

[10] ——, *Accessible segments of the fast growing hierarchy*, **Logic colloquium '95 (Haifa)**, Springer, Berlin, 1998, pp. 339–348.

[11] ANDREAS WEIERMANN, *How to characterize provably total functions by local predicativity*, **The Journal of Symbolic Logic**, vol. 61 (1996), no. 1, pp. 52–69.

[12] ——, *Sometimes slow growing is fast growing*, **Annals of Pure and Applied Logic**, vol. 90 (1997), no. 1–3, pp. 91–99.

[13] ——, *How is it that infinitary methods can be applied to finitary mathematics? Goedel's T: A case study*, **The Journal of Symbolic Logic**, vol. 63 (1998), no. 4, pp. 1348–1370.

[14] ——, *What makes a (pointwise) subrecursive hierarchy slow growing?*, **Sets and proofs** (Cambridge) (S. Barry Cooper et al., editors), London Mathematical Society Lecture Notes, vol. 258, Cambridge University Press, 1999, pp. 403–423.

[15] ——, *A very slow growing hierarchy for the Howard Bachmann ordinal*, preprint, Münster, 2000.

[16] ——, Γ_0 *may be subrecursively inaccessible*, **Mathematical Logic Quarterly**, vol. 47 (2001), pp. 397–408.

[17] ——, *Some interesting connections between the slow growing hierarchy and the Ackermann function*, **The Journal of Symbolic Logic**, vol. 66 (2001), pp. 609–628.

[18] ——, *Slow versus fast growing*, **Synthese**, vol. 133 (2002), pp. 13–19.

INSTITUT FÜR MATHEMATISCHE LOGIK UND GRUNDLAGENFORSCHUNG
DER WESTFÄLISCHEN WILHELMS-UNIVERSITÄT MÜNSTER
EINSTEINSTR. 62, D-48149 MÜNSTER, GERMANY
E-mail: weierma@math.uni-muenster.de
URL: http://wwwmath.uni-muenster.de/math/logik/org/staff/weiermann/

FIRST ORDER LOGICS OF INDIVIDUAL THEORIES

ROSTISLAV E. YAVORSKY

Abstract. A first order logic $\mathcal{L}(T)$ of a theory T is the set of predicate formulas provable in T under every interpretation of the predicate language in the language of T. Here we present recent results on first order logics of different classical first order theories.

§1. **Introduction.** The study of general logical laws was one of the fundamental purposes of mathematical logic. It was already mentioned by Aristotle that the following propositions

"A or not A,"
"if A implies B and B implies C then A implies C", etc.

are valid for any substitution of sentences for A, B, and C. Propositional logic describes all universal logical laws in the propositional language.

The shift from propositional to predicate language leads to essential increasing of expressive power. New principles which are universally valid appear, e.g.,

$$\forall x A(x) \to A(c), \quad \forall x A(x) \vee \exists x \neg A(x), \text{ etc.}$$

According to the Gödel completeness theorem the set of all first order formulas valid under any interpretation coincides with the first order calculus FO.

On the other hand, if we restrict ourselves by considering finite interpretations only, then the corresponding set of valid formulas is wider. The following formula is a well known example of a new principle:

$$\forall x R(x,x) \,\&\, \forall xyz(R(x,y) \,\&\, R(y,z) \to R(x,z))$$
$$\to \exists u \forall y(R(u,y) \to R(y,u)).$$

This formula is true in every finite model, but it could be easily refuted in an infinite one. By restricting the cardinal of considered finite models we come

---

Key words and phrases. Interpretation, first order logic.

The work is partially supported by the Russian Foundation for Basic Research, grants 98-01-00249, 99-01-01282, INTAS grant 97-1259, and grant DAAH04-96-1-0341, by DARPA under program LPE, project 34145.

Logic Colloquium '99
Edited by J. van Eijck, V. van Oostrom, and A. Visser
Lecture Notes in Logic, 17

to a family of different first order logics $\{\mathcal{L}_n\}_{n=1}^{\infty}$, where $\mathcal{L}_n$ is the set of first order formulas valid in all models with cardinal not greater that n.

Another way to obtain a new first order logic is to restrict the class of admitted interpretations of predicate symbols. In any case the following observation holds: the smaller is the class of considered models and interpretations, the wider is the corresponding first order logic.

In what follows a *predicate language* contains infinitely many predicate symbols of any arity but has no functional symbols, constants or special sign for equality. Formulas in this language are called *predicate formulas*. The first order calculus in this language is denoted by *FO*.

Let T be a first order theory. An *interpretation* f of the predicate language into the language of the theory T maps every atomic formula of the predicate language to a formula in the language of the theory T with the same set of free variables. It commutes with Boolean connectives, quantifiers and substitution of free variables, e.g., if $f(P(x)) = A(x)$, then $f(P(y)) = A(y)$ for any predicate symbol P and a formula A.

DEFINITION 1. *A first order logic $\mathcal{L}(T)$ of a theory T is the set of all predicate formulas provable in T under every interpretation of the predicate language into the language of the theory T.*

One can also consider a *first order logic of a given model* (or a family of models). It coincides with the first order logic of the elementary theory of this model (or the elementary theory of this family of models correspondingly).

It follows immediately from the definition that the first order logic $\mathcal{L}(T)$ of a theory T describes all logical laws valid in the framework of the theory T.

A similar notion was considered by V. A. Lifshits in [7]. A predicate formula φ is called T-valid if it is provable in T under every interpretation. A theory T is called Gödelian if every T-valid formula is provable in the first order calculus. In our terms T is Gödelian iff $\mathcal{L}(T) = FO$. It was established in this paper that every Gödelian theory is undecidable, and that Peano arithmetic *PA* and some of its subtheories are Gödelian theories.

First order logics of extensions of *PA* were studied by V. A. Vardanyan in [12]. In his paper the logic $\mathcal{L}(T)$ appeared as modal free fragment of the first order provability logic of the theory T. It was shown that for every degree of unsolvability d there exists an extension T of *PA* such that degrees of T and $\mathcal{L}(T)$ are equal to d.

First order logics of individual theories play a very important role in the constructive mathematics. Properties of first order logics of intuitionistic arithmetical theories were studied extensively by V. E. Plisko in [8, 9, 10]. In particular, it was proved that for any intuitionistic arithmetical theory T the set of theorems of T is 1-reducible to the logic $\mathcal{L}(T)$, and that the structure of intuitionistic arithmetical theories with the inclusion relation $\subseteq$

is isomorphic to the structure of first order logics of these theories with the inclusion relation $\subseteq$.

The definition of the first order logic of a theory T is a special case of the following general notion introduced by S. N. Artemov and F. Motagna in [1]. Let T be a theory, L be a language, and $\mathcal{F}$ be a set of interpretations of the language L into the language of the theory T. Then the logic $\mathcal{L}(T, L, \mathcal{F})$ is the set of all formulas of the language L provable in T under every interpretation from the set $\mathcal{F}$. If we take L to be the predicate language, and $\mathcal{F}$ to be the set of all interpretations of the predicate language into the language of the theory T, then the logic $\mathcal{L}(T, L, \mathcal{F})$ coincides with the first order logic $\mathcal{L}(T)$ of the theory T.

Our research was aimed at study of first order logics of different classical first order theories and properties of these logics. In particular, we studied decidability, finite axiomatizability, inclusion relations etc. In this paper we report recent results obtained in this area by the author.

§2. **General properties.** In this chapter we present some useful sufficient conditions for inclusion of the first order logics of different theories. Detailed proofs could be found in [13, 16].

PROPOSITION 1. *Let $\mathcal{L}_1$ denote the set of predicate formulas valid in all one-element models (singletons). Then for any theory T the following holds:*

$$FO \subseteq \mathcal{L}(T) \subseteq \mathcal{L}_1.$$

PROPOSITION 2. *Let a theory T_1 be an extension of a theory T_2 in the same language. Then $\mathcal{L}(T_1) \subseteq \mathcal{L}(T_2)$.*

PROPOSITION 3. *Let a theory T_2 be a conservative extension of a theory T_1. Then $\mathcal{L}(T_2) \subseteq \mathcal{L}(T_1)$.*

PROPOSITION 4. *Suppose there exists an interpretation g of the language of a theory T_1 into the language of a theory T_2 such that for any formula φ in the language of T_1 if $T_2 \vdash g(\varphi)$ then $T_1 \vdash \varphi$. Then $\mathcal{L}(T_2) \subseteq \mathcal{L}(T_1)$.*

PROPOSITION 5. *Suppose a theory T' is obtained from a theory T by the standard procedure of extension by definitions (cf. [6]). Then $\mathcal{L}(T') = \mathcal{L}(T)$.*

The following definition was formulated in [3].

DEFINITION 2. *Let $\mathcal{M}_1$ be a model of a signature $\sigma_1 = \langle P_1^{n_1}, \ldots, P_k^{n_k} \rangle$, which contains no functional symbols and no constants. Let $\mathcal{M}_2$ be a model of signature σ_2. The model $\mathcal{M}_1$ is called to be relatively definable in the model $\mathcal{M}_2$ if there are formulas $\delta(x)$, $P_1^*(x_1, \ldots, x_{n_1}), \ldots, P_k^*(x_1, \ldots, x_{n_k})$ in the signature σ_2 such that*

- *the set $K \rightleftharpoons \{a \mid \mathcal{M}_2 \models \delta(a)\}$ is not empty,*
- *a model K of the signature σ_1 with the domain K and predicates P_i defined by the formulas $P_i^*(x_1, \ldots, x_{n_i})$ is elementary equivalent to $\mathcal{M}_1$.*

PROPOSITION 6. *Suppose at least one constant is definable in a model $\mathcal{M}_1$, and the model $\mathcal{M}_1$ is relatively definable in a model $\mathcal{M}_2$. Then*

$$\mathcal{L}(\mathcal{M}_2) \subseteq \mathcal{L}(\mathcal{M}_1).$$

Let $\mathcal{L}_{fin}$ denote the set of predicate formulas valid in all finite models. The following facts are established in [13].

We say that a theory T is model-infinite if for any natural number n there exists a model M of T and formulas with one free variable $A_1(x), \ldots, A_n(x)$ in the language of T such that sets

$$M_i \rightleftharpoons \{a \in M \mid M \models A_i(a)\}$$

are pairwise distinct.

PROPOSITION 7. *Let T be a theory. Then*

$$\mathcal{L}(T) \subseteq \mathcal{L}_{fin} \text{ iff } T \text{ is model-infinite.}$$

PROPOSITION 8. *Suppose a theory T is complete with respect to a class of its finite models. Then*

$$\mathcal{L}_{fin} \subseteq \mathcal{L}(T).$$

§3. First order logics of expressively strong theories.

We show in this section that predicate logics of the following theories: the theory of rings, the theory of fields, the group theory, and all arithmetically correct theories, coincide with the predicate calculus *FO*.

In this section the following simple observation will be of use.

NOTE 1. *Since the inclusion $FO \subseteq \mathcal{L}(T)$ holds for every classical theory, in order to prove the equality $\mathcal{L}(T) = FO$ it is sufficient to show that $\mathcal{L}(T) \subseteq FO$.*

A first order theory T is called *arithmetically correct* if it is able to express all arithmetical relations, and all theorems of T is true in the standard model of arithmetic.

THEOREM 1. *For every arithmetically correct theory T the first order logic $\mathcal{L}(T)$ coincide with the predicate calculus FO.*

PROOF. It follows from the arithmetical formalization of the Gödel completeness theorem for *FO* (cf. [5, 4]). See [13, 16] for details. ⊣

COROLLARY 1. *The first order logic of the truth arithmetic TA and for all its subtheories in the full arithmetical language coincide with the predicate calculus FO. In particular, it is true for Peano arithmetic PA.*

THEOREM 2. (a) *The first order logic of the ring of integers coincides with the predicate calculus FO.*

(b) *The first order logic of the theory of rings coincides with FO.*

THEOREM 3. (a) *The first order logic of the field of rational numbers coincides with the predicate calculus FO.*

(b) *The first order logic of the theory of fields coincides with FO.*

PROOF. We use here a well known result of J. Robinson [11] about definability of the notion of integers and the notion of natural numbers in terms of rational numbers and operations of addition and multiplication.   ⊣

Let **G** denote the group theory in the language containing the unique non-logical symbol, namely, a binary functional symbol.

Let R be a ring with the unit. A square matrix over the ring R is called *an upper unitriangular matrix* if all of its elements below the main diagonal are equal to zero and all the diagonal elements are equal to 1. If the ring R is associative then the set of all upper unitriangular $n \times n$ matrices over R forms a group with respect to the operation of multiplication of matrices. This group is denoted by $UT_n(R)$.

THEOREM 4. (a) *The first order logic of the group $UT_3(\mathbf{Z})$ of upper unitriangular 3×3 matrices over the ring of integers coincides with the predicate calculus FO.*

(b) *The first order logic of the group theory coincides with the predicate calculus FO.*

PROOF. (a) The plan of our proof is the following. We consider two auxiliary models:

$$\mathcal{M}_1 = \langle \mathbf{N},\ P(x,y,z),\ x \cdot y = z \rangle,$$

where $P(x, y, z)$ denotes the predicate $(x + y = z \vee |x - y| = z)$, and

$$\mathcal{M}_2 = \langle \mathbf{Z},\ x + y = z,\ |x \cdot y| = |z| \rangle.$$

We will prove that $\mathcal{L}(\mathcal{M}_1) = FO$, $\mathcal{L}(\mathcal{M}_2) \subseteq \mathcal{L}(\mathcal{M}_1)$, and $\mathcal{L}(UT_3(\mathbf{Z})) \subseteq \mathcal{L}(\mathcal{M}_2)$. The complete proof is given in [16]. We present here the main three lemmas.

LEMMA 4.1. *The predicate of addition is definable in the model $\mathcal{M}_1$.*

In particular, it now follows that all arithmetical relations are definable in the model $\mathcal{M}_1$. Thus, the elementary theory of $\mathcal{M}_1$ is arithmetically correct in the sense of theorem 1. So, $\mathcal{L}(\mathcal{M}_1) = FO$.

Consider now the model $\mathcal{M}_2 = \langle \mathbf{Z},\ x + y = z,\ |x \cdot y| = |z| \rangle$.

LEMMA 4.2. *There exists an interpretation f of the language of $\mathcal{M}_1$ into the language of $\mathcal{M}_2$ such that for every closed formula φ in the language of $\mathcal{M}_1$ one has: $\mathcal{M}_1 \models \varphi \Leftrightarrow \mathcal{M}_2 \models f(\varphi)$.*

According to proposition 4 we obtain $\mathcal{L}(\mathcal{M}_2) \subseteq \mathcal{L}(\mathcal{M}_1)$ and $\mathcal{L}(\mathcal{M}_2) = FO$.

LEMMA 4.3. *The model $\mathcal{M}_2 = \langle \mathbf{Z},\ x + y = z,\ |x \cdot y| = |z| \rangle$ is relatively definable in the group $UT_3(\mathbf{Z})$.*

The proof of the last lemma immediately follows from recent results by O. V. Belegradek [2] about the group $UT_3(\mathbf{Z})$.   ⊣

§4. Other theories. Let *Pre* be Presburger's arithmetic of addition, *Sko* be Skolem's arithmetic of multiplication, and *DO* stand for the theory of discrete linear order with minimal element (the order type of natural numbers).

THEOREM 5. *The following proper inclusion holds*:

$$FO \subset \mathcal{L}(Sko) \subset \mathcal{L}(Pre) \subset \mathcal{L}(DO) \subset \mathcal{L}_{fin}.$$

PROOF. It follows from propositions 1–6. The most complicated parts are unequalities $\mathcal{L}(Sko) \neq \mathcal{L}(Pre)$ and $\mathcal{L}(Pre) \neq \mathcal{L}(DO)$. See [13, 14] for details.   ⊣

Let $\mathbf{Q}$ be the set of rationals.

THEOREM 6. *The first order logic $\mathcal{L}(Eq)$ of the theory of equality is decidable but can not be axiomatized by any set of schemes with restricted arity. The same holds for the first order logic $\mathcal{L}(\mathbf{Q}, \leq)$ of the theory of dense linear order without minimal and maximal elements.*

PROOF. See [13, 15].   ⊣

§5. Conclusion remarks. All the results mentioned above could be summarized on the following picture. This scheme describes the lattice of first order theories with respect to the inclusion relation.

First order logic is a first order theory in the predicate language which admits the substitution rule. One can easily see that for any first order logic F we have

$$\mathcal{L}(F) = F.$$

Thus, first order logics are exactly the first order logics of individual theories. This observation justifies the title of the scheme.

It was already mentioned above that *FO* is the minimal first order logic while the logic of one-element models $\mathcal{L}_1$ is the maximal element in this structure. The logic of finite models $\mathcal{L}_{fin}$ is situated in the middle.

The logic $\mathcal{L}(T)$ coincides with *FO* for the following expressively strong theories: the group theory, the theory of rings, the theory of fields, and all arithmetically correct theories.

First order logics of decidable fragments of arithmetic are proper extensions of the first order calculus. Moreover, the logic of Skolem's arithmetic of multiplication is properly included in the logic of Presburger's arithmetic of addition, while the later is a proper subtheory of the logic of the theory of discrete linear order with the minimal element (the ordering type of natural numbers). All first order logics situated below $\mathcal{L}_{fin}$ are undecidable because $\mathcal{L}_{fin}$ is known to be hereditary undecidable.

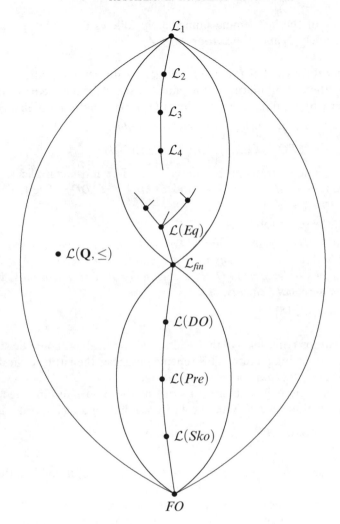

FIGURE 1. The structure of first order logics.

One can easily see that first order logics of the axiomatic theories of finite structures such as finite fields, finite rings, finite groups etc. coincide with $\mathcal{L}_{fin}$.

First order logic $\mathcal{L}(\mathbf{Q}, \leq)$ of the theory of dense linear order without minimal and maximal elements is incompatible with the logic of finite models. It is decidable.

First order logic $\mathcal{L}(Eq)$ of the theory of equality is a proper extension of $\mathcal{L}_{fin}$. It is decidable and for every decidable extension of the theory of equality the corresponding first order logic is decidable too. Another infinite family of decidable first order logics is $\{\mathcal{L}_n\}_{n=1}^{\infty}$. It is easy to see that for every natural number n the logic $\mathcal{L}_n$ of n-element models is decidable. Note that $\mathcal{L}_{fin}$ coincides with the limit of this family, i.e.,

$$\mathcal{L}_{fin} = \bigcap_{n=1}^{\infty} \mathcal{L}_n.$$

§6. Open problems. In this section the most interesting (from the author's point of view) open problems are listed. Most of them are formulated as directions for future research in this field.

6.1. Let us consider an extension of the predicate language by the equality sign. Stipulate that every interpretation f leaves equality unchanged. Let $\mathcal{L}_e(T)$ be the corresponding first order logic with equality, FO_e be the first order calculus in the extended language. A very interesting question in this direction is whether $\mathcal{L}_e(\mathbf{G}) = FO_e$ or not, where $\mathbf{G}$ denotes the group theory.

6.2. Consider a fragment of the predicate language which contains only binary predicate symbols. Let $\mathcal{L}_2(T)$ be the corresponding set of T-valid formulas. In a similar way one can define the logic $\mathcal{L}_n(T)$ for any natural number n. The author's conjecture is that for a wide class of theories one has $\mathcal{L}(\mathcal{L}_n(T)) = \mathcal{L}(T)$ for $n \geq 2$.

6.3. It would be nice to find the first order logics of other classical first order theories: the theory of abelian groups, the theory of the field of reals, the theory of fields of characteristics p, of a given finite algebraic structure etc.

6.4. Another examples of decidable first order logics, especially infinite families of decidable first order logic are welcomed.

6.5. It would be interesting to know properties of the structure of first order logics.

6.6. An interesting question is to find any natural necessary conditions for the inclusion $\mathcal{L}(T_1) \subseteq \mathcal{L}(T_2)$. The same question is open also for the equality $\mathcal{L}(T_1) = \mathcal{L}(T_2)$ and the inclusion $\mathcal{L}(M_1) \subseteq \mathcal{L}(M_2)$, where M_1 and M_2 are any algebraic structures.

§7. Acknowledgements. I am very grateful to my scientific advisor professor Sergei N. Artemov who advised me on this work. Also I am thankful to professor Oleg V. Belegradek for the proof of lemma 4.3.

REFERENCES

[1] SERGEI N. ARTEMOV and FRANCO MONTAGNA, *On first-order theories with provability operator*, **The Journal of Symbolic Logic**, vol. 59 (1994), no. 4, pp. 1139–1153.

[2] OLEG V. BELEGRADEK, *Teorija modelej unitreugol'nykh i ekzistencial'no zamknutykh grupp*, **Ph.D. thesis**, Novosibirsk State University, 1995.

[3] YURI L. ERSHOV, **Problemy razreshimosti i konstruktivnye modeli**, Nauka, Moscow, 1980, (in Russian).

[4] SOLOMON FEFERMAN, *Arithmetization of metamathematics in a general setting*, **Fundamenta Mathematicae**, vol. 49 (1960), pp. 35–92.

[5] DAVID HILBERT and PAUL BERNAYS, **Grundlagen der Mathematik**, vol. 2, Springer–Verlag, New York Berlin Heidelberg Tokio, 1968.

[6] S. KLEENE, **Introduction to metamathematics**, Inostrannaya Literatura, Moscow, 1957.

[7] VLADIMIR A. LIFSHITS, *Deduktivnaya obshcheznachimost' i klassy svedenija*, **Matematicheskie Zametki**, vol. 4 (1967), pp. 69–77, (in Russian).

[8] VALERI E. PLISKO, *Nekotorye varianty ponyatija realizuemosti dlya predikatnykh formul*, **Izvestiya Akademii Nauk SSSR**, vol. 42 (1978), no. 3, pp. 636–653, (in Russian).

[9] ——, *Konstruktivnaya formalizaciya teoremy Tennenbauma i ee primenenie*, **Matematicheskie Zametki**, vol. 48 (1990), no. 3, pp. 108–118.

[10] ——, *Ob arfmeticheskoj slozhnosti nekotorykh predikatnykh logik*, **Matematicheskie Zametki**, vol. 52 (1992), no. 1, pp. 94–104, (in Russian).

[11] JULIA ROBINSON, *Definability and decision problem in arithmetic*, **The Journal of Symbolic Logic**, vol. 14 (1949), no. 2, pp. 98–114.

[12] VALERI A. VARDANYAN, *Predikatnaya logika dokazuemosti bez dokazuemosti*, **Intensionalnye logiki i logicheskaya struktura teorii** (Tbilisi), 1988, (in Russian), pp. 65–69.

[13] ROSTISLAV E. YAVORSKY, *Logical schemes for first order theories*, **Logical foundations of computer science** (Sergei I. Adian and Anil Nerode, editors), Lecture Notes in Computer Science, vol. 1234, Springer-Verlag, Berlin Heidelberg, July 1997, pp. 410–418.

[14] ——, *Predikatnye logiki razreshimih fragmentov arifmetiki*, **Vestnik Moskovskogo Universiteta. Matematika. Mekhanika**, (1998), no. 2, pp. 12–16, (in Russian).

[15] ——, *Razreshimie logiki pervogo poryadka*, **Fundamental'naya i Prikladnaya Matematika**, vol. 4 (1998), no. 2, pp. 733–749, (in Russian).

[16] ——, *Predikatnye logiki vyrazitelno silnykh teorij*, **Matematicheskie Zametki**, vol. 66 (1999), no. 5, pp. 777–788, (in Russian).

STEKLOV MATHEMATICAL INSTITUTE
GSP-1, GUBKINA STR. 8, 117966, MOSCOW, RUSSIA
E-mail: rey@mi.ras.ru

LECTURE NOTES IN LOGIC

General Remarks

This series is intended to serve researchers, teachers, and students in the field of symbolic logic, broadly interpreted. The aim of the series is to bring publications to the logic community with the least possible delay and to provide rapid dissemination of the latest research. Scientific quality is the overriding criterion by which submissions are evaluated.

Books in the Lecture Notes in Logic series are printed by photo-offset from master copy prepared using LaTeX and the ASL style files. For this purpose the Association for Symbolic Logic provides technical instructions to authors. Careful preparation of manuscripts will help keep production time short, reduce costs, and ensure quality of appearance of the finished book. Authors receive 50 free copies of their book. No royalty is paid on LNL volumes.

Commitment to publish may be made by letter of intent rather than by signing a formal contract, at the discretion of the ASL Publisher. The Association for Symbolic Logic secures the copyright for each volume.

The editors prefer email contact and encourage electronic submissions.

Editorial Board

Editorial Policy

1. Submissions are invited in the following categories:
i) Research monographs   iii) Reports of meetings
ii) Lecture and seminar notes   iv) Texts which are out of print
Those considering a project which might be suitable for the series are strongly advised to contact the publisher or the series editors at an early stage.

2. Categories i) and ii). These categories will be emphasized by Lecture Notes in Logic and are normally reserved for works written by one or two authors. The goal is to report new developments quickly, informally, and in a way that will make them accessible to non-specialists. Books in these categories should include
– at least 100 pages of text;
– a table of contents and a subject index;
– an informative introduction, perhaps with some historical remarks, which should be accessible to readers unfamiliar with the topic treated;

In the evaluation of submissions, timeliness of the work is an important criterion. Texts should be well-rounded and reasonably self-contained. In most cases the work will contain results of others as well as those of the authors. In each case, the author(s) should provide sufficient motivation, examples, and applications. Ph.D. theses will be suitable for this series only when they are of exceptional interest and of high expository quality.

Proposals in these categories should be submitted (preferably in duplicate) to one of the series editors, and will be refereed. A provisional judgment on the acceptability of a project can be based on partial information about the work: a first draft, or a detailed outline describing the contents of each chapter, the estimated length, a bibliography, and one or two sample chapters. A final decision whether to accept will rest on an evaluation of the completed work.

3. Category iii). Reports of meetings will be considered for publication provided that they are of lasting interest. In exceptional cases, other multi-authored volumes may be considered in this category. One or more expert participant(s) will act as the scientific editor(s) of the volume. They select the papers which are suitable for inclusion and have them individually refereed as for a journal. Organizers should contact the Managing Editor of Lecture Notes in Logic in the early planning stages.

4. Category iv). This category provides an avenue to provide out-of-print books that are still in demand to a new generation of logicians.

5. Format. Works in English are preferred. After the manuscript is accepted in its final form, an electronic copy in LaTeX format will be appreciated and will advance considerably the publication date of the book. Authors are strongly urged to seek typesetting instructions from the Association for Symbolic Logic at an early stage of manuscript preparation.

LECTURE NOTES IN LOGIC

From 1993 to 1999 this series was published under an agreement between the Association for Symbolic Logic and Springer-Verlag. Since 1999 the ASL is Publisher and A K Peters, Ltd. is Co-publisher. The ASL is committed to keeping all books in the series in print.

Current information may be found at http://www.aslonline.org, the ASL Web site. Editorial and submission policies and the list of Editors may also be found above.

Previously published books in the *Lecture Notes in Logic* are:

1. *Recursion theory.* J. R. Shoenfield. (1993, reprinted 2001; 84 pp.)

2. *Logic Colloquium '90; Proceedings of the Annual European Summer Meeting of the Association for Symbolic Logic, held in Helsinki, Finland, July 15–22, 1990.* Eds. J. Oikkonen and J. Väänänen. (1993, reprinted 2001; 305 pp.)

3. *Fine structure and iteration trees.* W. Mitchell and J. Steel. (1994; 130 pp.)

4. *Descriptive set theory and forcing: how to prove theorems about Borel sets the hard way.* A. W. Miller. (1995; 130 pp.)

5. *Model theory of fields.* D. Marker, M. Messmer, and A. Pillay. (1996; 154 pp.)

6. *Gödel '96; Logical foundations of mathematics, computer science and physics; Kurt Gödel's legacy. Brno, Czech Republic, August 1996, Proceedings.* Ed. P. Hajek. (1996, reprinted 2001; 322 pp.)

7. *A general algebraic semantics for sentential objects.* J. M. Font and R. Jansana. (1996; 135 pp.)

8. *The core model iterability problem.* J. Steel. (1997; 112 pp.)

9. *Bounded variable logics and counting.* M. Otto. (1997; 183 pp.)

10. *Aspects of incompleteness.* P. Lindstrom. (1997, 2nd ed. 2003; 163 pp.)

11. *Logic Colloquium '95; Proceedings of the Annual European Summer Meeting of the Association for Symbolic Logic, held in Haifa, Israel, August 9–18, 1995.* Eds. J. A. Makowsky and E. V. Ravve. (1998; 364 pp.)

12. *Logic Colloquium '96; Proceedings of the Colloquium held in San Sebastian, Spain, July 9–15, 1996.* Eds. J. M. Larrazabal, D. Lascar, and G. Mints. (1998; 268 pp.)

13. *Logic Colloquium '98; Proceedings of the Annual European Summer Meeting of the Association for Symbolic Logic, held in Prague, Czech Republic, August 9–15, 1998.* Eds. S. R. Buss, P. Hájek, and P. Pudlák. (2000; 541 pp.)

14. *Model Theory of Stochastic Processes.* S. Fajardo and H. J. Keisler. (2002; 136 pp.)

15. *Reflections on the Foundations of Mathematics; Essays in honor of Solomon Feferman.* Eds. W. Seig, R. Sommer, and C. Talcott. (2002; 444 pp.)

16. *Inexhaustibility; a non-exhaustive treatment.* T. Franzén. (2004; 255 pp.)

17. *Logic Colloquium '99; Proceedings of the Annual European Summer Meeting of the Association for Symbolic Logic, held in Utrecht, Netherlands, August 1–6, 1999.* Eds. J. van Eijck, V. van Oostrom, and A. Visser. (2004; 208 pp.)